Mathematics UNLIMITED

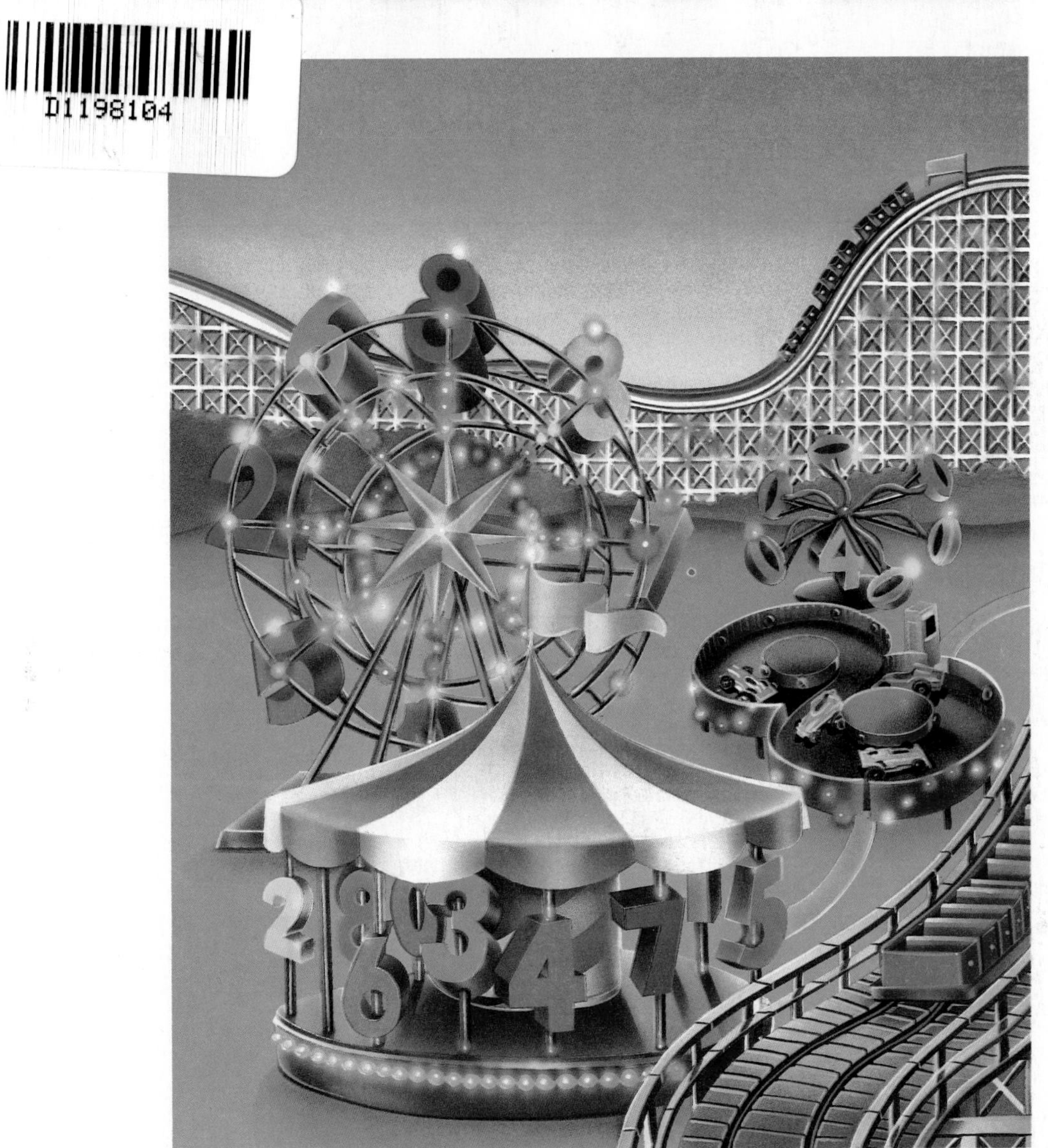

HOLT, RINEHART and WINSTON, Publishers
New York • Toronto • Mexico City • London • Sydney • Tokyo

AUTHORS

Francis "Skip" Fennell
Chairman, Education Department
Associate Professor of Education
Western Maryland College
Westminster, Maryland

Barbara J. Reys
Assistant Professor of Curriculum and Instruction
University of Missouri, Columbia, Missouri
Formerly Junior High Mathematics Teacher
Oakland Junior High, Columbia, Missouri

Robert E. Reys
Professor of Mathematics Education
University of Missouri
Columbia, Missouri

Arnold W. Webb
Senior Research Associate
Research for Better Schools
Philadelphia, Pennsylvania
Formerly Asst. Commissioner of Education
New Jersey State Education Department

ILLUSTRATION

Teresa Anderko: pp. 2, 3, 14, 15, 41, 44, 45, 153, 156, 157, 164, 168, 169, 182, 242, 243, 252, 261, 295, 316, 332, 333, 362, 402, 403 • Alex Block: pp. 192, 193, 372 • Janet Bohn: pp. 90, 117, 154, 183, 186, 228, 229, 250, 251, 273, 274, 284, 288, 289, 298, 306, 314, 315, 337, 339, 390, 391 • Penny Carter: pp. 8, 9, 38, 39, 50, 51, 198, 202, 203 • Donald Crews: p. 37 • Carolyn Croll: pp. 158, 170, 171, 176, 194, 195 • Ivan Dieruf: pp. 304, 305, 360 • Edwin Figueroa: pp. 87, 292, 361 • Judy Filippo: pp. 6, 13, 134, 135, 278, 279, 290, 291, 401 • Simon Galkin: pp. 98, 108, 132, 140, 190, 191, 200, 245, 310, 320, 321, 333 right, 335, 340, 344, 374, 375, 376 • Mark Giglio: pp. 16, 17, 282, 283 • Fred Harsh: pp. 352, 378, 379 • John Killgrew: pp. 7, 10, 11, 32, 46, 47, 83, 112, 116, 187, 206, 207, 212, 213, 254, 255, 260, 296, 382 • Elliot Kreloff: pp. 82, 166, 167, 174, 220, 276, 277 • Diana Magnuson: pp. 76, 217, 226, 227, 238, 326, 327, 350, 351, 356 • Jan Palmer: pp. 4, 5 • Beverly Pardee-Rich: pp. 55, 308, 322 • Karen Pellaton: pp. 18, 23, 54, 79, 138, 139, 146, 180, 181, 241, 270, 271, 275, 384, 385 • Jan Pyk: pp. 88, 89, 311 • Claudia Sargent: pp. 40, 42, 43, 102, 120, 208, 319, 387, 404 • Joel Snyder: pp. 64, 65, 236, 256, 257, 259 • Debbie Tilley: pp. 126, 142, 143, 160, 161 • George Ulrich: pp. 12, 20, 52, 53, 94, 95, 130, 137, 148, 163, 216, 246, 247, 342, 343, 354, 394 • Vantage Art Inc.: pp. 272, 420 • Nina Wallace: p. 325 • Jack Wallen: pp. 268, 269 • Fred Winkowski: pp. 26, 59, 62, 104, 105, 149, 230, 231, 330, 331, 363. **Chapter Opener Illustrations:** Bob Aiese: pp. 1, 29, 61, 91, 125, 155, 189, 225, 267, 303, 341, 371. **Cover Illustration:** Jeannette Adams.

PHOTOGRAPHY

George Ancona: p. 172 • California State Library: p. 48 bottom • Culver Pictures: p. 388 • DPI/Linda K. Moore: p. 178 bottom • DRK Photo/Len Rue, Jr.: p. 144 • DRK Photo/Stephen J. Krasemann: p. 136 • Exxon: p. 356 center, bottom • Focus on Sports/Warren Morgan: p. 272 • Focus on Sports: p. 258 • HRW Photo by Yoav Levy: p. 223 • Grant Heilman/Alan Pitcairn: pp. 80, 81 • Image Bank/G. Gscheidle: p. 40 • Library of Congress: p. 48 top • Lawrence Migdale 1983: p. 123 • Michal Heron: p. 68 • NASA: pp. 328, 329 • Odyssey Productions/Robert Frerck: pp. 34, 35; Walter Frerck: p. 73 • Omni-Photo Communications, Inc./Ken Karp: pp. 66, 67, 78, 158, 159, 174, 175, 194, 236, 246, 247, 248, 380, 381; John Lei: pp. 36, 37, 74, 75, 76, 79, 100, 101, 106, 107, 111, 164, 165, 210, 228, 234, 242, 287, 314, 315, 348, 349, 358, 359 • Photo Researchers, Inc./Patrick Grace: p. 96 • Science Museum, London: p. 114 • Stock Market/Jeffrey E. Blackman: p. 232; Viviane Holbrooke: p. 179 top; William Roy: p. 214 • Taurus Photos/Betsy Lee: p. 252 • Wheeler Pictures/Paul Solomon: p. 53 • Woodfin Camp & Associates/Craig Aurness: p. 356 top; Michal Heron: p. 368; William Hubbell: p. 346; Wally McNamee: p. 110.

Printed in the United States of America

ISBN 0-03-006427-9

6 7 8 9 0 032 9 8 7 6 5 4 3 2

CONTENTS

4 ADDITION OF WHOLE NUMBERS

5 SUBTRACTION OF WHOLE NUMBERS

6 MULTIPLICATION FACTS

7 DIVISION FACTS

8 FRACTIONS AND DECIMALS

9 MEASUREMENT

10 GEOMETRY AND GRAPHING

11 MULTIPLICATION

12 DIVISION

1 ADDITION AND SUBTRACTION FACTS

Can you name your favorite animal actor? Is the animal a TV star or a movie star, or is it a make-believe animal in a cartoon? Ask 5 friends which is their favorite animal actor. How would you decide which animal is the favorite?

Addition Facts to 10

A. When a trainer tossed a fish, 3 dolphins jumped up for it. Another 2 dolphins just swam in the pool. How many dolphins were there in all?

You can add to find how many dolphins there were.

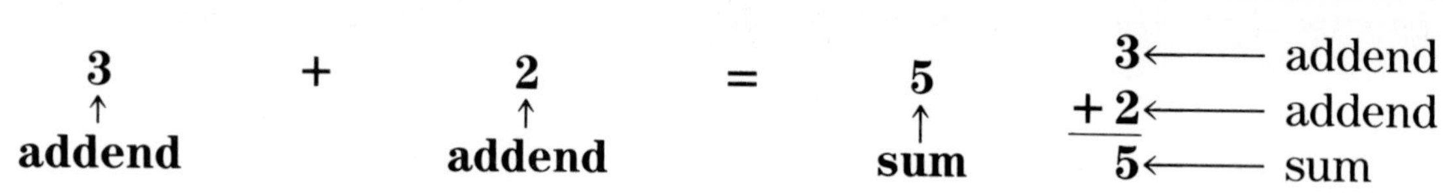

Read this fact: "Three plus two equals five."

There were 5 dolphins in all.

B. If one addend is 0, the sum equals the other addend.

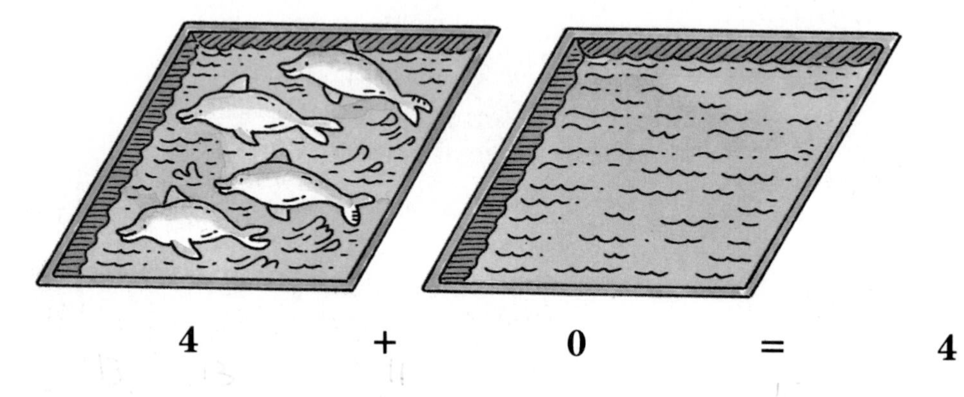

Add.

1.	**2.**	**3.**	**4.**	**5.**	**6.**	**7.**
$1 + 3$	$4 + 2$	$0 + 1$	$2 + 2$	$6 + 2$	$1 + 5$	$3 + 5$

8. $4 + 3 =$ ___ **9.** $3 + 3 =$ ___ **10.** $1 + 2 =$ ___ **11.** $3 + 4 =$ ___

12. $7 + 3 =$ ___ **13.** $5 + 0 =$ ___ **14.** $3 + 6 =$ ___ **15.** $0 + 7 =$ ___

Add.

16. $\begin{array}{r} 1 \\ +7 \\ \hline \end{array}$	17. $\begin{array}{r} 4 \\ +1 \\ \hline \end{array}$	18. $\begin{array}{r} 5 \\ +2 \\ \hline \end{array}$	19. $\begin{array}{r} 2 \\ +0 \\ \hline \end{array}$	20. $\begin{array}{r} 3 \\ +1 \\ \hline \end{array}$	21. $\begin{array}{r} 5 \\ +3 \\ \hline \end{array}$	22. $\begin{array}{r} 3 \\ +6 \\ \hline \end{array}$
23. $\begin{array}{r} 8 \\ +0 \\ \hline \end{array}$	24. $\begin{array}{r} 5 \\ +5 \\ \hline \end{array}$	25. $\begin{array}{r} 8 \\ +2 \\ \hline \end{array}$	26. $\begin{array}{r} 6 \\ +0 \\ \hline \end{array}$	27. $\begin{array}{r} 7 \\ +2 \\ \hline \end{array}$	28. $\begin{array}{r} 6 \\ +1 \\ \hline \end{array}$	29. $\begin{array}{r} 1 \\ +4 \\ \hline \end{array}$
30. $\begin{array}{r} 7 \\ +3 \\ \hline \end{array}$	31. $\begin{array}{r} 0 \\ +5 \\ \hline \end{array}$	32. $\begin{array}{r} 4 \\ +6 \\ \hline \end{array}$	33. $\begin{array}{r} 1 \\ +6 \\ \hline \end{array}$	34. $\begin{array}{r} 4 \\ +4 \\ \hline \end{array}$	35. $\begin{array}{r} 7 \\ +1 \\ \hline \end{array}$	36. $\begin{array}{r} 4 \\ +5 \\ \hline \end{array}$

37. $2 + 6 =$ ■	38. $9 + 1 =$ ■	39. $0 + 4 =$ ■	40. $3 + 7 =$ ■
41. $2 + 4 =$ ■	42. $6 + 2 =$ ■	43. $5 + 1 =$ ■	44. $2 + 8 =$ ■
45. $6 + 3 =$ ■	46. $2 + 7 =$ ■	47. $5 + 4 =$ ■	48. $2 + 5 =$ ■

Solve.

49. A dolphin show was filmed from land by 7 cameras. Another 2 cameras filmed it from underwater. How many cameras filmed the show?

50. During the show, 2 dolphins were noisy. Another 6 were quiet. How many dolphins took part in the show?

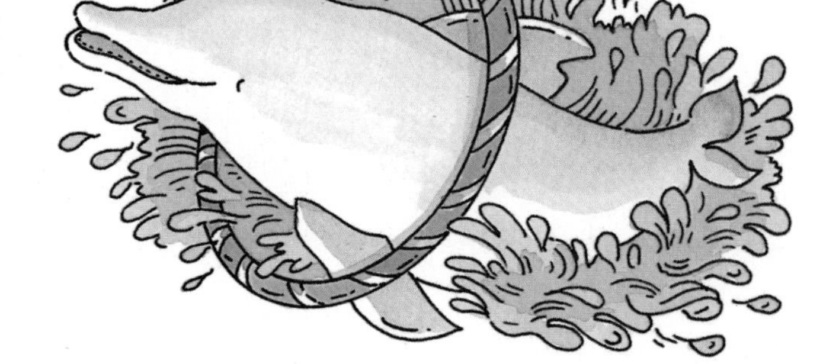

FOCUS: MENTAL MATH

Look for patterns.

$3 + 1 = 4$
$4 + 1 = 5$

Use the first fact to complete the second.

1. $2 + 1 = 3$ $3 + 1 =$ ■	2. $3 + 2 = 5$ $4 + 2 =$ ■	3. $5 + 1 = 6$ $5 + 0 =$ ■	4. $4 + 5 = 9$ $4 + 4 =$ ■
5. $2 + 2 = 4$ $2 + 3 =$ ■	6. $3 + 4 = 7$ $2 + 4 =$ ■	7. $5 + 3 = 8$ $5 + 2 =$ ■	8. $8 + 2 = 10$ $8 + 1 =$ ■

Addition Facts to 18

A. In a TV ad, 3 bulls ran down a street. Then 9 more bulls followed them. How many bulls were there in the ad?

You can add to find how many bulls there were in the TV ad.

3 + 9 = 12

$$\begin{array}{r} 3 \\ +9 \\ \hline 12 \end{array}$$

Altogether, there were 12 bulls in the TV ad.

B. You can add numbers in any order. The sum is always the same.

4 + 7 = 11

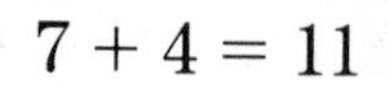

7 + 4 = 11

Think: The sum of 4 + 7 **is always** the same as the sum of 7 + 4.

Add.

1. $\begin{array}{r} 8 \\ +5 \\ \hline 13 \end{array}$ $\begin{array}{r} 5 \\ +8 \\ \hline 13 \end{array}$

2. $\begin{array}{r} 8 \\ +3 \\ \hline 11 \end{array}$ $\begin{array}{r} 3 \\ +8 \\ \hline 11 \end{array}$

3. $\begin{array}{r} 6 \\ +7 \\ \hline 13 \end{array}$ $\begin{array}{r} 7 \\ +6 \\ \hline 13 \end{array}$

4. $\begin{array}{r} 3 \\ +2 \\ \hline 5 \end{array}$ $\begin{array}{r} 2 \\ +3 \\ \hline 5 \end{array}$

5. $\begin{array}{r} 1 \\ +9 \\ \hline 10 \end{array}$ $\begin{array}{r} 9 \\ +1 \\ \hline 10 \end{array}$

6. $\begin{array}{r} 3 \\ +7 \\ \hline 10 \end{array}$ $\begin{array}{r} 7 \\ +3 \\ \hline 10 \end{array}$

7. $\begin{array}{r} 5 \\ +2 \\ \hline 7 \end{array}$ $\begin{array}{r} 2 \\ +5 \\ \hline 7 \end{array}$

8. $\begin{array}{r} 6 \\ +5 \\ \hline 11 \end{array}$ $\begin{array}{r} 5 \\ +6 \\ \hline 11 \end{array}$

9. $\begin{array}{r} 3 \\ +9 \\ \hline 12 \end{array}$ $\begin{array}{r} 9 \\ +3 \\ \hline 12 \end{array}$

Add.

10.	11.	12.	13.	14.	15.	16.
$8 + 6$	$6 + 8$	$4 + 9$	$9 + 4$	$6 + 5$	$5 + 6$	$9 + 8$

17.	18.	19.	20.	21.	22.	23.
$5 + 7$	$8 + 4$	$6 + 6$	$8 + 7$	$2 + 9$	$9 + 6$	$8 + 8$

24. $2 + 9 =$ ____ 25. $9 + 3 =$ ____ 26. $3 + 9 =$ ____ 27. $4 + 8 =$ ____

28. $7 + 4 =$ ____ 29. $5 + 6 =$ ____ 30. $9 + 9 =$ ____ 31. $8 + 7 =$ ____

32. $4 + 7 =$ ____ 33. $5 + 9 =$ ____ 34. $9 + 5 =$ ____ 35. $9 + 2 =$ ____

Solve.

36. Grecco the bull broke 9 fences in the morning. He broke 4 more fences in the afternoon. How many fences did Grecco break in all?

37. There were 3 trainers trying to chase Grecco out of the barn. Then 8 more people came to help. How many people were trying to get Grecco out?

38. Grecco the bull took 6 tries to do the first ad. He took 7 tries to do the second one. How many tries did it take Grecco to do the two ads?

39. Grecco received 4 bags of fan mail after his first TV show. He received 8 bags during the rest of the year. How many bags of mail did Grecco receive in all?

CHALLENGE

Find the number for each shape.

1. $2 + 1 =$ ■ ■ $+ 2 =$ ▲ ■ $+$ ▲ $=$ ●
2. $4 + 2 =$ ■ ■ $+ 3 =$ ▲ ■ $+$ ▲ $=$ ●
3. $5 + 4 =$ ■ ■ $+ 1 =$ ▲ ■ $+$ ▲ $=$ ●

More Practice, page 405

PROBLEM SOLVING

Using the Help File

Many problems seem too hard to solve. The Help File on page 397 can give you ideas about how to solve them. Here's how to use it.

Go to **Questions.** Go to **Tools.** Go to **Solutions.** Go to **Checks.**

1. Try to figure out why you are having trouble. Go to the part of the file that can help you.
2. Look over the ideas in the part of the file you choose. Find one that you think will help you. Remember that there is no rule about which idea you should select. Different people will choose different ways to help them solve problems.
3. Try to solve the problem by using the idea you choose.
4. If you still haven't solved the problem, look for another idea in the Help File.

Read the problem. Help each student below by deciding where in the Help File each can find ideas for solving the problem. Write the letter of the correct answer.

Sam has 3 fish in his fish tank. Mrs. Lee gives Sam 4 more fish. How many fish does Sam have in his fish tank now?

1. Jane read the problem. She read it again. She read it one more time. Then she said, "I don't even know what I am looking for." Where should Jane look in the Help File?

   a. Questions b. Tools
   c. Solutions d. Checks

2. John read the problem. "I get it," he said. "Sam had fish. He was given more fish. But how do I figure out how to find how many fish Sam has?" Where should John look in the Help File?

   a. Questions b. Tools
   c. Solutions d. Checks

3. Sara knew she had to add to solve the problem. But she had trouble adding. "What do I do now?" she wondered. Where should Sara look in the Help File?

   a. Questions b. Tools
   c. Solutions d. Checks

4. "Sam has 7 fish," answered Felix. "Now I should check the answer." Where should Felix look in the Help File?

   a. Questions b. Tools
   c. Solutions d. Checks

Subtraction Facts to 10

A. At the marine show, 10 seals sat on a rock. When the trainer blew her whistle, 2 seals jumped through a hoop. How many seals stayed on the rock?

You can subtract to find how many seals stayed on the rock.

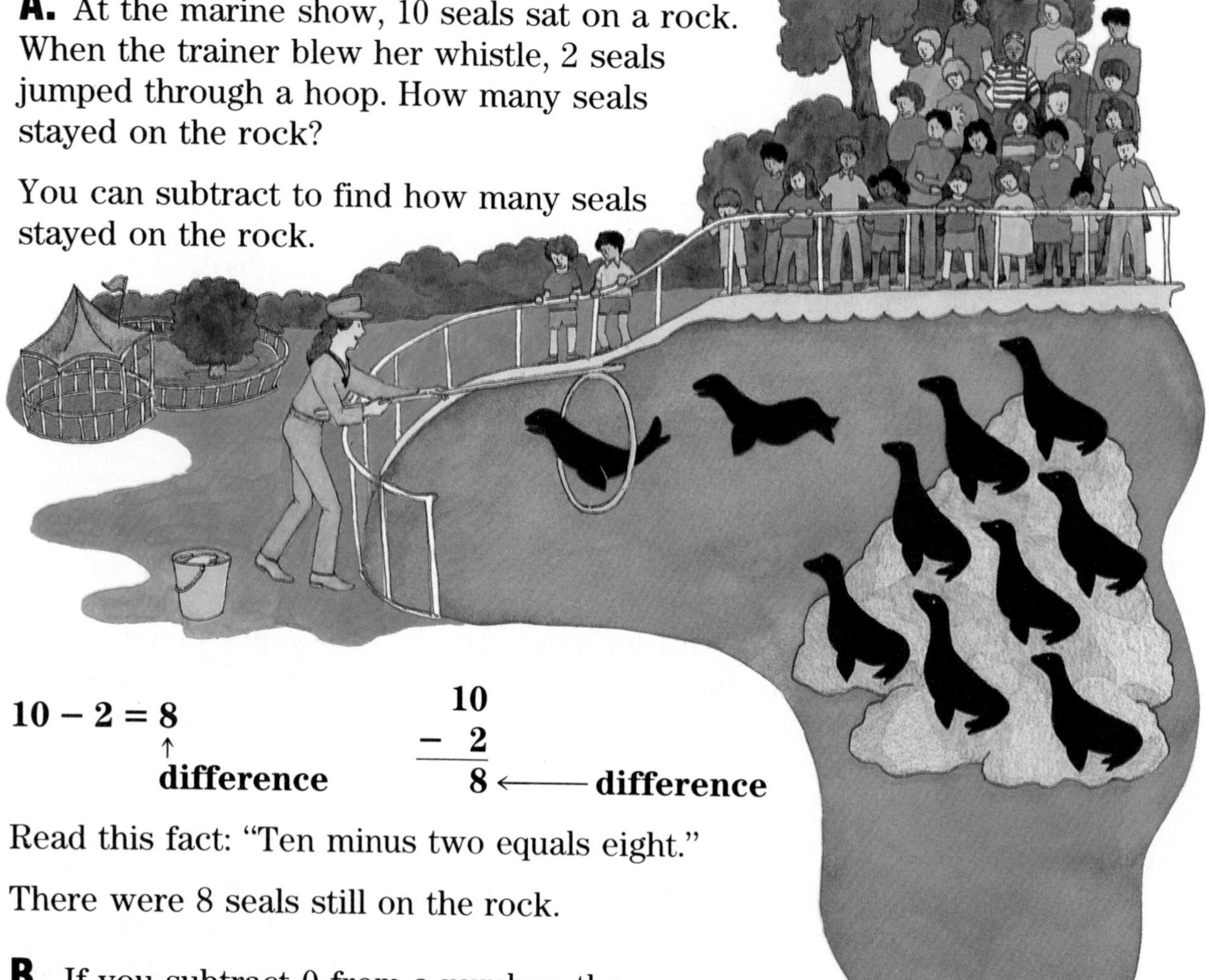

$10 - 2 = 8$ ← difference

$$\begin{array}{r} 10 \\ -\ 2 \\ \hline 8 \end{array}$$ ← difference

Read this fact: "Ten minus two equals eight."

There were 8 seals still on the rock.

B. If you subtract 0 from a number, the difference equals that number.

$9 - 0 = 9$

Subtract.

1. $\begin{array}{r} 6 \\ -2 \\ \hline \end{array}$
2. $\begin{array}{r} 8 \\ -3 \\ \hline \end{array}$
3. $\begin{array}{r} 2 \\ -1 \\ \hline \end{array}$
4. $\begin{array}{r} 3 \\ -0 \\ \hline \end{array}$
5. $\begin{array}{r} 10 \\ -\ 3 \\ \hline \end{array}$
6. $\begin{array}{r} 5 \\ -4 \\ \hline \end{array}$
7. $\begin{array}{r} 7 \\ -2 \\ \hline \end{array}$

8. $6 - 5 =$ ____
9. $5 - 0 =$ ____
10. $4 - 1 =$ ____
11. $7 - 4 =$ ____
12. $4 - 4 =$ ____
13. $8 - 5 =$ ____
14. $9 - 8 =$ ____
15. $6 - 2 =$ ____

Subtract.

16. $9 - 2$ 17. $3 - 1$ 18. $7 - 0$ 19. $9 - 3$ 20. $10 - 7$ 21. $9 - 5$ 22. $6 - 3$

23. $8 - 2$ 24. $5 - 2$ 25. $10 - 4$ 26. $8 - 5$ 27. $7 - 5$ 28. $10 - 5$ 29. $8 - 4$

30. $9 - 0$ 31. $7 - 6$ 32. $10 - 3$ 33. $6 - 4$ 34. $8 - 6$ 35. $9 - 6$ 36. $7 - 3$

37. $4 - 3 =$ ■ 38. $9 - 7 =$ ■ 39. $10 - 8 =$ ■ 40. $9 - 1 =$ ■

41. $7 - 2 =$ ■ 42. $10 - 6 =$ ■ 43. $5 - 5 =$ ■ 44. $5 - 4 =$ ■

45. $9 - 4 =$ ■ 46. $10 - 9 =$ ■ 47. $8 - 7 =$ ■ 48. $5 - 0 =$ ■

Solve.

49. There are 6 balls on a rock. A seal knocks 3 of them into the water. How many balls are still on the rock?

50. There are 5 seals playing in a band. They are joined by 3 more seals. How many seals are there now?

51. A trainer has 10 fish. She feeds some to the seals. Make up a word problem about the fish.

★52. Sally the seal had 6 rings on her neck. She tossed 3 to Sammy the seal. Then she tossed him 1 more. How many rings were left on her neck?

CHALLENGE

Complete the number sentences. Write + or − for ●.

1. 3 ● 7 = 10 2. 8 ● 4 = 4 3. 11 ● 6 = 5 4. 9 ● 8 = 17

5. 8 ● 5 = 13 6. 16 ● 8 = 8 7. 15 ● 9 = 6 8. 2 ● 9 = 11

Subtraction Facts to 18

A. There were 12 parrots on the stage. In the show 5 of them rode bicycles. How many parrots did not ride bicycles?

You can subtract to find how many parrots did not ride bicycles.

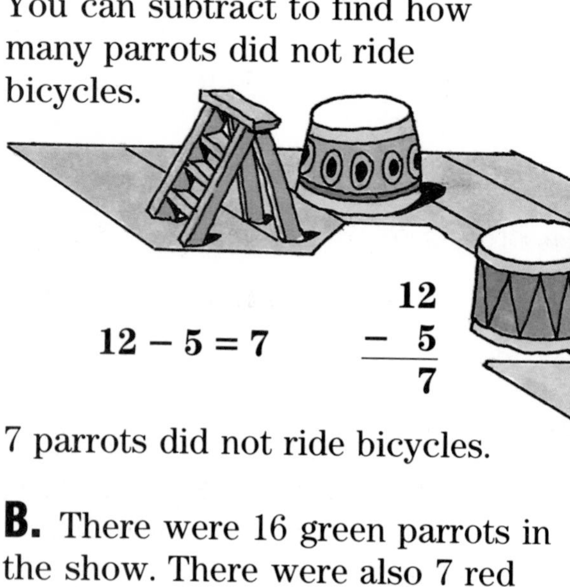

12 − 5 = 7

$$\begin{array}{r} \mathbf{12} \\ \mathbf{-\ 5} \\ \hline \mathbf{7} \end{array}$$

7 parrots did not ride bicycles.

B. There were 16 green parrots in the show. There were also 7 red parrots. How many more green parrots than red parrots were in the show?

You can subtract to compare green parrots to red parrots.

16 − 7 = 9

There were 9 more green parrots than red parrots.

Subtract.

1. $\begin{array}{r} 11 \\ -\ 8 \\ \hline \end{array}$ **2.** $\begin{array}{r} 12 \\ -\ 7 \\ \hline \end{array}$ **3.** $\begin{array}{r} 14 \\ -\ 5 \\ \hline \end{array}$ **4.** $\begin{array}{r} 13 \\ -\ 5 \\ \hline \end{array}$ **5.** $\begin{array}{r} 16 \\ -\ 9 \\ \hline \end{array}$ **6.** $\begin{array}{r} 14 \\ -\ 6 \\ \hline \end{array}$

7. 12 − 6 = ____ **8.** 14 − 8 = ____ **9.** 11 − 4 = ____ **10.** 13 − 7 = ____

11. 12 − 4 = ____ **12.** 16 − 9 = ____ **13.** 15 − 8 = ____ **14.** 17 − 9 = ____

Subtract.

15. $13 - 6$ 16. $13 - 4$ 17. $17 - 8$ 18. $15 - 8$ 19. $18 - 9$ 20. $11 - 9$

21. $16 - 8$ 22. $13 - 5$ 23. $13 - 8$ 24. $14 - 9$ 25. $15 - 7$ 26. $11 - 2$

27. $12 - 3$ 28. $11 - 4$ 29. $11 - 6$ 30. $13 - 9$ 31. $11 - 7$ 32. $14 - 7$

33. $11 - 3 =$ ____ 34. $15 - 9 =$ ____ 35. $11 - 5 =$ ____ 36. $16 - 7 =$ ____

37. $15 - 6 =$ ____ 38. $12 - 9 =$ ____ 39. $12 - 8 =$ ____ 40. $14 - 6 =$ ____

Solve.

41. The trainer had 12 buttons on his shirt. The parrot took 7 of them. How many buttons were left on the trainer's shirt?

42. The parrot knew 9 words. The trainer taught the parrot 7 new words. How many words did the parrot know?

43. A concert was given by 11 parrots. There were 5 parrots playing flute. The others played bells. How many parrots played bells?

44. There were 14 parrots in the first show. Then 6 of them flew away before the second show began. How many parrots were left?

MIDCHAPTER REVIEW

Add or subtract.

1. $16 - 8$ 2. $9 + 7$ 3. $14 - 5$ 4. $7 + 4$ 5. $15 - 8$ 6. $12 - 7$

7. $7 + 6 =$ ____ 8. $9 + 9 =$ ____ 9. $11 - 4 =$ ____ 10. $13 - 8 =$ ____

More Practice, page 405

PROBLEM SOLVING

Looking for a Pattern

Sometimes, you are given a group of numbers, shapes, or letters and asked to tell what comes next. You can often tell by looking at the pattern that is already there.

What should the last 2 elephants in the row be juggling? What would a fifteenth elephant be juggling?

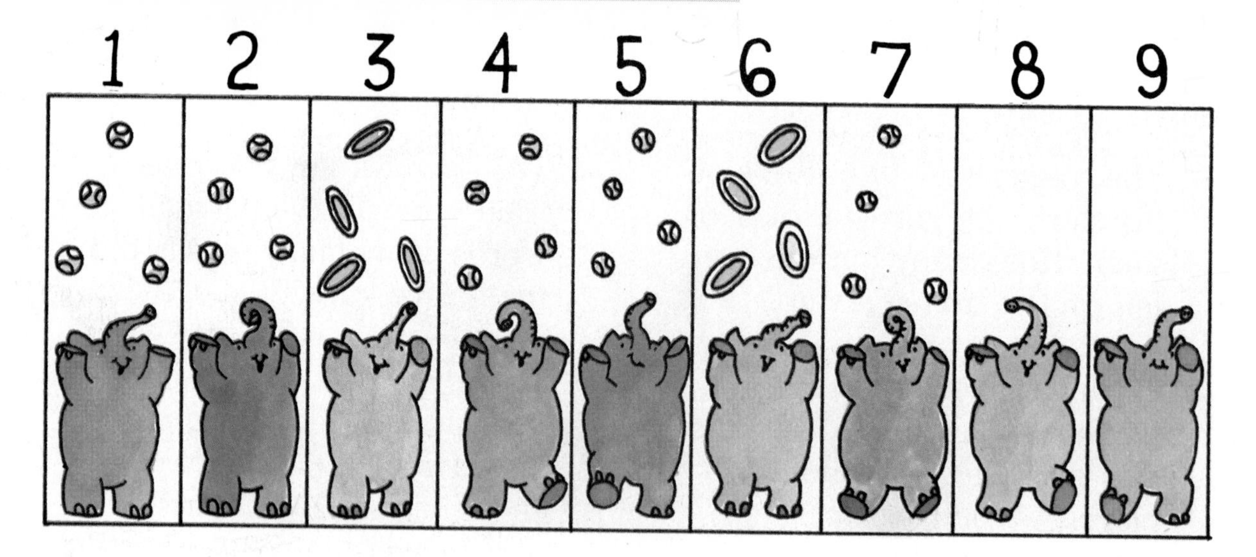

The first elephant is juggling baseballs. So is the second elephant. The third one is juggling plates. The fourth and fifth elephants are juggling baseballs. The sixth one juggles plates. The seventh elephant juggles baseballs.

The pattern is baseballs, baseballs, plates, baseballs, baseballs, plates, and so on.

The eighth elephant should be juggling baseballs. The ninth elephant should be juggling plates. If there were a fifteenth elephant, it would be juggling plates.

Find the pattern to solve.

1. The animals line up for the opening parade. How should the last 3 animals line up?

2. What kind of animal will the sixteenth animal in the row be?

3. The poodles do tricks on these stands. What should the next 2 stands look like?

? ?

★4. Some animals line up for the closing parade in this order: a goose, an ostrich, a lion, a swan, a duck, an alligator. In what order would a chicken, a turtle, and another duck join the end of this line? HINT: Count each animal's legs.

Fact Families

You can write four true number sentences that use the numbers 6, 3, and 9.

$6 + 3 = 9$ $\quad$ $3 + 6 = 9$

$9 - 3 = 6$ $\quad$ $9 - 6 = 3$

These four number sentences are called a **family of facts.**

Complete each family of facts.

1. (4 7 3)

$4 + 3 =$ 7
$3 + 4 =$ 7
$7 - 4 =$ 3
$7 - 3 =$ 4

2. (9 14 5)

$14 - 5 =$ 9
$14 - 9 =$ 5
$9 + 5 =$ 14
$5 + 9 =$ 14

3. (5 13 8)

$13 - 5 =$ 8
$13 - 8 =$ 5
$5 + 8 =$ 13
$8 + 5 =$ 13

4. (9 17 8)

$17 - 9 =$ 8
$17 - 8 =$ 9
$9 + 8 =$ 17
$8 + 9 =$ 17

5. (6 10 4)

$6 + 4 =$ 10
$4 + 6 =$ 10
$10 - 4 =$ 6
$10 - 6 =$ 4

6. (7 9 2)

$7 + 2 =$ 9
$2 + 7 =$ 9
$9 - 7 =$ 2
$9 - 2 =$ 7

7. (18 9)

$9 + 9 =$ 18
$18 - 9 =$ 9

8. (10 5)

$5 + 5 =$ 10
$10 - 5 =$ 5

9. (16 8)

$16 - 8 =$ 8
$8 + 8 =$ 16

10. (8 4)

$4 + 4 =$ 8
$8 - 4 =$ 4

11. (12 6)

$12 - 6 =$ 6
$6 + 6 =$ 12

12. (14 7)

$14 - 7 =$ 7
$7 + 7 =$ 14

Complete each family of facts.

13. (11 3 8)

$3 + 8 =$ ___

$8 +$ ___ $= 11$

$11 - 3 =$ ___

___ $- 8 = 3$

14. (2 5 7)

$2 +$ ___ $= 7$

$5 + 2 =$ ___

___ $- 2 = 5$

$7 - 5 =$ ___

15. (12 8 4)

___ $- 4 = 8$

$12 - 8 =$ ___

$4 +$ ___ $= 12$

$8 + 4 =$ ___

16. (11 4 7)

___ $+ 4 = 11$

$4 + 7 =$ ___

$11 -$ ___ $= 7$

$11 - 7 =$ ___

17. (11 6 5)

$6 + 5 =$ ___

$5 +$ ___ $= 11$

___ $- 6 = 5$

$11 - 5 =$ ___

18. (13 4 9)

$13 - 4 =$ ___

___ $- 9 = 4$

$9 + 4 =$ ___

$4 +$ ___ $= 13$

19. $3 + 2 =$ ___

$2 + 3 =$ ___

$5 - 3 =$ ___

___ $-$ ___ $=$ ___

20. $8 + 7 =$ ___

$7 + 8 =$ ___

$15 - 7 =$ ___

___ $-$ ___ $=$ ___

21. $11 - 9 =$ ___

$11 - 2 =$ ___

$9 + 2 =$ ___

___ $+$ ___ $=$ ___

Write the family of facts for each set of numbers.

22. 1, 8, 7 **23.** 6, 0, 6 **24.** 3, 10, 7 **25.** 5, 9, 4

26. 5, 8, 3 **27.** 9, 3, 12 **28.** 16, 7, 9 **29.** 8, 13, 5

CHALLENGE

Name the missing number in each fact family.

1. 14 — 9 — ■ ___ **2.** 7 — 11 — ■ ___ **3.** 17 — 9 — ■ ___

Missing Addends

Harry saw 3 chimps do trampoline tricks. There were 9 chimps in all. How many chimps were hiding behind the curtain?

You can find how many chimps were hiding behind the curtain by finding the missing addend.

Think: $3 + \blacksquare = 9$.
You know that $9 - 3 = 6$.
So, $3 + 6 = 9$.
There were 6 chimps hiding behind the curtain.

Find the missing addend.

1. 6 + ___ = 12	**2.** ___ + 4 = 12	**3.** 5 + ___ = 13	**4.** 6 + ___ = 9
5. ___ + 6 = 15	**6.** 8 + ___ = 15	**7.** ___ + 3 = 11	**8.** 9 + ___ = 14
9. 3 + ___ = 8	**10.** ___ + 4 = 10	**11.** 5 + ___ = 12	**12.** ___ + 9 = 18
13. 7 + ___ = 12	**14.** ___ + 3 = 9	**15.** 3 + ___ = 5	**16.** 6 + ___ = 13
17. ___ + 4 = 6	**18.** 8 + ___ = 14	**19.** ___ + 8 = 13	**20.** ___ + 8 = 17
21. 3 + ___ = 9	**22.** 5 + ___ = 14	**23.** ___ + 8 = 16	**24.** ___ + 5 = 11
25. 2 + ___ = 9	**26.** ___ + 4 = 8	**27.** 8 + ___ = 13	**28.** ___ + 7 = 15

Find the missing addend.

29. 7 + ___ = 9	**30.** 3 + ___ = 6	**31.** 5 + ___ = 5	**32.** 2 + ___ = 8
33. ___ + 6 = 7	**34.** ___ + 2 = 4	**35.** ___ + 3 = 3	**36.** ___ + 7 = 10
37. 9 + ___ = 15	**38.** 3 + ___ = 7	**39.** 8 + ___ = 8	**40.** 7 + ___ = 16
41. ___ + 9 = 11	**42.** ___ + 5 = 14	**43.** ___ + 7 = 13	**44.** ___ + 9 = 18
45. 3 + ___ = 5	**46.** ___ + 7 = 12	**47.** 9 + ___ = 9	**48.** 1 + ___ = 6
49. ___ + 7 = 14	**50.** 4 + ___ = 8	**51.** ___ + 7 = 11	**52.** ___ + 6 = 6
53. 2 + ___ = 7	**54.** 9 + ___ = 11	**55.** ___ + 8 = 15	**56.** 8 + ___ = 12

Solve.

57. The chimp threw the basketball through the hoop 7 times. Then he missed a few tries. He tried 12 times in all. How many times did he miss?

58. A chimp had 7 rings on his arm. The trainer threw some more rings to him. When he started to juggle, the chimp had 13 rings. How many rings did the trainer throw to him?

59. In the morning, there were 8 chimps in the tent. Then the trainer brought in some more. By lunch time, there were 15 chimps in the tent. How many chimps did the trainer bring in?

60. The chimp collected 6 toys from the tent. Then he collected some more toys from outside. He had 15 toys when he was done. How many toys did the chimp collect from outside?

FOCUS: MENTAL MATH

Tens are easy to find mentally. Find the missing numbers to make tens.

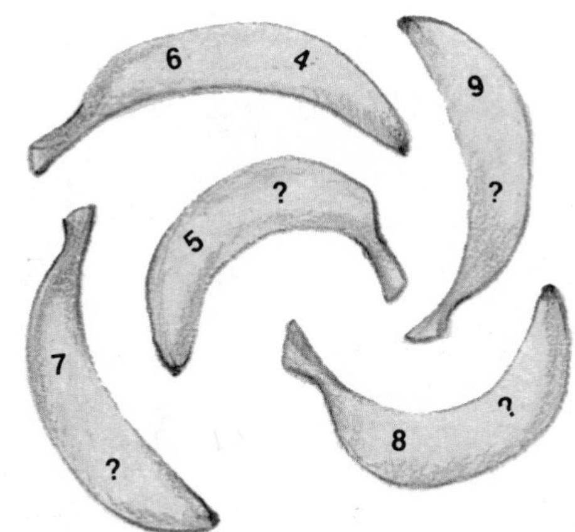

Three and Four Addends

A. A television ad for Yum-Yum Cat Food shows kittens at mealtime. There are 6 kittens eating from one bowl and 3 more eating from another bowl. Another kitten is not eating. How many kittens are there altogether?

You can add to find how many kittens there are.

You can change the grouping of the addends. The sum is always the same.

$6 + 3 + 1 = \square$		$6 + 3 + 1 = \square$
$(6 + 3) + 1 = \square$	OR	$6 + (3 + 1) = \square$
$9 + 1 = 10$		$6 + 4 = 10$

There are 10 kittens in all.

B. Grouping tens makes addition easier.

$$\begin{array}{r} 3 \\ 7 \\ +4 \\ \hline 14 \end{array} \quad \begin{array}{r} 5 \\ 2 \\ +5 \\ \hline 12 \end{array} \quad \begin{array}{r} 6 \\ 4 \\ 3 \\ +2 \\ \hline 15 \end{array} \quad \begin{array}{r} 2 \\ 6 \\ 1 \\ +8 \\ \hline 17 \end{array}$$

Add.

1. $\begin{array}{r} 3 \\ 4 \\ +1 \\ \hline \end{array}$ **2.** $\begin{array}{r} 4 \\ 5 \\ +1 \\ \hline \end{array}$ **3.** $\begin{array}{r} 2 \\ 1 \\ +2 \\ \hline \end{array}$ **4.** $\begin{array}{r} 4 \\ 2 \\ +3 \\ \hline \end{array}$ **5.** $\begin{array}{r} 6 \\ 3 \\ +2 \\ \hline \end{array}$ **6.** $\begin{array}{r} 2 \\ 6 \\ +1 \\ \hline \end{array}$ **7.** $\begin{array}{r} 4 \\ 1 \\ +4 \\ \hline \end{array}$

Add.

8.	9.	10.	11.	12.	13.	14.
3	5	3	1	3	1	1
2	1	1	3	3	2	7
1	2	4	4	3	4	1
+ 2	+ 2	+ 2	+ 2	+ 3	+ 2	+ 1

15. $2 + 7 + 2 =$ ___ 16. $6 + 2 + 1 =$ ___ 17. $5 + 1 + 1 =$ ___

18. $5 + 3 + 1 + 1 =$ ___ 19. $2 + 3 + 4 + 2 =$ ___ 20. $4 + 1 + 1 + 4 =$ ___

Group tens to help you find the sum.

21.	22.	23.	24.	25.	26.	27.
7	4	5	4	8	3	2
1	3	5	4	2	5	2
+ 3	+ 6	+ 1	6	2	1	6
			+ 1	+ 5	+ 5	+ 8

28. $3 + 1 + 7 =$ ___ 29. $6 + 3 + 7 =$ ___ 30. $3 + 5 + 2 + 8 =$ ___

Solve.

31. At first, 8 dogs are eating Brand X dog food. Then 6 switch to Zowie brand dog food. How many dogs are still eating Brand X?

32. An ad for a movie will show 15 animals. Choose 3 groups of animals. Write an addition problem.

FOCUS: REASONING

Find the number that completes each pattern.

1. 3, 5, 7, 9, ___
2. 1, 5, 9, ___
3. 11, 8, 5, ___
4. ___, 4, 6, 8
5. ___, 9, 6, 3
6. 12, ___, 4, 0
7. 1, 2, 4, 7, ___

★8. 15, 10, ___, 3, 1, 0

PROBLEM SOLVING

Choosing the Operation

Here are two hints to help you decide if you can add or subtract to solve a problem.

A. The Do-Si-Dogs is a square-dance team of 8 people and 8 dogs. How many dancers are there in all?

You know		You want to find	You can
how many in one group.	how many in another group.	how many in all.	ADD.

8	+	8	=	16
people		dogs		dancers

There are 16 dancers in all.

B. Each of the 16 dancers in the Do-Si-Dogs has 1 balloon. During the show, 7 balloons float up to the ceiling. How many balloons are left?

You know		You want to find	You can
how many in all.	how many taken away.	how many are left.	SUBTRACT.

16	–	7	=	9
balloons		float away		left

There are 9 balloons left.

Which operation would you choose to solve each problem? Write the letter of the correct answer.

1. One night, the Do-Si-Dogs performed 4 dances. The next night, they performed 6 dances. How many dances did they perform in the two nights?

   a. add
   b. subtract

2. There are 16 dancers in the Do-Si-Dogs. In one dance, 6 of the dancers leave the group and sit down. How many are left?

   a. add
   b. subtract

Solve.

3. Of the 8 dogs in the Do-Si-Dogs, 5 needed new collars and 4 needed new leashes. How many new things are needed?

4. The Do-Si-Dogs have 12 hats altogether. In one show, 7 dancers wear their hats. How many hats are not worn?

5. If 5 out of the 8 people in the Do-Si-Dogs need new costumes, how many of them do not need new costumes?

6. In one of the dances, 3 of the dogs wear blue ribbons and 3 of the dogs wear red ribbons. How many dogs wear ribbons?

7. Smokey the Bear receives 2 sacks of mail every week. His friends take 1 sack of mail to answer. How many sacks of mail are left?

8. In one week, Rin Tin Tin receives 4 sacks of mail. If 1 sack is answered on the first day, how many sacks remain?

★9. If Rin Tin Tin and Smokey the Bear shared the same mailbox, how many sacks of mail would they find in it every week?

★10. Before Lassie became a star, her name was Pal, and she chewed things. If she chewed 2 shoes from one pair and 1 shoe from another pair, how many pairs of shoes did she ruin?

CALCULATOR

You can use a calculator to explore the grouping property for addition. The numbers in parentheses ( ) are added first. Clear the display before starting a new calculation.

Example	Press these Keys	Display Shows
(23 + 42) + 14	2 3 + 4 2 =	65
	6 5 + 1 4 =	79
23 + (42 + 14)	4 2 + 1 4 =	56
	2 3 + 5 6 =	79

So, (23 + 42) + 14 = 79
23 + (42 + 14) = 79 The sums are equal.

Use your calculator to find the sum. Write the keys that you press and what the display shows. Are each pair of sums equal?

Exercise	Keys you Press	Display Shows
1. (15 + 51) + 22	1 5 + 5 1 =	66
	6 6 ______	______
15 + (51 + 22)	______	88
	______	79

2. (17 + 41) + 21 79
17 + (41 + 21)

3. (34 + 31) + 28 = 93
34 + (31 + 28)

4. (45 + 29) + 8 82
45 + (29 + 8)

GROUP PROJECT

The Fur and Fin Talent Show

The problem: Your class is having a pet talent show. The pets can be real or imaginary. Draw a picture of your pet, and be ready to tell a story about your pet's talents. Then the show can begin.

Here is the scorecard.

funny: add 2	scary: subtract 1
smart: add 2	noisy: subtract 1
brave: add 2	lazy: subtract 1
helpful: add 2	too big: subtract 1
friendly: add 2	too small: subtract 1

Here is how it works.
Everyone will take turns at being a judge. The first judge will decide whether the pet is funny. If the judge says yes, add 2.

The second judge will decide whether the pet is scary. If the judge says yes, subtract 1.

Everyone can keep score. The pet with the highest score wins. Good luck!

CHAPTER TEST

Add or subtract. (pages 2, 4, 8, and 10)

1. $\begin{array}{r} 6 \\ +8 \\ \hline \end{array}$	**2.** $\begin{array}{r} 7 \\ +2 \\ \hline \end{array}$	**3.** $\begin{array}{r} 8 \\ +8 \\ \hline \end{array}$	**4.** $\begin{array}{r} 10 \\ -5 \\ \hline \end{array}$	**5.** $\begin{array}{r} 12 \\ -6 \\ \hline \end{array}$
6. $\begin{array}{r} 9 \\ -0 \\ \hline \end{array}$	**7.** $\begin{array}{r} 9 \\ -6 \\ \hline \end{array}$	**8.** $\begin{array}{r} 8 \\ +2 \\ \hline \end{array}$	**9.** $\begin{array}{r} 18 \\ -9 \\ \hline \end{array}$	**10.** $\begin{array}{r} 2 \\ +4 \\ \hline \end{array}$

11. $8 + 9 =$ ___ **12.** $9 + 7 =$ ___ **13.** $9 - 5 =$ ___

14. $6 + 7 =$ ___ **15.** $12 - 8 =$ ___ **16.** $7 + 3 =$ ___

17. $7 + 7 =$ ___ **18.** $16 - 8 =$ ___ **19.** $8 - 4 =$ ___

Complete each family of facts. (page 14)

20. (3 9 6)
$3 + 6 =$ ___
$6 + 3 =$ ___
$9 - 3 =$ ___
$9 - 6 =$ ___

21. (16 7 9)
$9 +$ ___ $= 16$
$7 + 9 =$ ___
___ $- 7 = 9$
$16 -$ ___ $= 7$

22. (6 3)
$3 + 3 =$ ___
$6 - 3 =$ ___

23. (8 16)
$8 + 8 =$ ___
___ $- 8 = 8$

Solve. (page 20)

24. There are 5 eggs in a basket. Ellen adds 7 more eggs to the basket. How many eggs are there now?

25. Jeb has 17 balloons. He gives away 8 of them. How many balloons does Jeb have left?

26. Blake took 11 books from the library. Later he returned 3 books. How many books does he still have?

27. Bar-X Ranch has 7 white horses. Then they buy 7 black horses. How many horses do they have now?

Complete and write the missing number sentence. (page 14)

28. $5 + 6 =$ ___
$6 +$ ___ $= 11$
$11 - 5 =$ ___
___ $-$ ___ $=$ ___

29. $9 + 8 =$ ___
$8 +$ ___ $= 17$
$17 - 9 =$ ___
___ $-$ ___ $=$ ___

Find the missing addends. (page 16)

30. $4 +$ ___ $= 8$ **31.** $8 +$ ___ $= 12$ **32.** $6 +$ ___ $= 11$ **33.** ___ $+ 4 = 9$

34. $7 +$ ___ $= 7$ **35.** ___ $+ 6 = 10$ **36.** $6 +$ ___ $= 13$ **37.** ___ $+ 9 = 18$

38. ___ $+ 8 = 14$ **39.** $3 +$ ___ $= 12$ **40.** ___ $+ 9 = 16$ **41.** $7 +$ ___ $= 15$

Add. (page 18)

42. $3 + 4 + 2$

43. $8 + 1 + 6 + 1$

44. $3 + 5 + 4$

45. $6 + 7 + 1 + 4$

46. $9 + 1 + 1$

47. $3 + 2 + 8 =$ ___

48. $5 + 3 + 5 + 2 =$ ___

49. $2 + 1 + 3 + 1 =$ ___

50. $1 + 3 + 5 + 0 =$ ___

BONUS

Complete each number sentence. Write + or − for ●. The first one is done for you.

1. $7 \ominus 2 \oplus 5 = 10$

2. 6 ● 2 ● 4 = 8

3. 9 ● 5 ● 3 = 7

4. 14 ● 9 ● 3 = 8

5. 6 ● 7 ● 6 = 7

6. 8 ● 3 ● 6 = 5

7. 11 ● 8 ● 1 = 2

8. 8 ● 8 ● 9 = 7

9. 7 ● 6 ● 9 = 4

RETEACHING

You can use what you know about addition and subtraction to find missing addends.

$3 + ? = 7$

$3 + \underline{\quad} = 7$

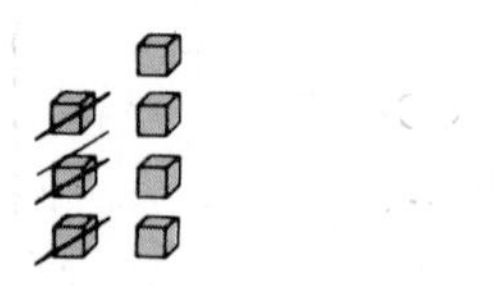

You know that $7 - 3 = 4$.

So, $3 + 4 = 7$.

Find the missing addend.

1. $5 + \underline{\quad} = 9$
2. $\underline{\quad} + 7 = 15$
3. $4 + \underline{\quad} = 12$
4. $7 + \underline{\quad} = 13$
5. $9 + \underline{\quad} = 12$
6. $\underline{\quad} + 8 = 16$
7. $3 + \underline{\quad} = 9$
8. $\underline{\quad} + 4 = 11$
9. $\underline{\quad} + 4 = 10$
10. $8 + \underline{\quad} = 13$
11. $\underline{\quad} + 3 = 11$
12. $2 + \underline{\quad} = 8$
13. $\underline{\quad} + 5 = 8$
14. $\underline{\quad} + 7 = 16$
15. $6 + \underline{\quad} = 14$
16. $6 + \underline{\quad} = 12$
17. $7 + \underline{\quad} = 14$
18. $4 + \underline{\quad} = 6$
19. $\underline{\quad} + 5 = 7$
20. $\underline{\quad} + 8 = 17$
21. $9 + \underline{\quad} = 13$
22. $2 + \underline{\quad} = 10$
23. $9 + \underline{\quad} = 14$
24. $9 + \underline{\quad} = 11$
25. $\underline{\quad} + 7 = 9$
26. $\underline{\quad} + 4 = 8$
27. $7 + \underline{\quad} = 12$
28. $3 + \underline{\quad} = 10$
29. $9 + \underline{\quad} = 15$
30. $5 + \underline{\quad} = 10$
31. $9 + \underline{\quad} = 18$
32. $8 + \underline{\quad} = 16$
33. $6 + \underline{\quad} = 11$

ENRICHMENT

Logic

The magician keeps 7 doves in his coat pocket and some rabbits in his top hat. He has a poodle that rides a tricycle and 3 goldfish that live in his bathtub. He has 15 animals in all.

Solve.

1. How many rabbits does the magician have?
2. How many more doves than rabbits are there?
3. How many more goldfish than poodles are there?
4. How many rabbits and poodles are there?
5. How many more doves than goldfish are there?
6. How many doves and rabbits are there?

The Red Tent Circus has 18 animals. There are some seals that juggle red and blue beach balls. There are 3 white horses. Each horse is ridden by a chimpanzee in a gold hat. Ramu the lion tamer has a lion named Tess. Tess likes to watch the 5 elephants squirt each other with their trunks.

Solve.

7. How many chimpanzees are there in the circus?
8. How many chimpanzees and horses are there?
9. If Tess and the elephants share a ring, how many animals are there in the ring?
10. If the chimpanzees and seals share a ring, how many animals are there in the rest of the tent?
11. How many seals are there in the circus?
12. Each seal has 2 beach balls. How many beach balls are there?

CUMULATIVE REVIEW

Write the letter of the correct answer.

1. $7 + 5 =$ 12

a. 2
b. 12
c. 75
d. not given

2. $9 + 3$ = 12

a. 3
b. 6
c. 9
d. not given

3. $14 - 6 =$ 8

a. 6
b. 8
c. 20
d. not given

4. $7 - 2$ = 5

a. 5
b. 9
c. 72
d. not given

5. Which sentence completes the family of facts?

$4 + 6 = 10$
$6 + 4 = 10$
$10 - 6 = 4$

a. $4 + 10 = 14$
b. $10 - 4 = 4$
c. $10 - 4 = 6$
d. not given

6. Which sentence belongs to the same family of facts as $14 - 7 = 7$?

a. $14 - 14 = 0$
b. $7 + 7 = 14$
c. $14 - 0 = 14$
d. not given

7. $5 +$ 3 $= 8$

a. 3
b. 4
c. 13
d. not given

8. 8 $+ 9 = 17$

a. 1
b. 8
c. 12
d. not given

9. $4 + 5 + 2$ = 11

a. 7
b. 11
c. 92
d. not given

10. $3 + 6 + 9 =$ 18

a. 9
b. 18
c. 360
d. not given

11. Eva has 6 glass birds and 7 glass fish. She places them on a shelf. How many glass animals are there on the shelf?

a. 1 animal
b. 13 animals
c. 42 animals
d. not given

12. Dan has 17 pumpkins. He sells 9 of them. How many pumpkins does he have left?

a. 6 pumpkins
b. 7 pumpkins
c. 8 pumpkins
d. not given

2 PLACE VALUE

Do you think there are more people, cows, cars, trees, or buildings on your street? How could you find this out?

Tens and Ones

Our numeration system is based on groups of ten.

10 ones = 1 ten

We use ten digits and place value to write numbers.

0 1 2 3 4 5 6 7 8 9

Examples:

1 ten 4 ones = 14
Read: fourteen.

4 tens 1 one = 41
Read: forty-one.

Checkpoint Write the letter of the correct answer.

Choose the correct number.

1.

a. 6 **b.** 15
c. 51 **d.** 105

2.

a. 5 **b.** 23
c. 32 **d.** 203

3. 3 tens 7 ones

a. 10 **b.** 30
c. 37 **d.** 73

4. ninety-one

a. 10 **b.** 19
c. 91 **d.** 901

Write the number.

1.

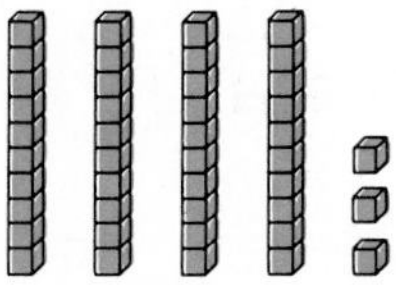

2.

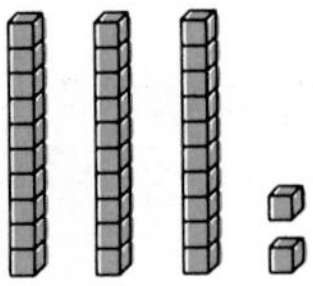

3.

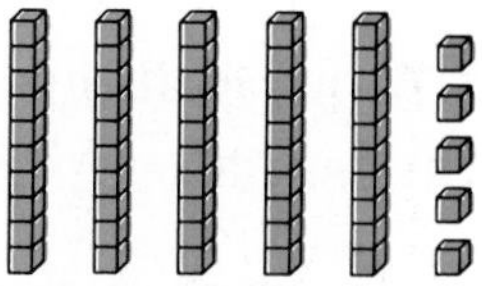

4.

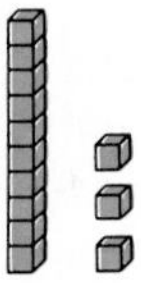

5.

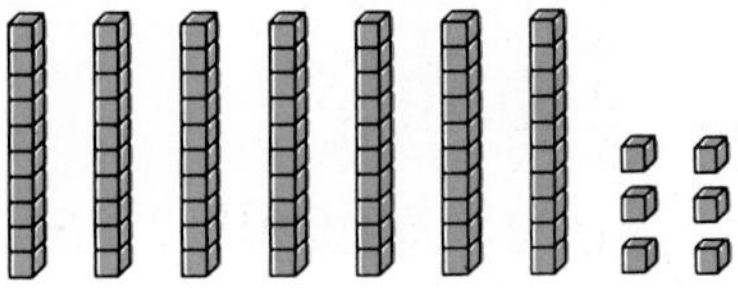

6.

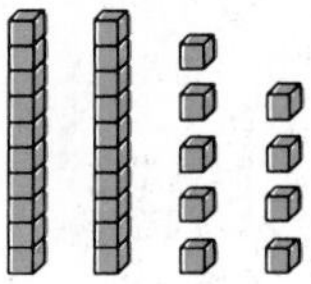

7. 4 tens 5 ones

8. 3 tens 0 ones

9. 6 tens 1 one

10. 2 tens 8 ones

11. 5 tens 6 ones

12. 4 tens 7 ones

13. fifty-six

14. seventy-two

15. thirty-one

16. ninety-eight

17. eighty-nine

18. forty-six

Write the digit that is in the ones place.

19. 15 20. 70 21. 55 22. 42 23. 68 24. 87 25. 18

26. 17 27. 59 28. 84 29. 44 30. 92 31. 37 32. 46

Write the digit that is in the tens place.

33. 62 34. 49 35. 81 36. 19 37. 43 38. 90 39. 74

40. 11 41. 76 42. 53 43. 98 44. 27 45. 83 46. 20

Solve.

47. San Francisco is built on 40 hills. How many tens and ones are there in the number 40?

48. Cable cars go up the hills. You can ride 10 miles on the cars. How many tens and ones are in the number 10?

49. In San Francisco, you can ride seventy-one miles on trains called BART. Write the number seventy-one.

★50. A train car has eighty seats. Write the number eighty. How many tens and ones are there?

Hundreds

A. In our numeration system, ten groups of ten equal one hundred.

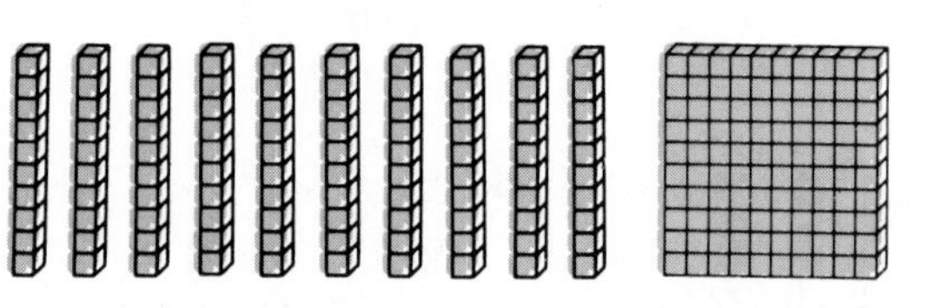

10 tens = 1 hundred

B. Numbers can be shown in many ways.

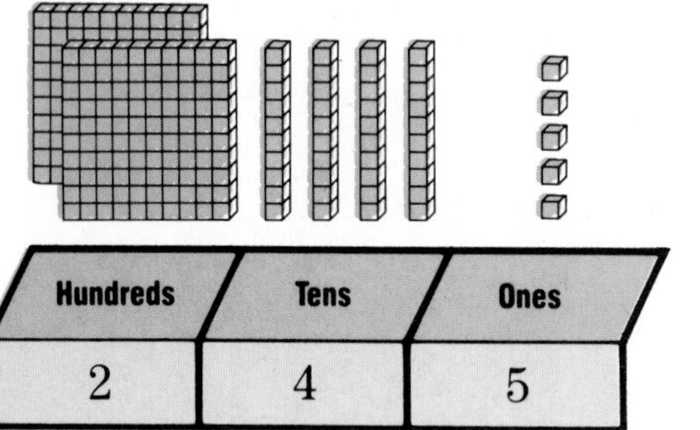

Hundreds	Tens	Ones
2	4	5

2 hundreds 4 tens 5 ones ← Expanded form

200 + 40 + 5 ← Expanded form

245 ← Standard form

two hundred forty-five

Another example:

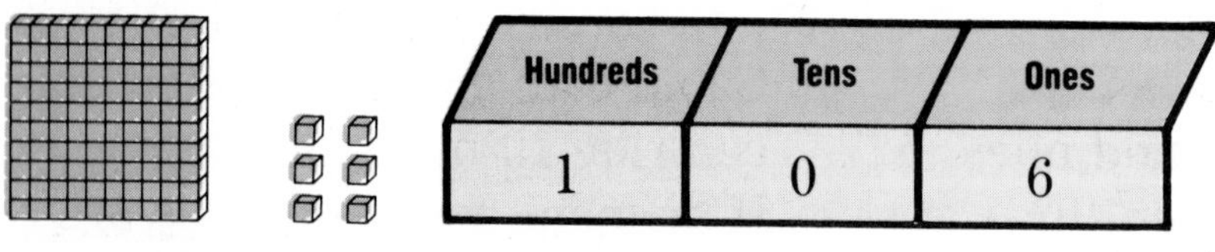

Hundreds	Tens	Ones
1	0	6

1 hundred 6 ones ← Expanded form

100 + 6 ← Expanded form

106 ← Standard form

one hundred six

Write the number.

1.

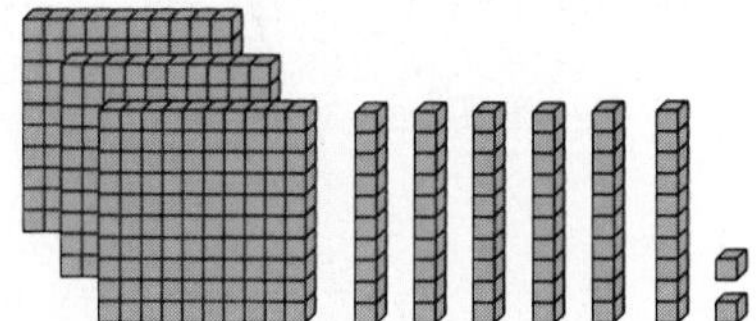

2.

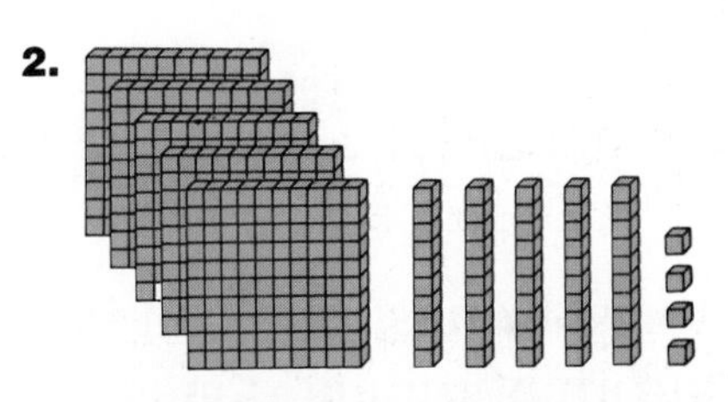

3.

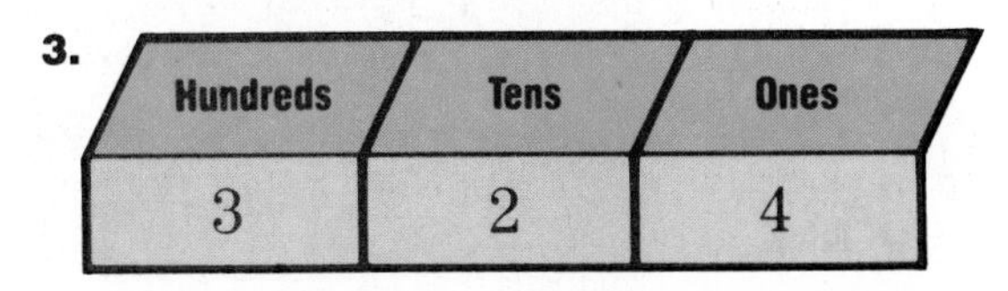

4.

Hundreds	Tens	Ones
4	3	7

5. 700 + 10 + 7 = ____ **6.** 200 + 40 + 1 = ____ **7.** 300 + 60 + 3 = ____

8. 800 + 8 = ____ **9.** 400 + 20 = ____ **10.** 500 + 30 + 6 = ____

11. 600 + 10 = ____ **12.** 100 + 30 + 4 = ____ **13.** 900 + 7 = ____

14. seven hundred twenty-nine **15.** five hundred eighty

16. nine hundred three **17.** three hundred eighty-two

Write the digit that is in the hundreds place.

18. 679 **19.** 357 **20.** 486 **21.** 267 **22.** 986

Write the digit that is in the tens place.

23. 854 **24.** 105 **25.** 546 **26.** 790 **27.** 662

Write the number.

28. When you walk across San Francisco's Golden Gate Bridge, you are two hundred twenty feet above the water.

29. Many people come to see the towers of the Golden Gate Bridge. These towers are seven hundred forty-six feet high.

FOCUS: MENTAL MATH

Study the chart; then cover it up.

What number is above 45?

What number is below 56?

30	31	32	33	34	35	36	37	38	39
40	41	42	43	44	45	46	47	48	49
50	51	52	53	54	55	56	57	58	59
60	61	62	63	64	65	66	67	68	69

PROBLEM SOLVING
Choosing/Writing a Sensible Question

Asking the right questions can help you organize information and make decisions.

Theo lives in San Francisco. He and his sister, Betsy, want to take their grandmother out to dinner. They must choose a restaurant. Which of these questions will help them choose?

- "How much money have we saved?"
  It is important to know this information. This amount is what they can afford to spend on dinner. They need enough money for their grandmother and themselves.
- "How much time should we spend eating dinner?"
  This information is not important unless they have plans for after dinner. Time won't change the cost of the dinner.
- "How much does each restaurant charge for dinner?"
  It is important to know how much each restaurant charges. They cannot go to a restaurant that charges more than they can afford.

Read each sentence or sentence group. For each, write two questions that Theo's family should answer before making a decision.

1. Theo's family and other relatives plan a party for Grandmother's birthday. The family must decide where to have the party.

2. The family needs to decide what food to serve at the party.

3. Some relatives are cooking food at their homes. They will bring it to the party.

4. Some relatives will bring pots, plates, and chairs.

5. Some out-of-town relatives will stay overnight in the city. Theo and his family have room in their home for 6 relatives. The family wants to help other relatives find places to stay.

6. All the relatives want to chip in to buy Grandmother a gift.

7. The family must decide who will make and put up decorations.

8. The family must decide who will take pictures at the party.

Thousands

A. Ten hundreds equal one thousand.

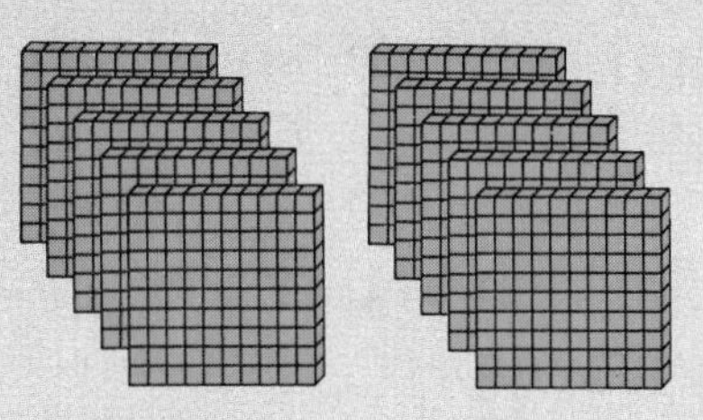
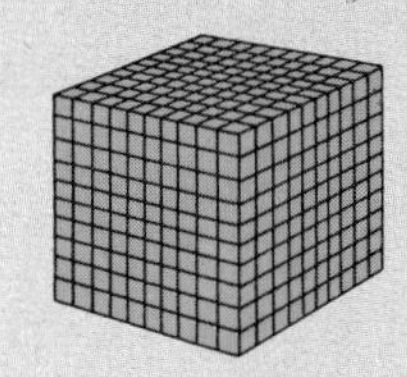

10 hundreds = **1 thousand**

B. Write a comma to separate thousands from hundreds.

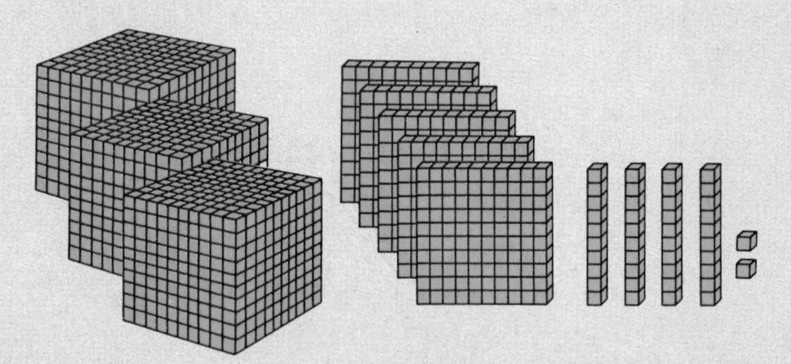

3,542

three thousand, five hundred forty-two

C. A chart shows the place value of each digit.

Thousands	Hundreds	Tens	Ones
6	4	3	8

6 is in the thousands place. 4 is in the hundreds place.
3 is in the tens place. 8 is in the ones place.

Checkpoint Write the letter of the correct answer.

1. In which place is the digit 6 in 9,681?
   - a. ones
   - b. tens
   - c. hundreds
   - d. thousands

2. two thousand, forty
   - a. 2000,40
   - b. 2,004
   - c. 2,040
   - d. 2,400

Write the number.

1.

2.

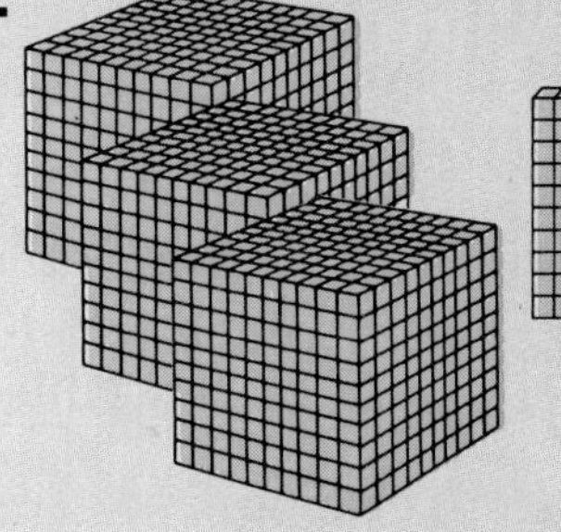

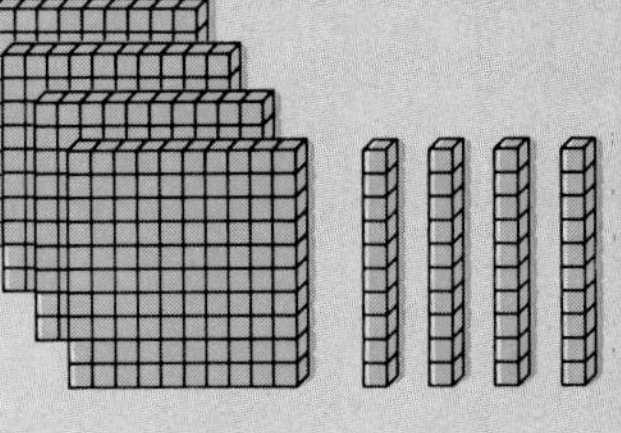

3.

Thousands	Hundreds	Tens	Ones
8	4	6	7

4.

Thousands	Hundreds	Tens	Ones
3	7	0	3

In which place is the digit 7 in each number?
Write *ones, tens, hundreds,* or *thousands.*

5. 4,670 **6.** 7,821 **7.** 3,765 **8.** 7,682

9. 5,678 **10.** 1,107 **11.** 8,872 **12.** 6,657

Use the picture to solve.

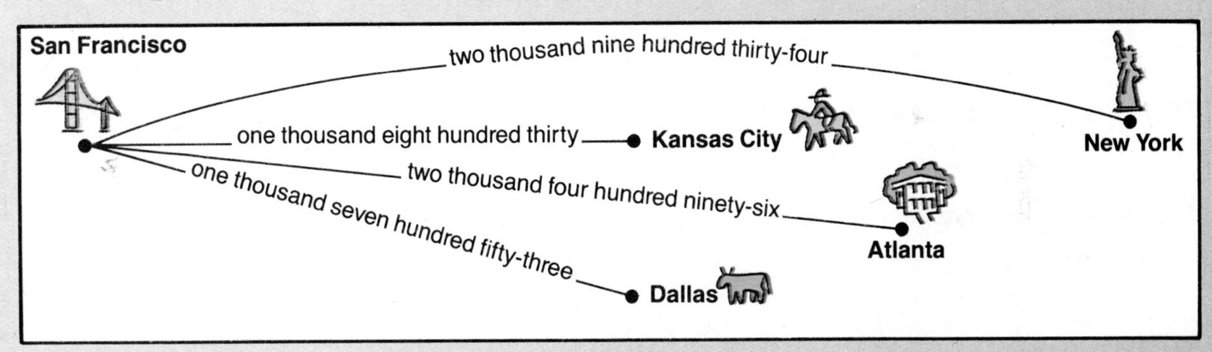

How many miles is it from San Francisco to each city? Write the number.

13. New York **14.** Kansas City **15.** Atlanta **16.** Dallas

ANOTHER LOOK

Add or subtract.

1. $5 + 7$ **2.** $9 - 5$ **3.** $15 - 7$ **4.** $8 + 7$ **5.** $15 - 6$ **6.** $8 + 9$

Order

A. At a town meeting, people who wish to speak are given a number. They are called to speak in the **order** of their numbers. If your number is 16, do you speak before or after the person who has number 17?

You can use the number line to find the order of numbers.

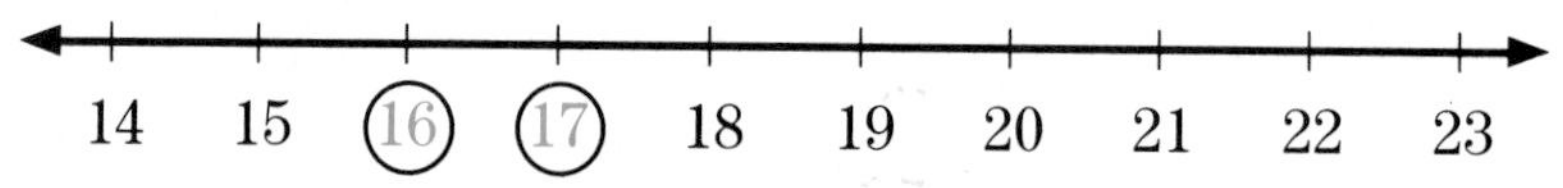

Number 16 comes *before* number 17.
Number 17 comes *after* number 16.

You will speak before the person holding number 17.

B. You can use the number line to find the number that comes *between* two other numbers.

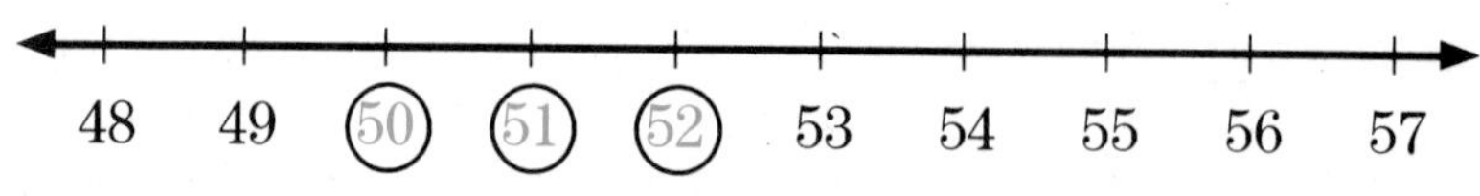

Number 51 comes between 50 and 52.

Checkpoint

Write the letter of the correct answer.

1. ___ comes just before 300.

a. 200 **b.** 301
c. 299 **d.** 400

2. ___ is between 119 and 121.

a. 110 **b.** 120
c. 118 **d.** 122

Write each missing number.

1. 7 8 9 10 ___ 12 13 14 15

2.

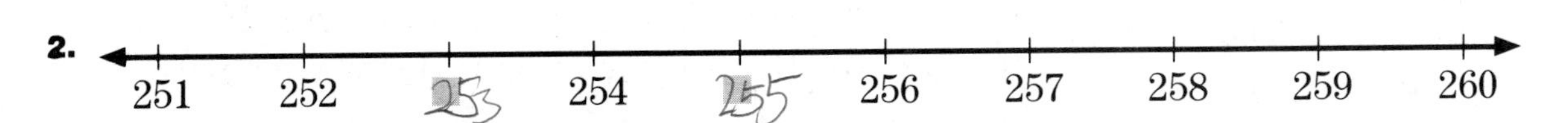

Write the number that comes just before.

3. 9	**4.** 7	**5.** 4	**6.** 3	**7.** 5
8. 27	**9.** 92	**10.** 74	**11.** 19	**12.** 88
13. 206	**14.** 730	**15.** 100	**16.** 429	**17.** 880

Write the number that comes just after.

18. 5	**19.** 8	**20.** 9	**21.** 2	**22.** 4
23. 64	**24.** 17	**25.** 99	**26.** 32	**27.** 43
28. 206	**29.** 721	**30.** 422	**31.** 699	**32.** 520

Write the number that is between.

33. 5, ___, 7	**34.** 1, ___, 3	**35.** 8, ___, 10	**36.** 6, ___, 8
37. 45, ___, 47	**38.** 69, ___, 71	**39.** 12, ___, 14	**40.** 90, ___, 92
41. 209, ___, 211	**42.** 744, ___, 746	**43.** 167, ___, 169	**44.** 934, ___, 936

Complete each sentence.
Write *more than* or *less than.*

1. There are ________ 20 students in my class.
2. There are ________ 200 students in my school.
3. There are ________ 2,000 students in my town.

PROBLEM SOLVING
Number Sense

Sometimes, you don't have to find the exact amount to solve a problem. Instead, you can estimate.

A. Ling is planning to visit her grandmother in San Francisco for the Chinese New Year in February. She is allowed to carry up to 10 pounds of luggage onto the plane. She has 3 presents weighing 7 pounds, 3 pounds, and 2 pounds. Can she carry them all onto the plane?

Without finding the exact sum, you can solve this problem. You know that 7 pounds + 3 pounds = 10 pounds. Ling has met the weight limit with only 2 presents. She will not be able to take all 3 presents onto the plane.

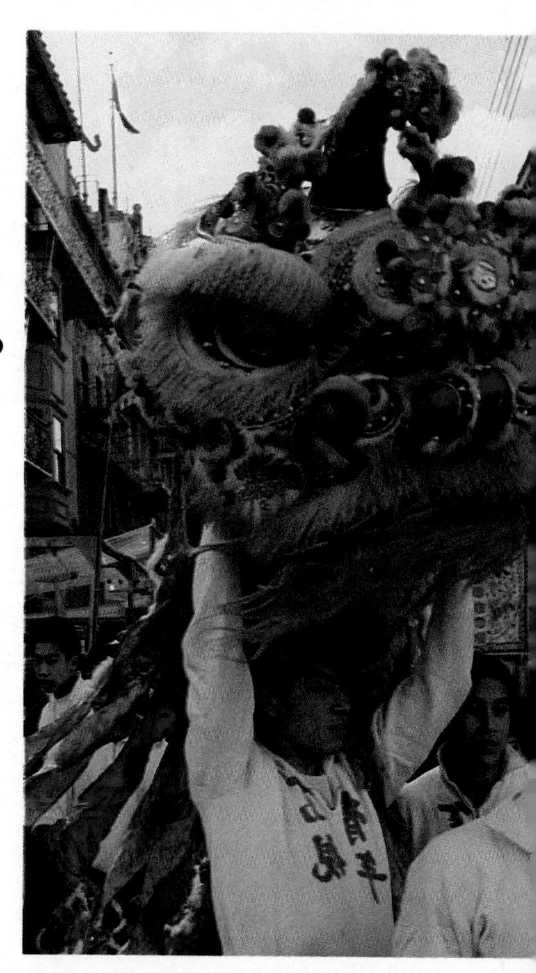

B. Ling buys picture postcards. She buys 9 cards that have pictures of the city, 6 cards that show animals, and 8 cards that have flower pictures. Does she have enough cards to send to 16 friends? Can you answer the question without using all 3 numbers?

You can add the 2 greatest amounts to get an answer.

$9 + 8 = 17 \qquad 17 > 16$

Ling has enough cards.

Write the letter of the best answer.

1. Ling's aunt came to San Francisco from San Diego. Ling came from Santa Barbara. Look at the map. San Diego is about 450 miles from San Francisco. How far is Santa Barbara from San Francisco?

   a. about 30 miles b. about 300 miles
   c. about 3,000 miles

2. The picture shows the Great Dragon at the Chinese New Year parade. The dragon on the left is about 50 feet long. How long is the Lion Dragon next to it?

   a. about 25 feet b. about 75 feet
   c. about 100 feet

Solve without finding the exact sum.

3. Ling has 2 pieces of paper with which to wrap 5 books that are the same size. One piece of paper wraps 3 books. Will she have enough paper with which to wrap all 5?

4. The last time Ling saw her cousin Chou, they were the same height. If Ling grew 2 inches and Chou grew 1 inch more than that, who is taller now?

5. Ling's grandmother wants to be sure there is enough rice for the New Year dinner. There will be 4 people from one family, 6 from another, and 9 from a third at the dinner. If each person eats more than 1 bowl, will 15 bowls of rice be enough?

Comparing and Ordering Numbers

A. The tallest tree on Earth grows near San Francisco. It is a redwood tree that is 370 feet high. A 36-story building is 368 feet high. Which is taller?

You can compare numbers to find which is taller. Begin to compare at the left.

Compare hundreds. The digits are the same.	Compare tens. 7 is greater than 6.
370	**370**
368	**368**
370 is greater than **368.**	**368** is less than **370.**
$370 > 368$	$368 < 370$

The arrow points to the lesser number.

The redwood tree is taller.

B. You can write numbers in order from the least to the greatest: 112, 89, 101, 97.

Line up the digits.	Compare to find the least number.	Compare the remaining numbers.	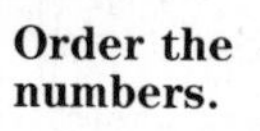 Order the numbers.
112	$89 < 112$	$101 < 112$	89
89	$89 < 101$	$97 < 101$	97
101	$89 < 97$	$97 < 112$	101
97	So, 89 is the least number.		112

To order from the greatest to the least, start with the greatest number.

Write $>$ or $<$ for ●.

1. 13 ● 31
2. 103 ● 96
3. 98 ● 11
4. 78 ● 70
5. 59 ● 95
6. 509 ● 95
7. 995 ● 905
8. 3,500 ● 9,050

Order the numbers from the least to the greatest.

9. 9, 8, 5, 4
10. 67, 78, 87, 86
11. 560; 650; 6,005; 600
12. 2,321; 3,211; 1,213
13. 8,909; 8,998; 9,008

Order the numbers from the greatest to the least.

14. 263, 362, 632
15. 496, 499, 604
16. 702, 207, 770
17. 3,462; 3,642; 3,264
18. 7,082; 7,280; 7,802
19. Look at the chart. The buildings are listed in alphabetical order. Make your own chart. Put the buildings on your chart in order from the tallest to the shortest.

SOME TALL BUILDINGS IN THE U.S.

Name of building	Height (in feet)
Empire State Building	1,250
Sears Tower	1,454
Transamerica Pyramid	853
United California Bank	858
World Trade Center	1,377

MIDCHAPTER REVIEW

Write the number.

1. 7 tens 6 ones
2. 3 hundreds 5 tens 2 ones
3. 9 tens 9 ones

In which place is the digit 6? Write *ones*, *tens*, *hundreds*, or *thousands*.

4. 3,467
5. 9,632
6. 6,501
★7. 2,696

More Practice, page 406

Even and Odd Numbers

A. To find the even numbers, you can skip-count by twos beginning with 0.

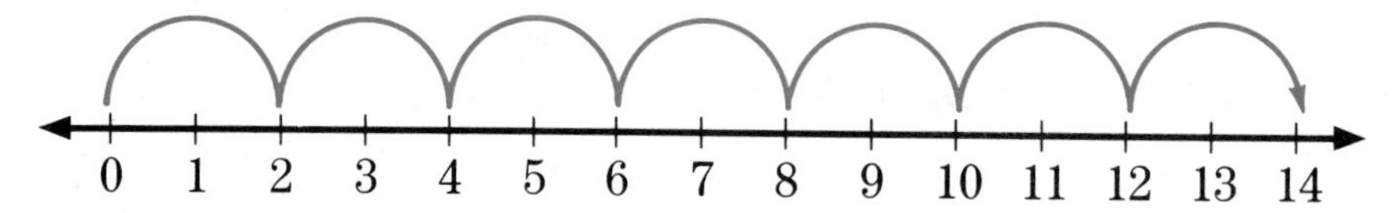

Even numbers end in 0, 2, 4, 6, or 8.

B. To find the odd numbers, you can skip-count by twos beginning with 1.

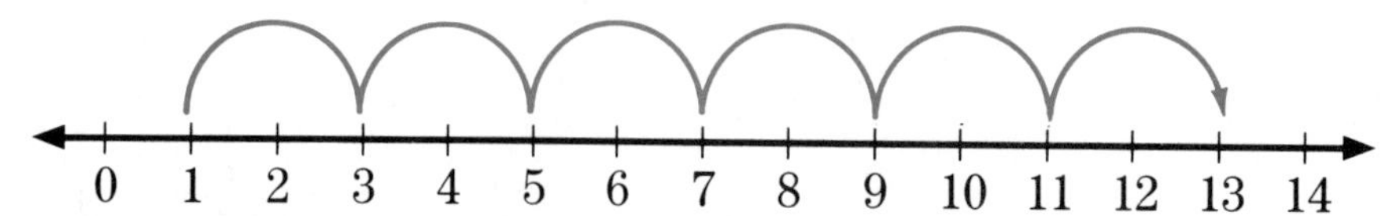

Odd numbers end in 1, 3, 5, 7, or 9.

Write *even* or *odd.*

1. 22 **2.** 15 **3.** 30 **4.** 126 **5.** 75 **6.** 29

7. 42 **8.** 104 **9.** 448 **10.** 1,338 **11.** 700 **12.** 4,027

Write the even numbers that come between.

★13. 9 and 17 **★14.** 47 and 55 **★15.** 83 and 91 **★16.** 161 and 169

Write the odd numbers that come between.

★17. 42 and 52 **★18.** 105 and 115 **★19.** 171 and 181 **★20.** 367 and 377

Solve.

21. Does the green house have an odd or an even number?

22. Are there more odd-numbered or even-numbered houses?

Counting by Fives and Tens

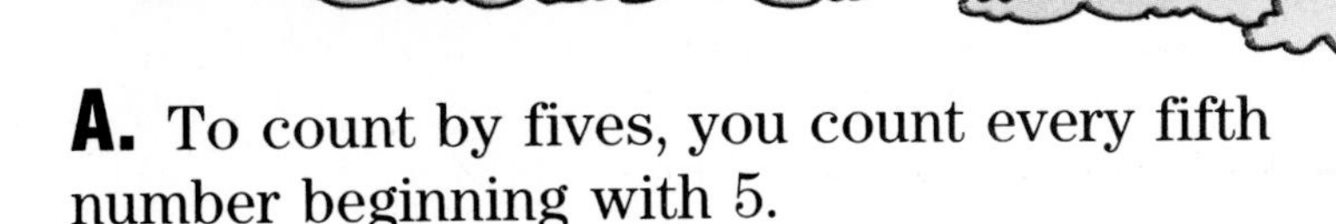

A. To count by fives, you count every fifth number beginning with 5.

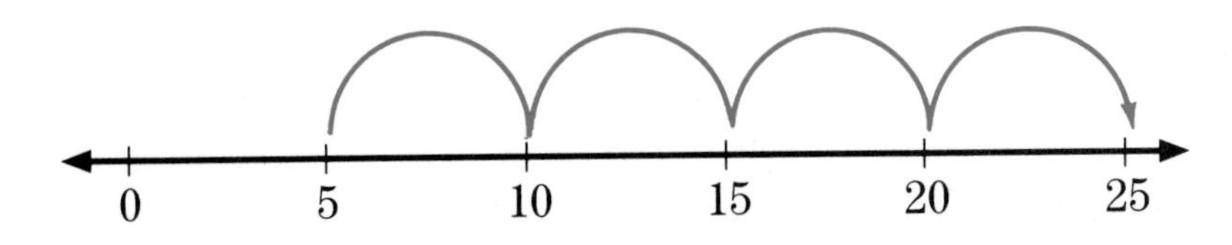

B. To count by tens, you count every tenth number beginning with 10.

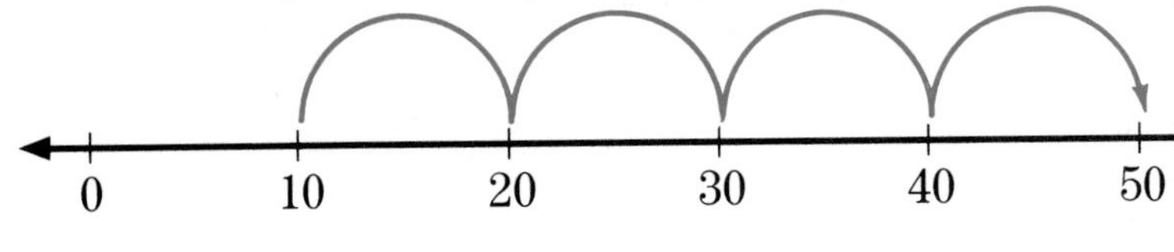

Count by fives. Write each missing number.

1.

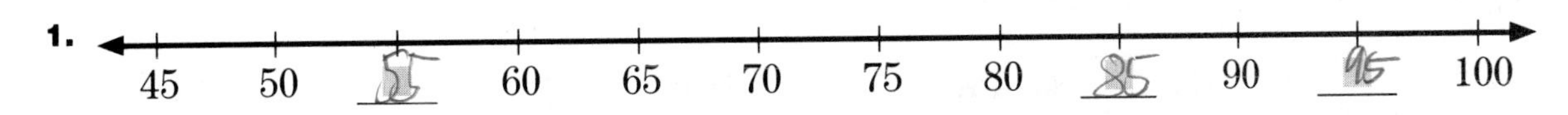

2. Begin at 30. Stop at 60.

3. Begin at 105. Stop at 135.

4. Begin at 200. Stop at 230.

5. Begin at 15. Stop at 45.

6. Begin at 85. Stop at 125.

7. Begin at 150. Stop at 180.

Count by tens. Write each missing number.

8.

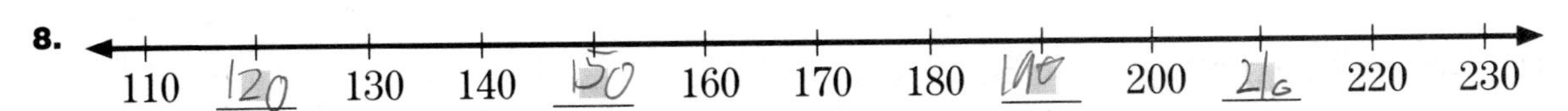

9. Begin at 10. Stop at 70.

10. Begin at 50. Stop at 110.

11. Begin at 280. Stop at 340.

12. Begin at 150. Stop at 220.

13. Begin at 170. Stop at 230.

14. Begin at 400. Stop at 460.

Rounding

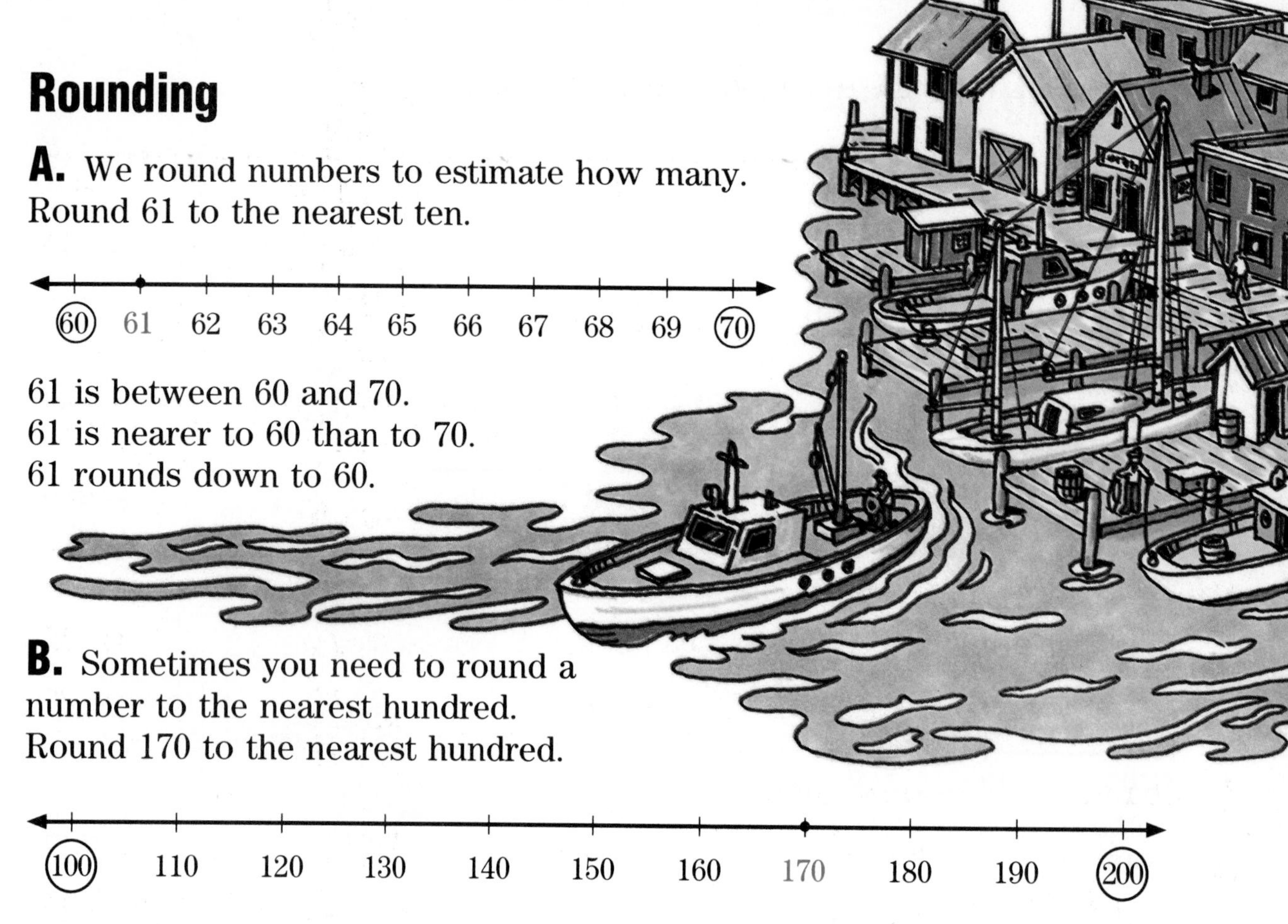

A. We round numbers to estimate how many.
Round 61 to the nearest ten.

(60) 61 62 63 64 65 66 67 68 69 (70)

61 is between 60 and 70.
61 is nearer to 60 than to 70.
61 rounds down to 60.

B. Sometimes you need to round a number to the nearest hundred.
Round 170 to the nearest hundred.

(100) 110 120 130 140 150 160 170 180 190 (200)

170 is between 100 and 200.
170 is nearer to 200 than to 100.
170 rounds up to 200.

C. When a number is halfway between two numbers, you round to the larger number.

Round 25 to the nearest ten.
25 is halfway between 20 and 30.
25 rounds up to 30.

Round 350 to the nearest hundred.
350 is halfway between 300 and 400.
350 rounds up to 400.

Use the number line to round to the nearest ten.

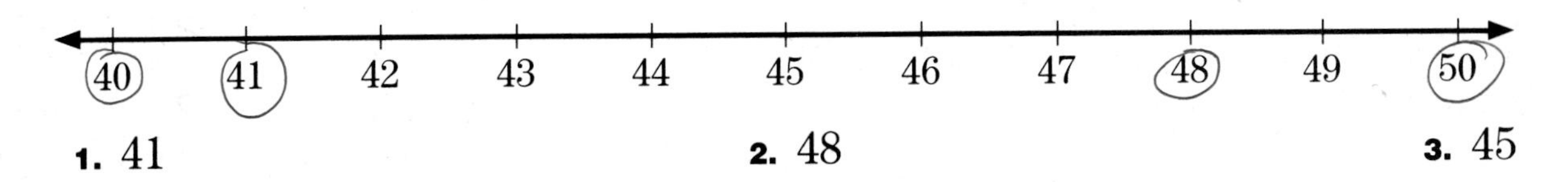

1. 41

2. 48

3. 45

Use the number line to round to the nearest hundred.

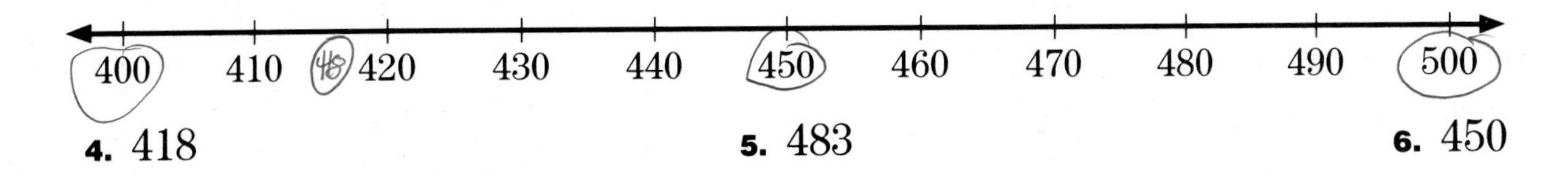

4. 418

5. 483

6. 450

Round to the nearest ten.

7. 9
8. 21
9. 44
10. 16
11. 45
12. 85
13. 63
14. 12
15. 92
16. 37

Round to the nearest hundred.

17. 430
18. 580
19. 246
20. 318
21. 847
22. 212
23. 185
24. 940
25. 470
26. 625

Solve.

27. Many people in San Francisco fish for a living. One fisher caught 45 pounds of shrimp. How many pounds is that to the nearest ten?

28. Other people fish for fun. One group of people had a fishing party. They caught 87 fish. How many fish is that to the nearest ten?

29. Some scientists study the fish and the water. They work on a boat that is 126 feet long. About how long is that to the nearest hundred?

★30. Use this list to write a question. Then solve the problem by rounding to the nearest hundred.

Fish Caught Today
138 bass
245 rock cod
350 kingfish

More Practice, page 406

Ten Thousands and Hundred Thousands

A. A long time ago, gold was found near San Francisco. In less than four years 210,856 people came looking for gold.

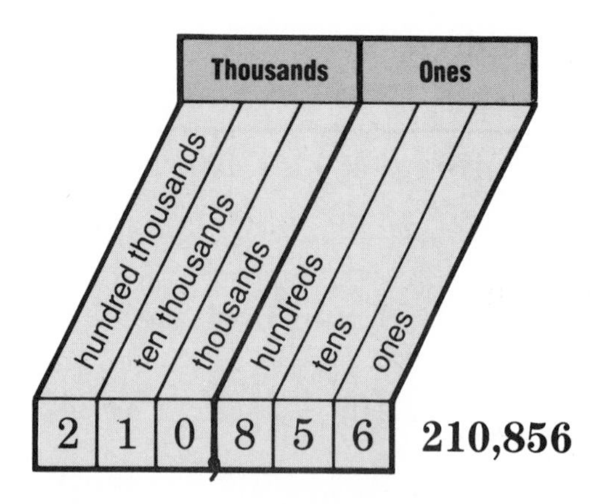

210,856

Read: two hundred ten thousand, eight hundred fifty-six.
Write a comma to separate the thousands from the hundreds.
Commas make large numbers easier to read.

B. You can show the value of each digit in a number.

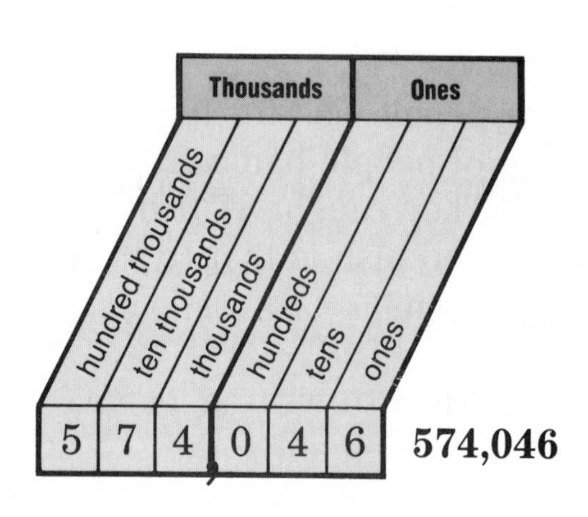

574,046

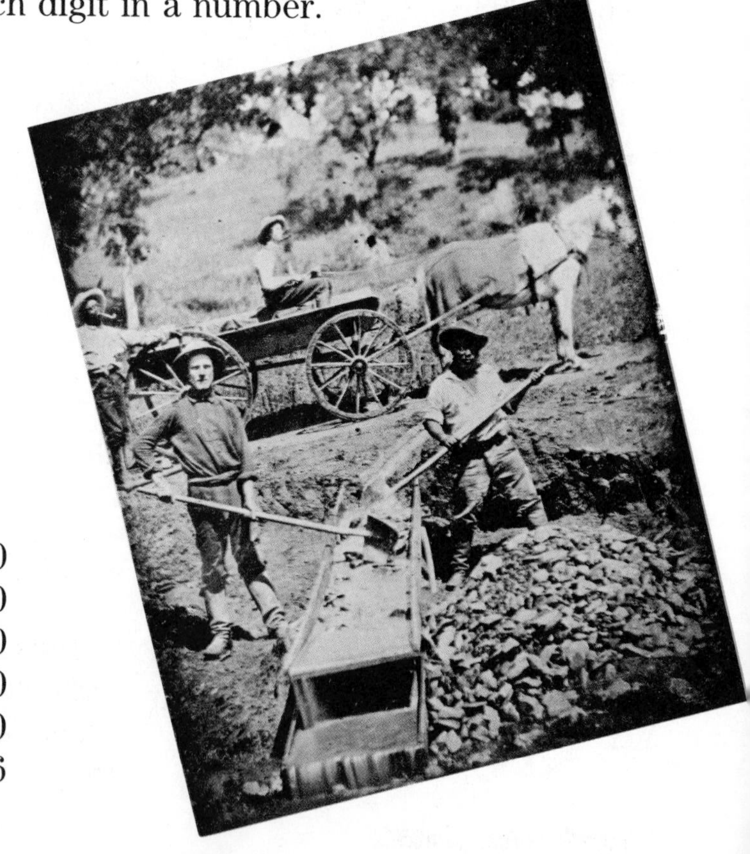

5 hundred thousands	500,000
7 ten thousands	70,000
4 thousands	4,000
0 hundreds	0
4 tens	40
6 ones	6

Read the number.

1. 68,000	2. 43,984	3. 161,000	4. 810,029
5. 45,200	6. 246,000	7. 732,921	8. 15,986
9. 100,634	10. 82,405	11. 780,980	12. 52,700

Write the number in standard form.

13. forty-two thousand, two hundred sixty-six
14. eight hundred seventy-five thousand, two hundred thirty-six
15. fifty-eight thousand, nine hundred ninety-nine
16. one hundred fifty-five thousand
17. eighty-eight thousand, two hundred twenty-two
18. twelve thousand, five hundred
19. sixty-nine thousand, two hundred eleven
20. seven hundred eighty-one thousand, forty-four
21. two hundred forty-six thousand, one hundred ninety-three
22. three hundred thousand eight

Write the value of the blue digit in each number.

23. 363	24. 1,203	25. 6,832	26. 82,045
27. 7,930	28. 37,438	29. 100,746	30. 670,012
31. 726	32. 52,993	33. 878,200	34. 200,372
35. 79,822	36. 72,000	37. 106,046	38. 36,607

CHALLENGE

Choose three different digits. Use them to write all the numbers you can. Then order the numbers from the least to the greatest.

More Practice, p. 406

Ordinal Words

You can use numbers to show order.

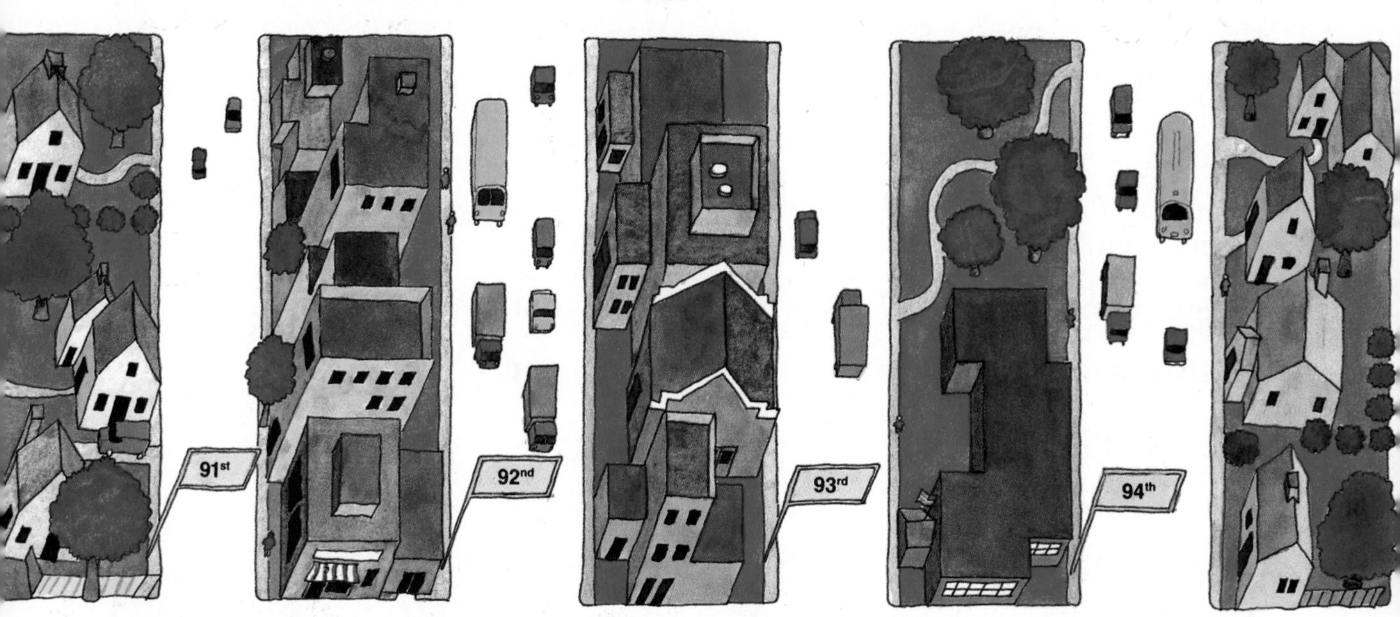

Ninety-first **Ninety-second** **Ninety-third** **Ninety-fourth**

In many towns, streets are numbered in order to help people find them easily.

Look at the street signs and think about the pattern.

Ninety-first Street comes before Ninety-second Street.

Ninety-third Street comes after Ninety-second Street.

Write the ordinal word that comes before

1. ninety-first.
2. seventieth.
3. thirty-sixth.
4. sixteenth.
5. forty-ninth.
6. sixty-second.

Write the ordinal word that comes after

7. fourth.
8. nineteenth.
9. eighty-sixth.
10. twenty-fourth.
11. thirty-third.
12. fifty-eighth.

Solve.

13. Don is thirty-sixth in a line. How many people are there ahead of him?

14. Fay is behind sixty-nine people in a line. What is her position in the line?

15. What position in the word *subtract* does the letter *a* have?

16. What positions in the word *there* does the letter *e* have?

17. In a five-letter word, *u* is the third letter, *e* is the fifth letter, *o* is the second, and *s* is fourth. The word also has an *h*. What is the word? house

18. In a five-letter word, *p* is the second letter, *t* is fourth, and *s* is both first and fifth. What position does *o* hold? What is the word? spots

CHALLENGE

Use the code to help you answer the riddle. Copy the answer spaces and ordinal words below. Write the letter that is in that position in the letter line.

Letter Line

What is black and white and has 16 wheels?

E	B	G	C	F	E
fifth	second	seventh	third	sixth	fifth

D	I
fourth	ninth

F	D	J	J	G	F
sixth	fourth	tenth	tenth	seventh	sixth

h	A	E	K	G	h
eighth	first	fifth	eleventh	seventh	eighth

PROBLEM SOLVING
Using a Pictograph

A **pictograph** gives information in pictures instead of just with words.

Use the pictograph to answer this question. How many cable cars travel Route 59?

NUMBER OF CABLE CARS IN SAN FRANCISCO

= 1 cable car

- The title tells you that this pictograph is about the number of cable cars in San Francisco.
- The labels at the left tell you which routes the cable cars travel. Find Route 59. It is the first label.
- The key tells you that each symbol stands for 1 cable car. You can find the number of cable cars that travel a route by counting the symbols in that row. Since each symbol equals 1 cable car, 9 cable cars travel Route 59.

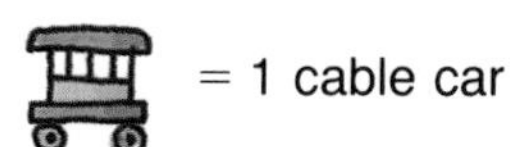

Use the pictograph on page 52 to answer each question.

1. How many cable cars travel Route 61?

2. Which route has the most cable cars?

3. How many cable cars are shown on this graph?

4. Suppose another route that has 10 cable cars is added. Would this route have more cars than Route 59?

5. A cable car travels at a speed of 9 miles per hour. How fast do 2 cable cars travel?

One day, Kathy counted the cars that passed her house. Use the pictograph she made to answer each question.

CARS THAT PASSED KATHY'S HOUSE IN ONE DAY

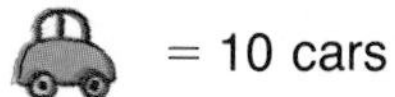

6. How many red cars did Kathy see?

★7. How many more blue cars than white cars did she see?

★8. If 20 of the red cars Kathy saw had 4 doors, how many red cars had 2 doors?

READING MATH

$$\begin{array}{rl} 5 & \text{addend} \\ +8 & \text{addend} \\ \hline 13 & \text{sum} \end{array}$$

Math has its own special words. The words *addends* and *sum* are math words.

Read the problem below. Then, answer each question.

Baxter, the talking dog, barked 7 times to tell his age. Then he barked 4 times to tell how many brothers he has. How many times did he bark?

Write the letter of the correct answer.

1. In this problem, what is the number 7 called?

a. *addend* **b.** *sum*
c. *top* **d.** *bottom*

2. When you solve this problem, what will you find?

a. *sum* **b.** *bottom*
c. *top* **d.** *addend*

3. When two or more numbers are added to find a total, what is each number called?

a. *sum* **b.** *addend*
c. *top* **d.** *bottom*

GROUP PROJECT

Planning a Playground

The problem: Your community is building a park. You and your classmates have been asked to plan the playground. On a separate sheet of paper, draw plans of what the playground will look like. Show where you would put each piece of playground equipment.

Key Questions

- How much room is needed for the playground?
- How much space is needed for your favorite games?
- Is there room for a seesaw?
- Where will the sandbox for the smaller children go?
- Do you want swings? Should they be fenced?
- What other areas should be fenced for safety?

CHAPTER TEST

Write the number. (pages 30, 32, 36, and 48)

1. forty-three
2. seven hundred thirty-six
3. three thousand fifty-nine
4. two thousand six
5. twenty-one thousand, nine hundred one
6. six hundred eighty-two thousand, four hundred twenty-eight
7. $600 + 20 + 5$
8. $100 + 2$
9. $700 + 20$
10. $200 + 70 + 3$
11. $800 + 8$
12. $900 + 40 + 4$

In which place is the digit 5 in each number? (pages 30, 32, and 36)
Write *ones*, *tens*, *hundreds*, or *thousands*.

13. 8,735
14. 2,537
15. 3,452
16. 5,621

Write $<$ or $>$ for ●. (page 42)

17. 39 ● 58
18. 932 ● 845
19. 99 ● 201
20. 4,809 ● 7,600

Order the numbers from the least to the greatest. (page 42)

21. 17, 22, 5, 3
22. 81, 75, 4, 101
23. 632; 27; 1,000; 582

Complete the pattern. (page 45)

24. 2, 4, ____, 8, 10
25. 10, 20, 30, ____, 50
26. 314, 316, ____, 320, 322
27. 45, ____, 55, 60, 65
28. 200, 205, ____, 215, 220
29. 410, ____, 430, 440, 450

Round to the nearest ten. (page 46)

30. 11
31. 29
32. 32
33. 57

Round to the nearest hundred. (page 46)

34. 309
35. 211
36. 589
37. 867

Write the value of the blue digit in each number. (pages 30, 32, 34, 36, and 48)

38. 278 **39.** 9,464 **40.** 81,382 **41.** 609,457 **42.** 17,846

Write the ordinal word that comes after. (page 50)

43. twenty-first **44.** ninety-eighth **45.** sixtieth **46.** forty-first

Solve. Use the pictograph. (page 52)

47. How many baseballs did Kenny hit?

48. Did Cora hit more baseballs on her first or her third try?

49. Who hit the greatest number of baseballs?

HITS			
Kenny	⚾	⚾⚾	⚾
Cora	⚾⚾	⚾	⚾⚾⚾
Jim	⚾⚾⚾	⚾⚾	⚾⚾
	First try	Second try	Third try
⚾ = 5 baseballs			

50. Jane is planning a trip to an ice-skating rink. Choose the letter of the question that is not needed for her plans. (page 34)

a. How much money does it cost to get in?

b. Will Jane need to rent a pair of ice skates?

c. Will there be music there?

d. How will Jane get there?

BONUS

Write > or < for ●.

1. 57,826 ● 62,578 **2.** 13,478 ● 13,495 **3.** 80,514 ● 80,451

4. 176,340 ● 167,950 **5.** 315,682 ● 315,689 **6.** 707,466 ● 622,317

RETEACHING

A. Which is greater, 437 or 433?

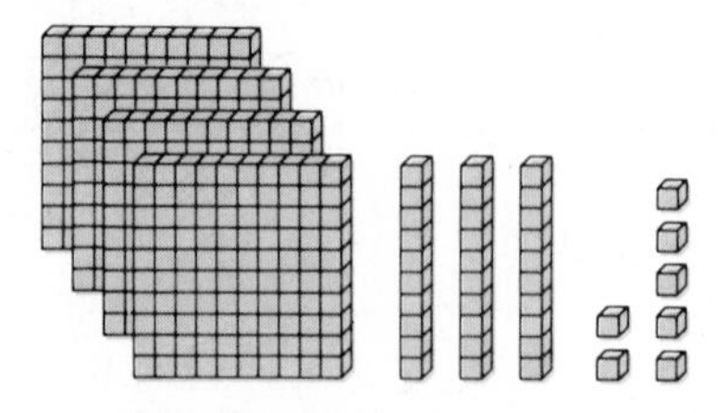

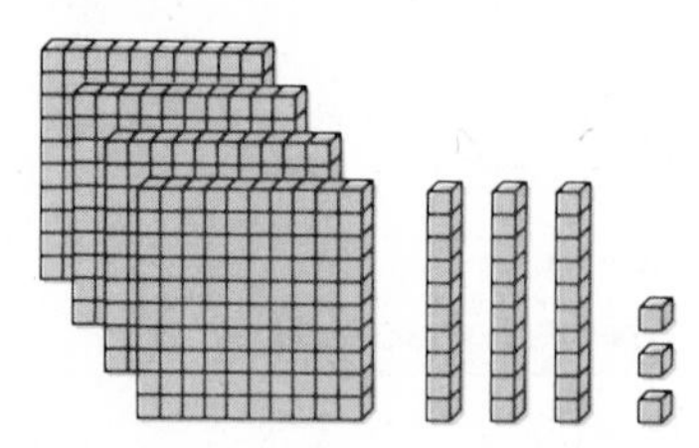

Compare the hundreds. The hundreds are the same. Compare the tens. The tens are the same. Compare the ones. 7 is greater than 3.

Therefore, 437 is greater than 433. Write: 437 > 433.

> The arrow always points to the lesser number.

B. You can write numbers in order from the least to the greatest. Write in order from the least to the greatest: 436, 71, 209, 58.

Line up the digits.	Compare to find the least number.	Compare the remaining numbers.	Order the numbers.
436	71 < 436	209 < 436	58, 71, 209, 436
71	71 < 209	71 < 209	
209	58 < 71		
58	So, 58 is the least number.		

Write > or < for ●.

1. 57 ● 51 **2.** 402 ● 482 **3.** 376 ● 367 **4.** 1,967 ● 1,847

Order from the least to the greatest.

5. 103, 86, 410, 87 **6.** 962, 937, 973, 692

7. 2,880; 8,280; 2,088; 8,208

Order from the greatest to the least.

8. 486, 991, 116, 345 **9.** 93, 39, 66, 69

10. 6,226; 7,136; 5,366; 6,267

ENRICHMENT

Roman Numerals

Many years ago, the Romans used letters to name numbers.

Use this key to read the first twenty Roman numerals.

I	II	III	IV	V	VI	VII	VIII	IX	X
1	2	3	4	5	6	7	8	9	10

IV is 5 − 1.

XI	XII	XIII	XIV	XV	XVI	XVII	XVIII	XIX	XX
11	12	13	14	15	16	17	18	19	20

XIV is 10 + 4.

When **I** is on the left of a letter, subtract 1.
When **I** is on the right of a letter, add 1.

Think Roman! Write the number for each Roman numeral.

1. You are asked to a feast on Floor XX. 20
2. Your Math class meets on Floor XIV. 14
3. Chariots are sold on Floor I. 1
4. You can buy sandals on Floor III. 3
5. Shields are made on Floor XIX. 19
6. There is a robe shop on Floor VI. 6
7. A senator rents space on Floor X. 10
8. The library is on Floor XV. 15
9. The emperor lives on Floor VIII. 8

★10. There is a party on Floor XXVI. 26

CUMULATIVE REVIEW

Write the letter of the correct answer.

1. $8 + 8 = \underline{16}$

a. 0 **b.** 6
c. 16 **d.** not given

2.
$$\begin{array}{r} 4 \\ +6 \\ \hline 10 \end{array}$$

a. 2
b. 10
c. 24
d. not given

3. $11 - 3 = \underline{8}$

a. 5 **b.** 12
c. 113 **d.** not given

4.
$$\begin{array}{r} 15 \\ -\ 6 \\ \hline 9 \end{array}$$

a. 9
b. 19
c. 21
d. not given

5. Which number completes the family of facts?

$4 + 7 = \underline{11}$
$7 + 4 = \underline{11}$
$\underline{11} - 4 = 7$
$\underline{11} - 7 = 4$

a. 3
b. 11
c. 12
d. not given

6. Which sentence completes the family of facts?

$3 + 2 = 5$
$5 - 2 = 3$
$5 - 3 = 2$

a. $3 - 2 = 1$
b. $2 + 3 = 5$
c. $2 \times 3 = 6$
d. not given

7. $\underline{3} + 9 = 12$

a. 3 **b.** 4
c. 21 **d.** not given

8. $8 + \underline{9} = 17$

a. 9 **b.** 11
c. 71 **d.** not given

9.
$$\begin{array}{r} 2 \\ 5 \\ +7 \\ \hline 14 \end{array}$$

a. 15
b. 77
c. 257
d. not given

10. $1 + 3 + 5 = \underline{9}$

a. 1 **b.** 4
c. 9 **d.** not given

11. In her dream, Sylvia wears a crown with 7 pearls. Her shoes have 5 pearls each. How many pearls does she dream of in all?

a. 2 pearls **b.** 12 pearls
c. 75 pearls **d.** not given

12. Stan picks 9 baskets of apples. He uses 5 to make cider. How many baskets does he have left?

a. 4 baskets **b.** 14 baskets
c. 45 baskets **d.** not given

3 TIME AND MONEY

Your class has won a trip. You can go anywhere in the mainland United States. Where will you go? List all the places that interest you and your classmates. What would be a fair way for the class to decide where to go?

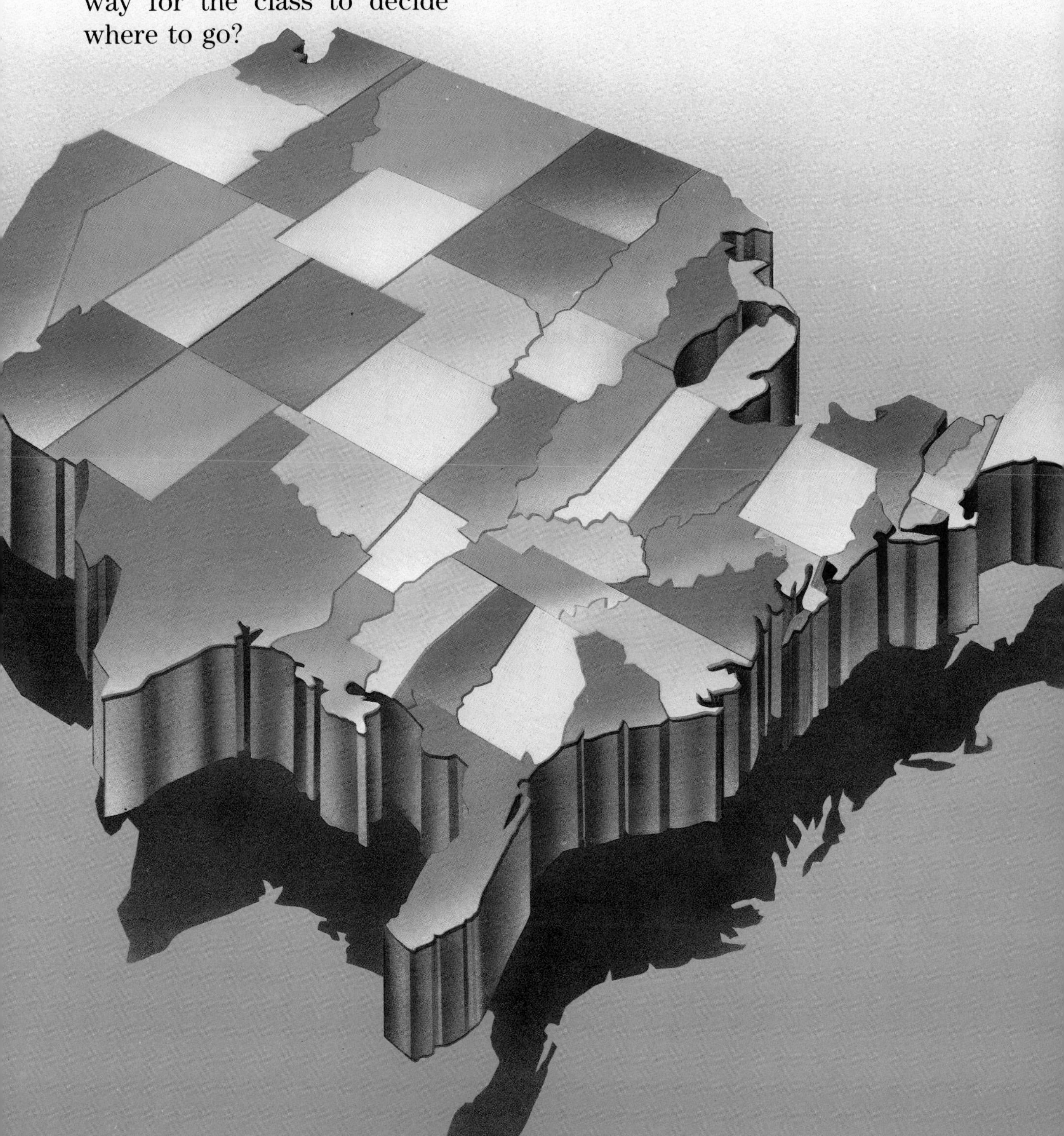

Hour, Half Hour, and Quarter Hour

A. Joan sets her alarm for 8:00 in the morning. At 8:30, she is eating breakfast. At 9:00, she is riding on her school bus.

Write: 8:00.
Read: eight o'clock.

Write: 8:30.
Read: eight-thirty, half past eight, or 30 minutes after 8.

Write: 9:00.
Read: nine o'clock.

There are 30 minutes in a half hour.
There are 60 minutes in an hour.
The hour hand moves from one number to the next in 1 hour.
The minute hand moves all the way around the clock in 1 hour.

The **minute hand** is the long hand.

The **hour hand** is the short hand.

B. Joan gets dressed by 8:15. At 8:45 she is walking to the bus stop.

Write: 8:15.
Read: eight-fifteen, quarter after eight, or 15 minutes after 8.

Write: 8:45.
Read: eight-forty-five, quarter to nine, 45 minutes after 8, or 15 minutes to 9.

There are 15 minutes in a quarter hour.

Write the time.

1. 
2. 
3. 
4. 
5. 
6. 
7. 
8. 
9. 
10. 
11. 
12. 

Complete each sentence. Write the time in words.

13. Rosa looks at this clock when she leaves her house. Rosa leaves her house at ■.

★14. At ten-fifteen, Brenda starts out for Pine Park. She stops every 45 minutes to rest. She reaches the park at one-fifteen. Brenda stops at ■, ■, and ■.

ANOTHER LOOK

Write > or < for ●.

1. 279 ● 301
2. 45 ● 54
3. 151 ● 143
4. 36 ● 30
5. 9 ● 7
6. 83 ● 67
7. 509 ● 515
8. 28 ● 17
9. 93 ● 77

Minutes

A. Tony timed the class trip to the museum. They left at 9:23 in the morning. They came back to the school at 4:37 in the afternoon.

9:23

Write: 9:23.
Read: nine-twenty-three,
or 23 minutes after 9.

If the minute hand is between the 12 and 6, you can read the time as "after" the hour.

4:37

Write: 4:37.
Read: four-thirty-seven,
or 23 minutes to 5.

If the minute hand is between the 6 and 12, you can read the time as "to" the hour.

B. You use **A.M.** and **P.M.** to show what part of the day it is.

Use A.M. for time from 12 midnight to 12 noon.
Write 9:23 in the morning as 9:23 A.M.

Use P.M. for time from 12 noon to 12 midnight.
Write 4:37 in the afternoon as 4:37 P.M.

Checkpoint

Write the letter of the correct answer.

What time is it?

1.

a. 10 minutes after 3
b. 20 minutes after 10
c. 10 minutes to 4
d. 4 minutes after 10

2.

a. 11 minutes to 6
b. 6 minutes to 11
c. 11 minutes after 6
d. 49 minutes to 6

Write the time.

1. 

2. 

3. 

4. 

5. 

6. 

Write the letter of the clock that shows the time.

7. 22 minutes after 3

8. 13 minutes to 7

9. 6 minutes to 1

10. 5 minutes after 1

a. 

b. 

c. 

d. 

Complete each sentence with A.M. or P.M.

11. The race began at noon and ended at 1:15 ____.

12. The school bus picks Ed up at home at 8:00 ____.

Estimate how many minutes it takes you to do each.

1. brush your teeth

2. say the alphabet

3. travel to school

4. eat lunch

5. count to 100

6. clean your room

More Practice, page 407

Elapsed Time

A. Dee's class rides a bus to the aquarium. The bus leaves the school at 9:05. It arrives at the aquarium at 9:30. How many minutes long is the bus ride?

The bus leaves. The bus arrives.

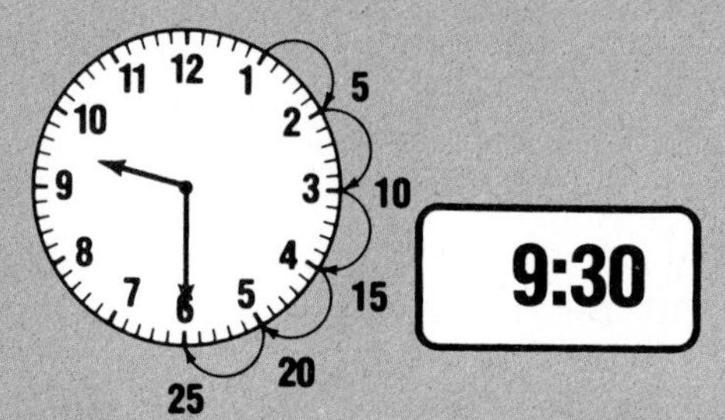

9:30

You can count by fives to find how many minutes long the bus ride is.

The bus ride is 25 minutes long.

B. A movie about whales begins at 9:55 A.M. It is 18 minutes long. When does the movie end?

The movie begins. The movie ends.

9:55

10:13

You can count by fives and ones to find when the movie ends.
5, 10, 15, 16, 17, 18

The movie ends at 10:13.

How many minutes are between

1.

1:05 and 1:50?

2.

5:45 and 6:10?

3.

7:15 and 7:35?

How many minutes are between

4. **1:15**

1:15 and 1:53?

5. 

10:12 and 11:00?

6. **9:47**

9:47 and 10:15?

What time will it be

7. **3:30**

in 45 minutes?

8. 

in 25 minutes?

9. **12:30**

in 30 minutes?

10. 

in 37 minutes?

11. 

in 12 minutes?

12. **12:21**

in 31 minutes?

MIDCHAPTER REVIEW

Write the time.

1. 

2. 

3. 

4. 

Complete each sentence with A.M. or P.M.

5. Jack eats dinner at 6:30 ___.

6. Mary has breakfast at 7:30 ___.

7. You should be asleep at 11:55 ___.

More Practice, page 407

PROBLEM SOLVING
Using a Schedule

When you plan to travel by bus, you will find that a schedule is helpful. A bus schedule tells you when a bus leaves each stop on its route.

This schedule is for the bus that goes from Eastern Avenue to Laurel Avenue.

Bus Stop	Time
Eastern Avenue	6:31 P.M.
Bennett Street	6:35 P.M.
Fall Street	6:40 P.M.
Grant Street	6:58 P.M.
Laurel Avenue	7:02 P.M.

The names of the streets where the bus stops are listed at the left. The time at which the bus leaves each stop is at the right.

Use the schedule to answer this question.

Mary lives in a house on Bennett Street. At what time must she be at the bus stop at Bennett Street to take the bus to Grant Street?

Find the name of Bennett Street on the schedule. The time listed is the time the bus leaves Bennett Street. The bus leaves Bennett Street at 6:35 P.M.

Mary should be at the bus stop at 6:35 P.M.

Use the schedule to answer each question.

Bus Stop	Time
Flynn Street	8:15 A.M.
Amber Street	8:18 A.M.
Tulip Avenue	8:30 A.M.
Sixth Street	8:45 A.M.
Bliss Avenue	8:57 A.M.
Fox Street	9:03 A.M.
Mill Street	9:15 A.M.
Dale Highway	9:20 A.M.

1. What is the stop after Bliss Avenue?
2. At what time does the bus stop at Fox Street?
3. How many stops does the bus make between Mill Street and Dale Highway?
4. Does the bus stop first at Amber Street or at Mill Street?

Solve.

5. How long does it take to go from Flynn Street to Sixth Street if the bus is on time?
6. Between which two stops does the bus take the longest time to travel?
7. Dan boarded the bus at Flynn Street. He left the bus at Mill Street. The bus ran on time. How much time did Dan spend on the bus?
8. Darcy reached the Bliss Avenue stop. The bus arrived 4 minutes later. When did Darcy reach the stop?

★9. It takes Jim 40 minutes to dress and eat and 10 minutes to walk to the Mill Street stop. At what time should he start to dress in order to take the bus?

★10. The bus leaves 2 minutes late from Amber Street and 5 minutes late from Tulip Avenue. What is the earliest time at which it could leave Sixth Street?

Reading a Calendar

Erica is planning a trip. She marked a calendar to show what state she will be in each day. Her birthday is June 15. What day of the week is that? Where will Erica be?

JUNE						
Sunday	Monday	Tuesday	Wednesday	Thursday	Friday	Saturday
			Vermont 1	Vermont 2	Vermont 3	Vermont 4
Vermont 5	Vermont 6	Vermont 7	Illinois 8	Illinois 9	Illinois 10	Illinois 11
Illinois 12	Texas 13	Texas 14	Texas 15	Texas 16	Texas 17	Texas 18
Texas 19	Texas 20	Texas 21	Texas 22	Nevada 23	Nevada 24	Nevada 25
Nevada 26	Nevada 27	Nevada 28	Nevada 29	Nevada 30		

June 15 is the third Wednesday of June. Erica will be in Texas.

Use the calendar to complete.

On which day of the week

1. does she leave Vermont?
2. is her second day in Nevada?
3. is her ninth day in Texas?
4. is her last day in Illinois?

Where is Erica

5. on the third Tuesday?
6. on June 7?
7. on June 11?
8. on the fourth Saturday?

What is the date

9. on which Erica leaves Illinois?
10. of the third Friday?

This calendar shows 1 year. There are 12 months in 1 year.

CALENDAR

JANUARY

S	M	T	W	T	F	S
					1	2
3	4	5	6	7	8	9
10	11	12	13	14	15	16
17	18	19	20	21	22	23
24	25	26	27	28	29	30
31						

FEBRUARY

S	M	T	W	T	F	S
	1	2	3	4	5	6
7	8	9	10	11	12	13
14	15	16	17	18	19	20
21	22	23	24	25	26	27
28	29					

MARCH

S	M	T	W	T	F	S
		1	2	3	4	5
6	7	8	9	10	11	12
13	14	15	16	17	18	19
20	21	22	23	24	25	26
27	28	29	30	31		

APRIL

S	M	T	W	T	F	S
					1	2
3	4	5	6	7	8	9
10	11	12	13	14	15	16
17	18	19	20	21	22	23
24	25	26	27	28	29	30

MAY

S	M	T	W	T	F	S
1	2	3	4	5	6	7
8	9	10	11	12	13	14
15	16	17	18	19	20	21
22	23	24	25	26	27	28
29	30	31				

JUNE

S	M	T	W	T	F	S
			1	2	3	4
5	6	7	8	9	10	11
12	13	14	15	16	17	18
19	20	21	22	23	24	25
26	27	28	29	30		

JULY

S	M	T	W	T	F	S
					1	2
3	4	5	6	7	8	9
10	11	12	13	14	15	16
17	18	19	20	21	22	23
24	25	26	27	28	29	30
31						

AUGUST

S	M	T	W	T	F	S
	1	2	3	4	5	6
7	8	9	10	11	12	13
14	15	16	17	18	19	20
21	22	23	24	25	26	27
28	29	30	31			

SEPTEMBER

S	M	T	W	T	F	S
				1	2	3
4	5	6	7	8	9	10
11	12	13	14	15	16	17
18	19	20	21	22	23	24
25	26	27	28	29	30	

OCTOBER

S	M	T	W	T	F	S
						1
2	3	4	5	6	7	8
9	10	11	12	13	14	15
16	17	18	19	20	21	22
23	24	25	26	27	28	29
30	31					

NOVEMBER

S	M	T	W	T	F	S
		1	2	3	4	5
6	7	8	9	10	11	12
13	14	15	16	17	18	19
20	21	22	23	24	25	26
27	28	29	30			

DECEMBER

S	M	T	W	T	F	S
				1	2	3
4	5	6	7	8	9	10
11	12	13	14	15	16	17
18	19	20	21	22	23	24
25	26	27	28	29	30	31

Use this calendar to complete.

11. Which months of the year have 30 days?

12. Which months of the year have 31 days?

13. What is the eleventh month of the year?

14. What is the third month of the year?

Give the date of

15. the first Sunday in March.

16. the third Wednesday in May.

17. the fourth Friday in July.

18. the second Tuesday in June.

What day of the week is

19. January 23?

20. November 24?

21. May 11?

PROBLEM SOLVING

Choosing the Operation

Here are three more hints to help you decide whether you should add or subtract.

A. One morning, 8 students visited the Space Center. Later, 2 more students arrived. How many students visited that day?

8	+	2	=	10
in the morning		later		in all

That day, 10 students visited.

You know	how many in one group. how many more joined the first group.
You want to find	how many altogether.
You can	ADD.

B. The center had 8 visitors in the morning. It had 10 visitors later. How many more people visited later?

10	−	8	=	2
later visitors		morning visitors		more

There were 2 more people who visited later.

You know	how many in one group. how many in another group.
You want to find	how many more in one group than another.
You can	SUBTRACT.

C. The center had 9 visitors. Of these, 4 had visited before. How many had not visited before?

9	−	4	=	5
visitors		visited before		new visitors

There were 5 people who had not visited before.

You know	how many in all. how many in part of the group.
You want to find	how many in the rest of the group.
You can	SUBTRACT.

Choose the operation you would use to solve each problem. Write the letter of the correct answer.

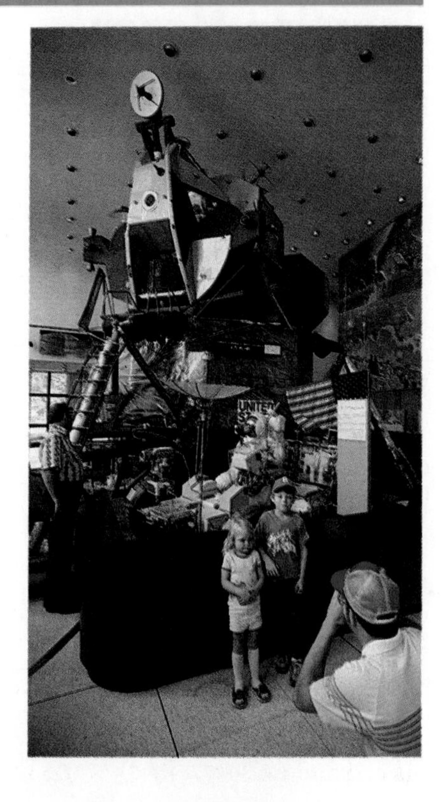

1. Stephen went on the Space Shuttle Ride for 10 minutes. Ashley went on the Weightless Machine for 5 minutes. How much longer was Stephen's ride?

   a. addition b. subtraction

2. At the Space Center shop, Jackie bought 7 postcards. Later, she bought 5 more postcards. How many postcards did Jackie buy?

   a. addition b. subtraction

Solve.

3. Eric spent 5 hours building a model rocket. Danny spent 3 hours building a model rocket. How much more time did Eric spend building a rocket than Danny spent?

4. In the morning, 6 students visited the model of the *Apollo* spacecraft. In the afternoon, 7 students visited the model. How many students visited the model that day?

5. During Space Food Class, Joe ate 7 tubes of space food. Karen ate only 3 tubes. How many more tubes of space food did Joe eat than Karen ate?

6. Ed's class buys space suits. There are 11 students who buy silver suits. There are 7 students who buy blue suits. How many more students buy silver suits?

7. Edna buys 3 Apollo mission patches. Jon buys 5 Spacelab patches. How many patches did they buy altogether?

8. Sol bought 6 astronaut action figures. He gives 4 of them to his friends. How many action figures does he have left?

Counting Money

Here are some coins and their values.

penny
1¢
one cent

nickel
5¢
five cents

dime
10¢
ten cents

quarter
25¢
twenty-five cents

half dollar
50¢
fifty cents

Look at the coins. How much money is there?

Count: 50¢ 75¢ 85¢ 90¢ 95¢ 96¢ 97¢ 98¢ 99¢

The amount of money is 99¢.

Write the amount.

1.

3.

2.

4.

6.

5.

Match the values that are the same.
Write the letter of the correct answer.

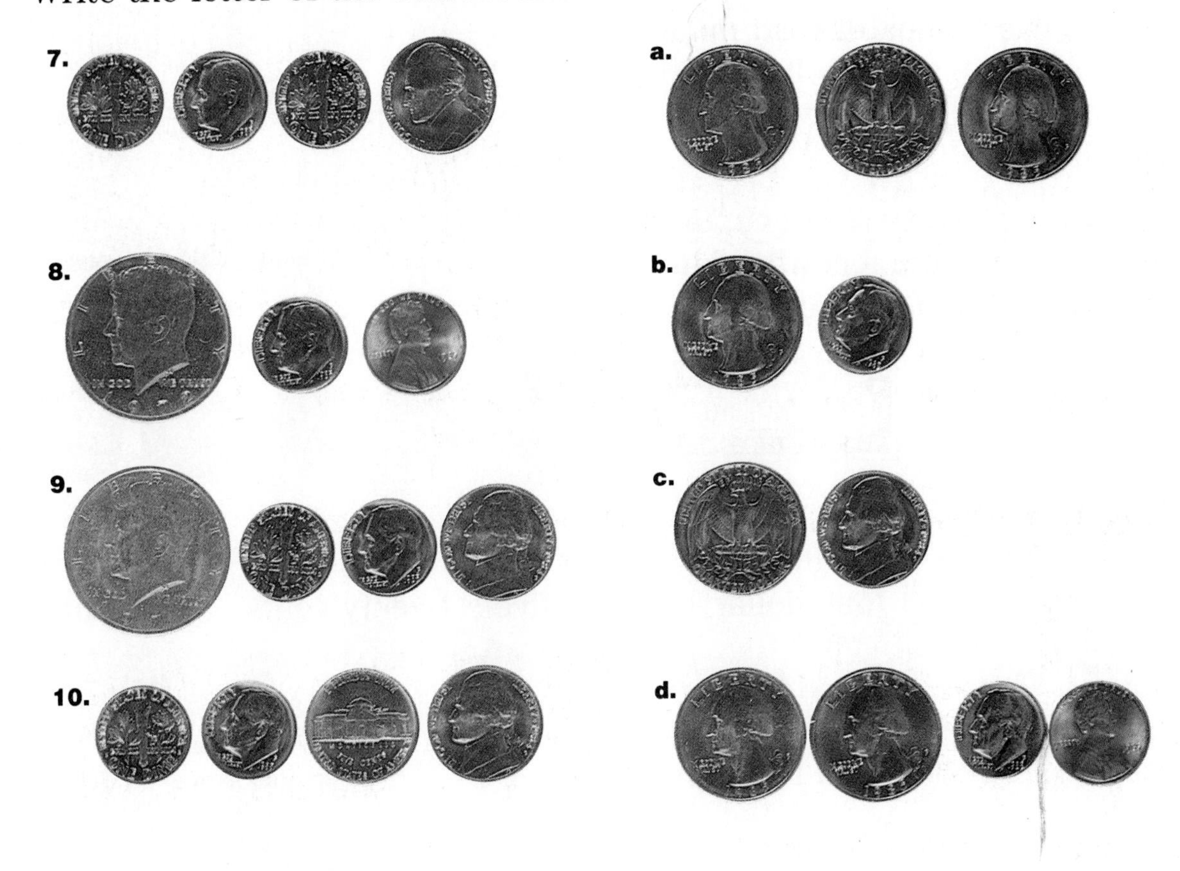

Write the amount.

11. 2 dimes
 1 nickel
 4 pennies

12. 1 half dollar
 6 nickels
 3 pennies

13. 1 quarter
 3 nickels
 2 pennies

14. 2 quarters
 3 dimes
 2 pennies

15. 3 dimes
 5 nickels
 1 penny

16. 1 half dollar
 1 dime
 4 nickels

CHALLENGE

List five ways to give exactly eighty-five cents in change.
You can use as many as 10 coins.

Dollars and Cents

A. Matt buys a book of road maps. He uses one 5-dollar bill, one 1-dollar bill, 1 quarter, 2 dimes, and 2 pennies. How much does he pay for the book?

You can use a **cents sign (¢)**, or a **dollar sign ($)** and a **cents point (.)** to write amounts of money.

Write:	$5.00	$1.00	$0.25	$0.20	$0.02
	500¢	100¢	25¢	20¢	2¢
	five dollars	one dollar	twenty-five cents	twenty cents	two cents
Count:	$5.00	$6.00	$6.25	$6.45	$6.47
Write:	$6.47				
Read:	six dollars and forty-seven cents.				

Matt pays $6.47 for the book.

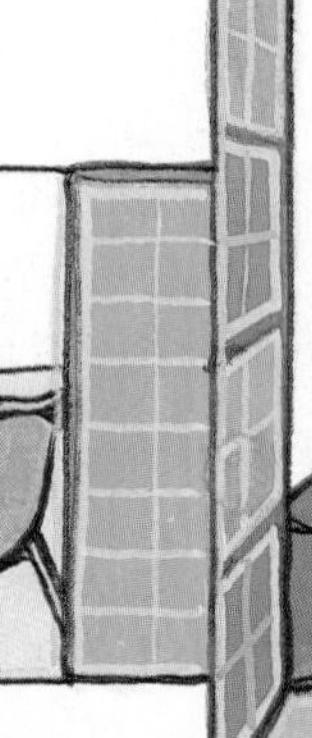

B. Keiko buys a bus ticket to Omaha for $7.75. What is the fewest number of bills and coins that she can use?

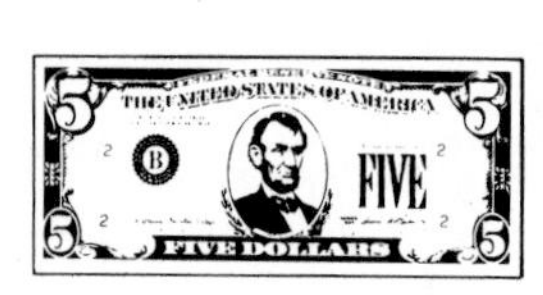

Count:	$5.00	$7.00	$7.50	$7.75

Keiko can use three bills and two coins.

Write the amount. Write the dollar sign ($) and the cents point (.).

1. 

2. 

3. 6 dollars and 75 cents

4. 4 dollars and 33 cents

5. 9 dollars and 2 cents

6. 3 dollars and 95 cents

7. 5 dollars and 20 cents

8. 8 dollars and 8 cents

9. twenty-seven cents

10. ninety-six cents

11. four dollars and one cent

12. one dollar and eight cents

List the fewest number of bills and coins for each amount.

	Amount							
13.	$1.26	0	1	0	1	0	0	1
14.	$8.30	1	3	0	1	0	1	0
15.	$0.99	0	0	0	3	0	3	0
16.	$5.39	1	0	0	1	1	0	4

CHALLENGE

Write *add* or *subtract* to solve the problem.

1. $5.00 earned
$2.50 spent
How much is left?

2. 124 quarters
300 pennies
How many coins?

3. $0.92 earned
$1.88 earned
How much is earned?

More Practice, page 407

Comparing Money

A bicycle bell costs $7.35. Tony has one 5-dollar bill, two 1-dollar bills, 3 quarters, and 1 dime. Does he have enough money to buy the bell?

Compare the amount of money Tony has with the cost of the bell to find if he has enough money.

Count: $5.00 $7.00 $7.75 $7.85

Compare $7.85 to $7.35.
Think: 785 > 735.
So, $7.85 > $7.35.

> You can compare money the same way you compare whole numbers.

Tony has enough money to buy the bell.

Compare. Is there enough money to buy the ticket? Write *yes* or *no*.

1.

2.

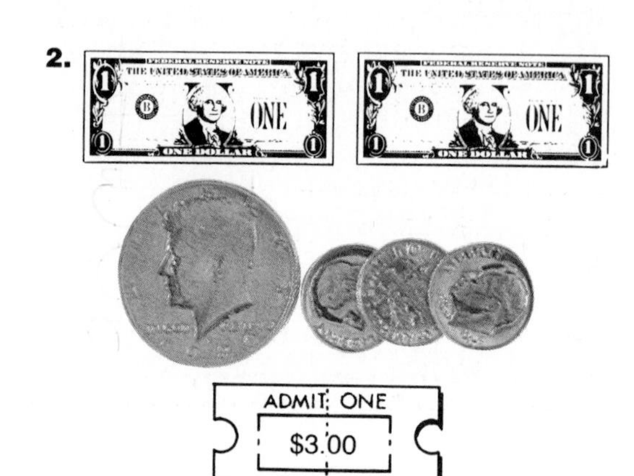

Solve.

3. Dana wants to buy a Giant Redwood button for 57¢. She has 2 quarters and 3 pennies. Does she have enough money to buy the button?

★4. Jason wants to buy two road maps. One map costs $2.19. The other map costs $2.00. He has a 5-dollar bill. Does he have enough money to buy both maps?

Making Change

Jack buys a book about the Redwood National Park for $2.39. He pays with a 5-dollar bill. How much change does Jack receive?

Start with $2.39.

Count up to find the change.

$2.40

$2.50

$2.75

$3.00

$4.00

$5.00

Jack receives $2.61 change.

Find the correct change. The first one is done for you.

The fare is	You give	Count up to find the change.							Your change is
		1¢	5¢	10¢	25¢	50¢	$1.00	$5.00	
1. 60¢	$1.00		1	1	1				40¢
2. $1.59	$2.00								■
3. $2.65	$5.00								■
4. 48¢	$1.00								■
5. $4.44	$5.00								■

FOCUS: MENTAL MATH

Write > or < for ●.

1. 1 half-dollar ● 3 quarters

2. 2 quarters ● 3 dimes

3. 1 half-dollar ● 15 nickels

4. 5 dimes ● 45 pennies

PROBLEM SOLVING
Making a Table To Find a Pattern

Making a table can sometimes help you find a pattern to solve a problem.

> The bus fare in Hilltown is based on how far a person rides. A ride of up to 5 blocks costs $0.10. A 10-block ride costs $0.20. A 15-block ride costs $0.30. How much would it cost to ride 25 blocks?

Make a table like this one.

BUS FARES BY BLOCK

Blocks	5	10	15	___	___	___
Cost	$0.10	$0.20	$0.30	___	___	___

The table makes it easier to see patterns. There is a pattern to the bus fare.

The number of blocks increases by 5 each time. The bus fare increases by $0.10 each time.

Use the pattern. Complete your table by writing the missing amounts.

It would cost $0.50 to ride 25 blocks.

Skip-count to find the pattern. Copy and complete the table to help you.

1. One bus goes 3 miles every 4 minutes. How long does it take to go 24 miles?

Miles	3	6	9	12	15	■	■	■
Minutes	4	8	12	16	20	■	■	■

2. How far will the bus go in 28 minutes?

Make a table. Solve.

3. The children are packing lunches for their trip. They put 3 pieces of fruit and 2 sandwiches into each lunch bag. They have packed 12 sandwiches. How many pieces of fruit have they packed?

4. Kitty and Ron each have $0.75. They are both trying to save money. Kitty plans to save $0.25 each week. Ron plans to save $0.50 each week. By the end of the first week, Kitty has $1.00. Ron has $1.25. How much has each saved by the end of the fourth week?

5. Kitty wants to buy a bus ticket that costs $2.75. For how many weeks must she save to have enough money to buy the ticket?

★6. The Parrs drive 10 miles every 15 minutes. They leave home at 9:00 A.M. At what time will they have traveled 30 miles?

CALCULATOR

You can use your calculator to help you figure change. Find the cents point on your calculator. Use it to do this problem: Fred buys a robot for \$3.24. He gives the clerk a five-dollar bill. How much change should he receive?

Think: \$5.00 − \$3.24 = ■

There is no [\$] key. Press these keys:

The display should show 1.76. You write \$1.76. Fred should receive \$1.76 in change.

Use your calculator to find the exact change.

Item	Cost	Amount Given	Change
1. Soap	\$0.79	\$1.00	■
2. Toothbrush	\$0.89	\$5.00	■
3. Shirt	\$4.03	\$5.00	■
4. Gloves	\$3.75	\$5.00	■
5. Hat	\$3.17	\$4.00	■
6. Socks	\$2.90	\$5.00	■
7. Hair curler	\$7.92	\$10.00	■
8. Tie	\$5.43	\$10.00	■
★9. pen and paper	\$4.86 and \$2.49	\$10.35	■

GROUP PROJECT

Travel Tips

The problem: You and your family are going from Los Angeles to Albuquerque. You need to arrive in time for your aunt's wedding, which will take place in 3 days. How will you travel there?

The options: You might drive a car or take a train.

Key Facts

Car

- You have a small car.
- There are 5 people in your family.
- You will need a car once you get there.
- It will take 2 days of driving 10 hours per day to get there.
- You could visit the Grand Canyon on the way, but it would mean an extra 4 hours of driving.
- You will have to spend a night in a motel.

Train

- Your aunt could use your help if you arrive early.
- You won't have a car once you get there.
- The trip will take 17 hours nonstop.
- You will spend the night on the train traveling 47 miles every hour while you are sleeping.
- You will buy your meals in the train's dining car.
- Everyone will be able to see the sights along the way.

CHAPTER TEST

Write the time. (page 62 and 64)

1.

2. 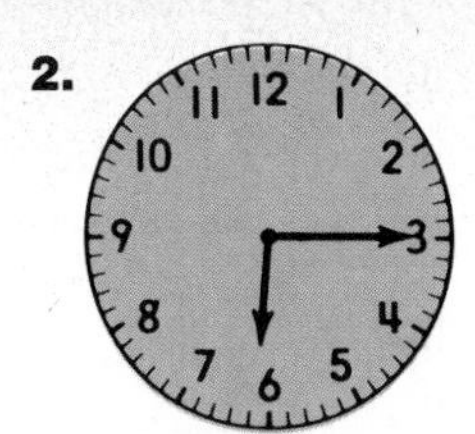

3. 

4. 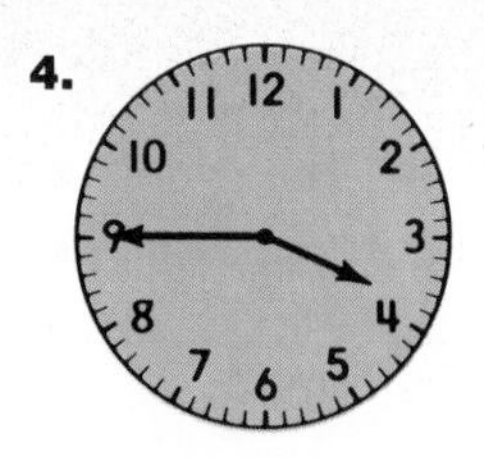

How many minutes are there between (page 66)

5. 

1:25 and 1:44?

6. 

11:55 and 12:23?

7. 

3:32 and 4:07?

Use the calendar to complete. (page 70)

8. What is the date of the third Sunday in October?

9. What day of the week is October 20?

10. What day of the week is the last day of October?

OCTOBER

S	M	T	W	T	F	S
						1
2	3	4	5	6	7	8
9	10	11	12	13	14	15
16	17	18	19	20	21	22
23	24	25	26	27	28	29
30	31					

Write the amount. (page 74 and 76)

11.    

12.   

Write the amount. Write the dollar sign ($) and the cents point (.).

13. 

14. 

15. 4 dollars and 4 cents

16. 9 dollars and 90 cents

Solve. (page 78)

17. A bus ticket costs $4.05. Jim has 11 quarters and two $1-bills. Does he have enough?

18. A hat costs $6.98. Sal has one $5-bill and 4 half dollars. Does she have enough?

19. A game costs $2.98. Joe has two $1-bills, 1 half dollar, 4 dimes, and 4 pennies. Does he have enough?

20. Cindy has 5 quarters, 3 dimes, 3 nickels, and 6 pennies. Can she buy a book that costs $1.75?

Use the schedule to solve. (page 68)

21. How long is the bus ride from Branton to Granville?

22. When does the bus arrive in Lockton?

23. Does the bus arrive in Granville before 12 noon?

BUS SCHEDULE			
Branton	Granville		Lockton
10:15 A.M.	12:20 P.M.	1:10 P.M.	5:30 P.M.
leave	arrive	leave	arrive

Solve. (page 72)

24. Lil has 9 pears. She gets 8 more. How many pears does she have?

25. Jay has 14 toy cars. He gives 5 cars to Al. How many cars does Jay have?

RETEACHING

It is 10:35.
What time will it be
in 27 minutes?

It will be 11:02.

You can count by fives and ones to find what time it will be.

What time will it be

1. 

in 25 minutes?

2. 

in 15 minutes?

3. 

in 20 minutes?

You can count by fives to find what time it will be.

4. 

in 53 minutes?

5.

in 22 minutes?

6. 

in 58 minutes?

You can count by fives and ones to find what time it will be.

7. 

in 21 minutes?

8. 

in 34 minutes?

9. 

in 48 minutes?

ENRICHMENT

Time Zones

Earth is divided into different time zones. When it is 2:00 P.M. in Chicago, it is 1:00 P.M. in Denver. When it is noon in Los Angeles, it is 3:00 P.M. in New York. The map shows the time zones in the United States.

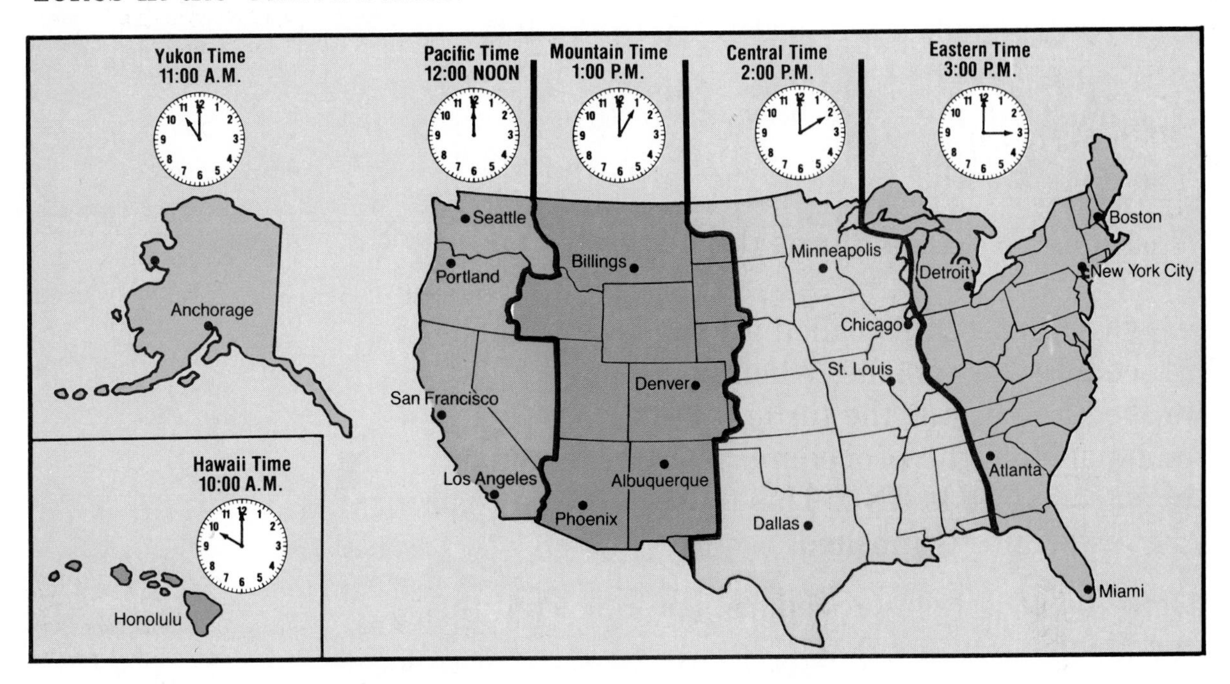

Use the map to solve.

When it is 8:00 A.M. in Honolulu, what time is it in

1. Anchorage? ____
2. Portland? ____
3. Minneapolis? ____

When it is 3:00 P.M. in Dallas, what time is it in

4. Boston? ____
5. Albuquerque? ____
★6. Honolulu? ____

7. A baseball game begins at 8:00 P.M. in Atlanta. It lasts two hours. When it ends, what time is it in Billings?

★8. An airplane leaves Los Angeles at 11:00 P.M. Six hours later, it arrives in New York. What time is it in New York?

TECHNOLOGY

Here are some LOGO commands.

FD	This makes the turtle move forward the number of steps shown.
FD10	This makes the turtle move 10 steps.
RT	This makes the turtle turn to the right.
LT	This makes the turtle turn to the left.

You must tell the turtle how far to turn.

The command RT 90 turns the turtle this far.

The command LT 45 turns the turtle this far.

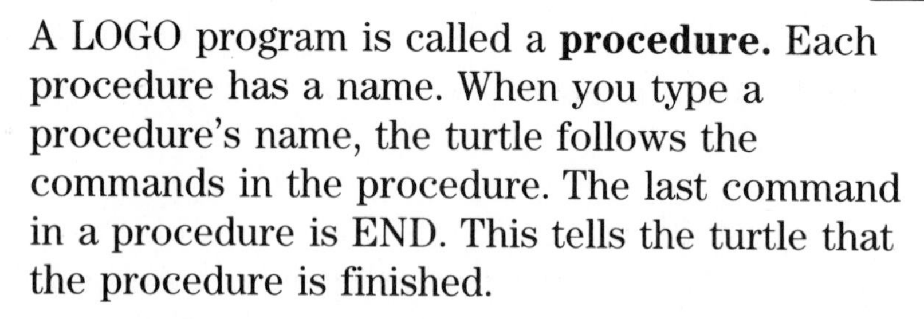

A LOGO program is called a **procedure.** Each procedure has a name. When you type a procedure's name, the turtle follows the commands in the procedure. The last command in a procedure is END. This tells the turtle that the procedure is finished.

Before you write a procedure, you must tell the turtle what you are doing. You give the procedure a name by typing the word TO and the procedure's name. For example:

TO LINE
FD 80 The computer would draw ________________
END

1. Follow the procedure step by step. Complete the blanks to draw the shape.

TO TURN
FD 40
RT 90
■ 80
■ 90
END

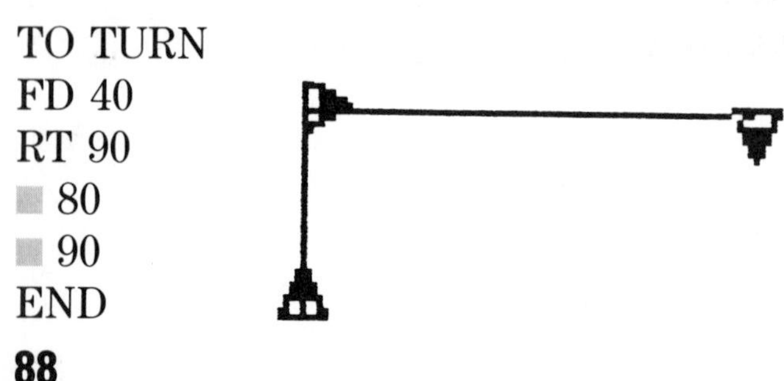

2. Start in the lower-left corner.
Write a procedure to draw this figure.

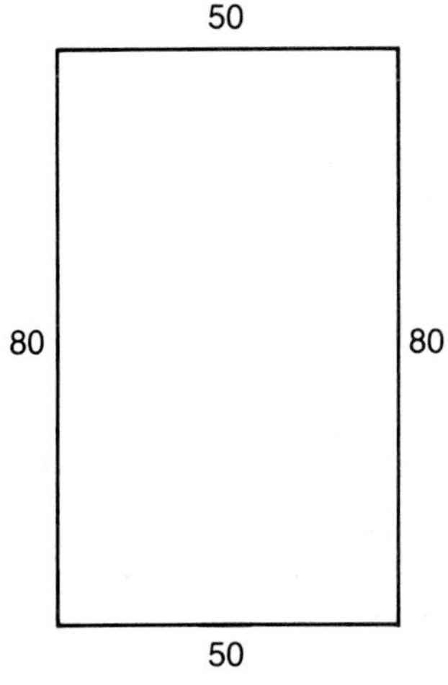

3. Change your procedure so that it draws a square. Each side should be 50 turtle steps long.

4. Write a procedure called STEP3 to draw the shape. The steps are 20 steps high and 20 steps wide.

★5. Change your procedure for TO STEP3 so that you use LT 45 and RT 45 instead of LT 90 and RT 90. How does this change the shape?

CUMULATIVE REVIEW

Write the letter of the correct answer.

1. $7 + 3 = \underline{\blacksquare}$

a. 4 **b.** 10
c. 21 **d.** not given

2. $\begin{array}{r} 16 \\ -\ 9 \\ \hline \end{array}$

a. 7 **b.** 25
c. 27 **d.** not given

3. Which sentence completes the family of facts?

$4 + 8 = 12$
$12 - 8 = 4$
$12 - 4 = 8$

a. $8 \div 4 = 2$
b. $8 + 4 = 12$
c. $8 \times 4 = 32$
d. not given

4. $\underline{\blacksquare} + 6 = 15$

a. 9 **b.** 10
c. 21 **d.** not given

5. $6 + 2 + 7 = \underline{\blacksquare}$

a. 8 **b.** 9
c. 14 **d.** not given

6. What is the number?
five hundred forty

a. 504 **b.** 514
c. 540 **d.** not given

7. What digit is in the ten thousands place in 439,721?

a. 3 **b.** 4
c. 9 **d.** not given

8. Write the number:
$700 + 20 + 3$

a. 12 **b.** 723
c. 700,203 **d.** not given

9. Compare. 1,190 ● 1,265

a. $<$ **b.** $>$
c. $=$ **d.** not given

10. Round 329 to the nearest hundred.

a. 300 **b.** 330
c. 400 **d.** not given

11. Roy caught 8 fish. Jay caught 7 fish. How many fish did they catch in all?

a. 1 fish **b.** 15 fish
c. 87 fish **d.** not given

12. How many cars park on Sunday?

WEEKEND PARKING

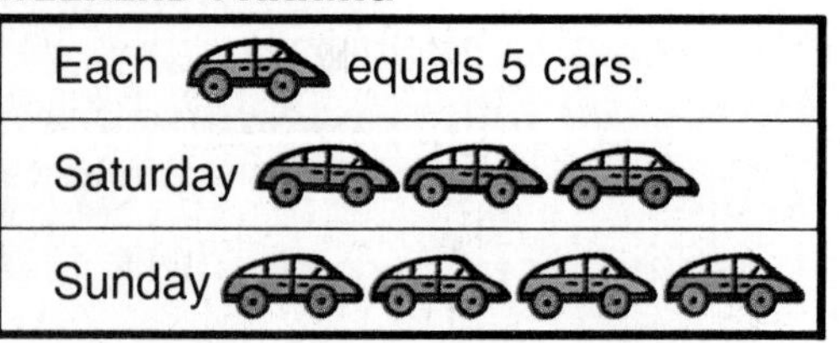

a. 4 cars **b.** 8 cars
c. 20 cars **d.** not given

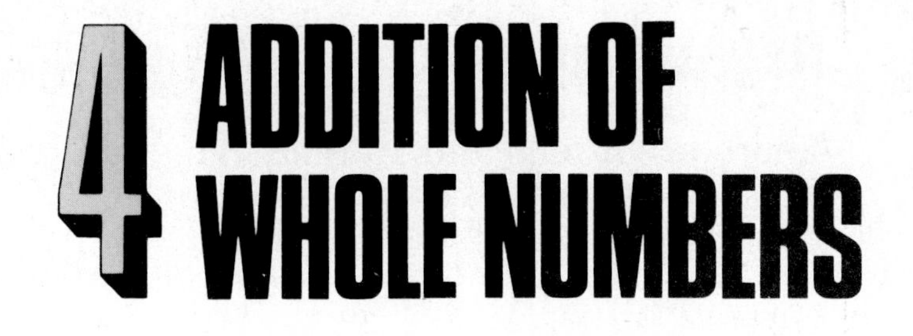

4 ADDITION OF WHOLE NUMBERS

How many pounds of garbage does your family throw away each week? You may be amazed! How could you find out?

Regrouping Ones

There are 3 tens and 16 ones. You can regroup 10 ones as 1 ten.

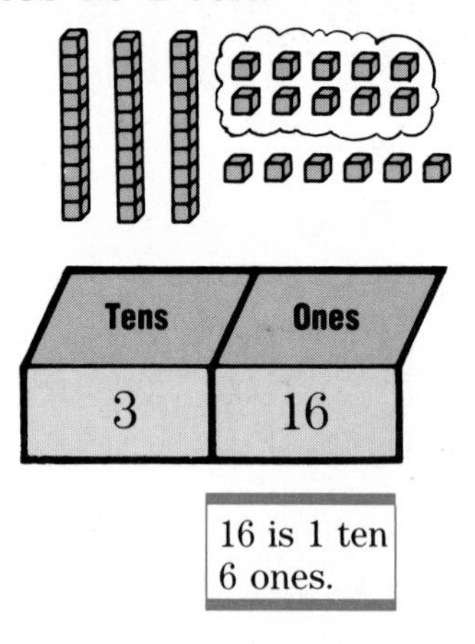

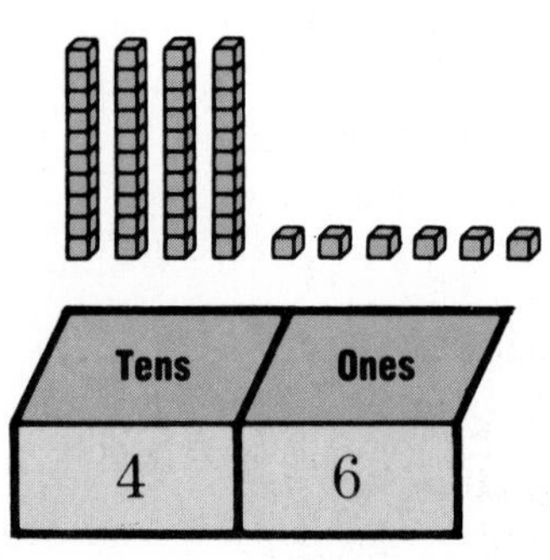

There are 4 tens and 6 ones.

Another example:

There are 3 tens and 10 ones. You can regroup 10 ones as 1 ten.

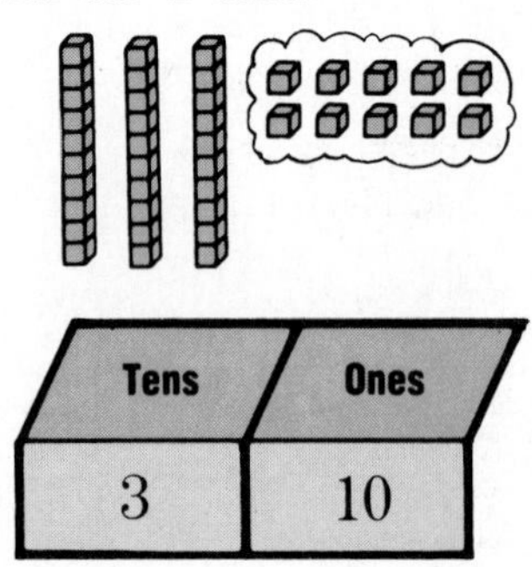

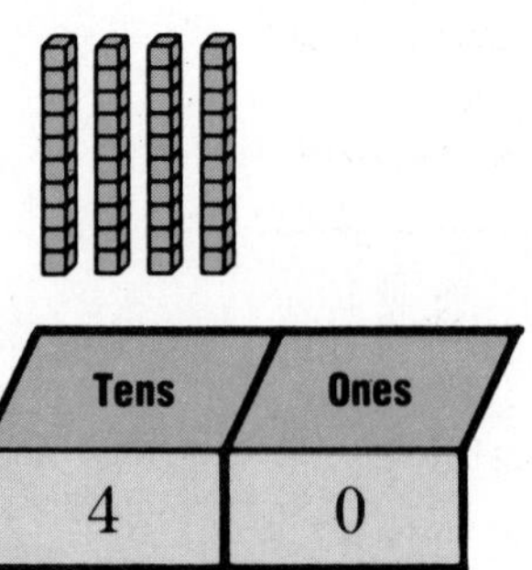

There are 4 tens and 0 ones.

Checkpoint Write the letter of the correct answer.

Regroup.

1.

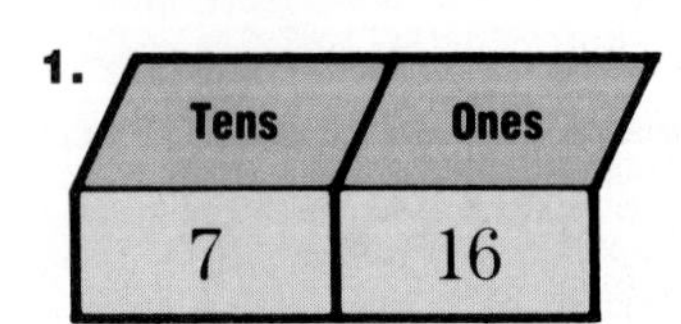

a. 7 tens 6 ones
b. 8 tens 6 ones
c. 8 tens 15 ones
d. 9 tens 6 ones

2.

Tens	Ones
4	10

a. 4 tens 0 ones
b. 5 tens 0 ones
c. 5 tens 9 ones
d. 50 tens 0 ones

Regroup. Write the new number of ones and tens.

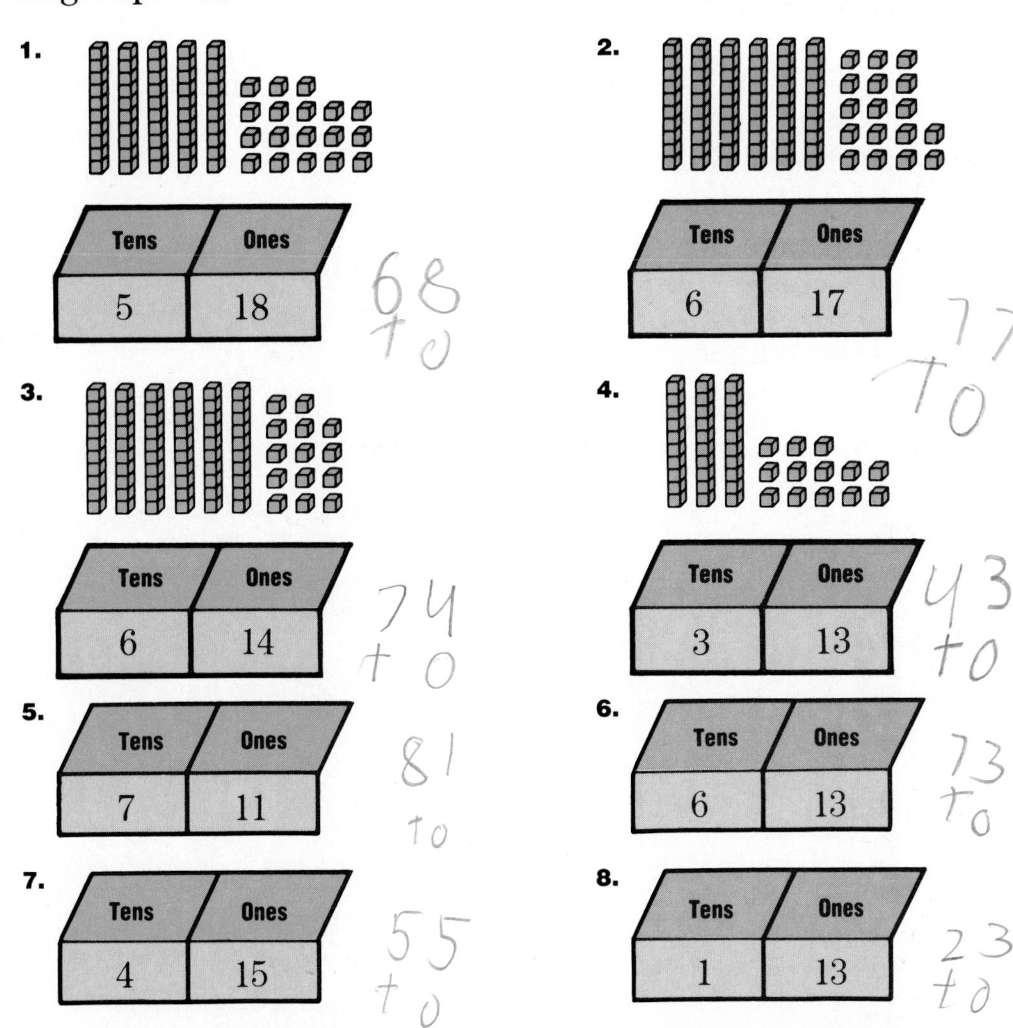

ANOTHER LOOK

Write the value of the underlined digit in each number.

1. 335,462   **2.** 525,873   **3.** 679,420

4. 207,661   **5.** 118,434   **6.** 841,911

Adding 2-Digit Numbers: Regrouping Ones

Antonio Albertando was the first person to swim across the English Channel and back again. He swam from England to France in 19 hours. He swam back to England in 24 hours. How many hours did it take Mr. Albertando to swim both ways?

You can add to find the number of hours he spent swimming.

Add 19 + 24.

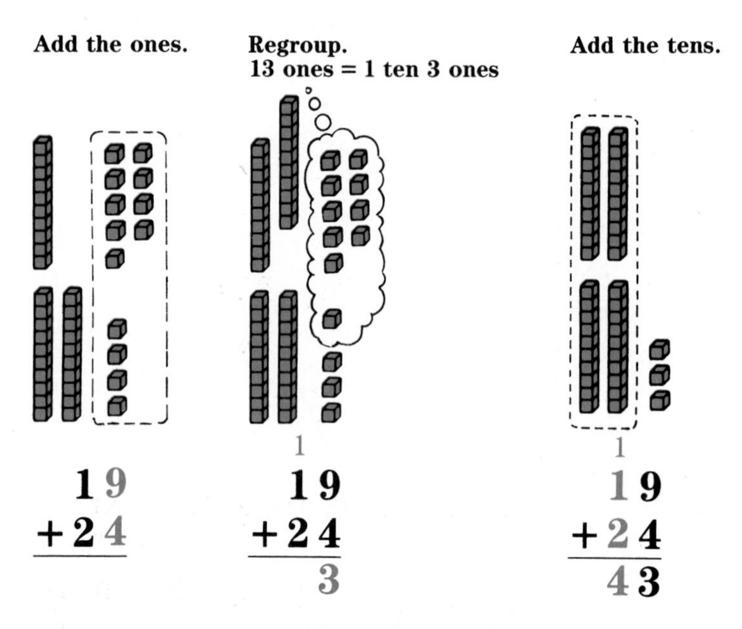

$$\begin{array}{r} 19 \\ +24 \\ \hline \end{array} \qquad \begin{array}{r} {}^{1} \\ 19 \\ +24 \\ \hline 3 \end{array} \qquad \begin{array}{r} {}^{1} \\ 19 \\ +24 \\ \hline 43 \end{array}$$

Mr. Albertando swam both ways in 43 hours.

Checkpoint Write the letter of the correct answer.

Add.

1. $\begin{array}{r} 35 \\ +47 \\ \hline \end{array}$

a. 12
b. 72
c. 82
d. 712

2. $\begin{array}{r} 16 \\ +\ 9 \\ \hline \end{array}$

a. 15
b. 25
c. 16
d. 115

3. 65 + 28 = ____

a. 21
b. 93
c. 6,528
d. 83

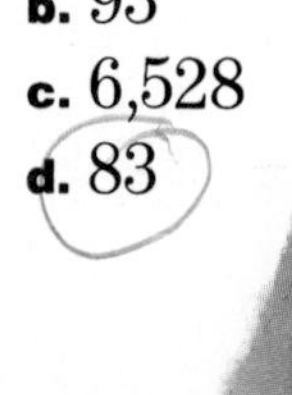

Add.

1. $36 + 52$	**2.** $42 + 5$	**3.** $74 + 25$	**4.** $83 + 12$	**5.** $62 + 3$	**6.** $27 + 10$
7. $19 + 54$	**8.** $7 + 36$	**9.** $42 + 38$	**10.** $15 + 6$	**11.** $67 + 24$	**12.** $21 + 49$
13. $54 + 18$	**14.** $23 + 67$	**15.** $45 + 3$	**16.** $34 + 29$	**17.** $86 + 6$	**18.** $46 + 22$
19. $31 + 18$	**20.** $63 + 29$	**21.** $12 + 36$	**22.** $24 + 7$	**23.** $53 + 15$	**24.** $44 + 38$

25. 65 + 13 = ____ **26.** 6 + 43 = ____ **27.** 34 + 14 = ____

28. 75 + 5 = ____ **29.** 77 + 18 = ____ **30.** 57 + 25 = ____

31. 54 + 24 = ____ **32.** 52 + 9 = ____ **33.** 33 + 26 = ____

34. 43 + 8 = ____ **35.** 27 + 14 = ____ **36.** 61 + 17 = ____

Solve.

37. A ferry crossed the channel 43 times in one week. It crossed 27 times the next week. How many times did it cross the channel in the two weeks?

38. A ferry carried 38 large cars and 15 small cars across the channel in one trip. How many cars did it carry?

CHALLENGE

Find two ways to make each sum.

1. ____ + ____ = 33 **2.** ____ + ____ = 21 **3.** ____ + ____ = 80

4. ____ + ____ = 42 **5.** ____ + ____ = 54 **6.** ____ + ____ = 36

More Practice, page 408

Adding 2-Digit Numbers: Regrouping Ones and Tens

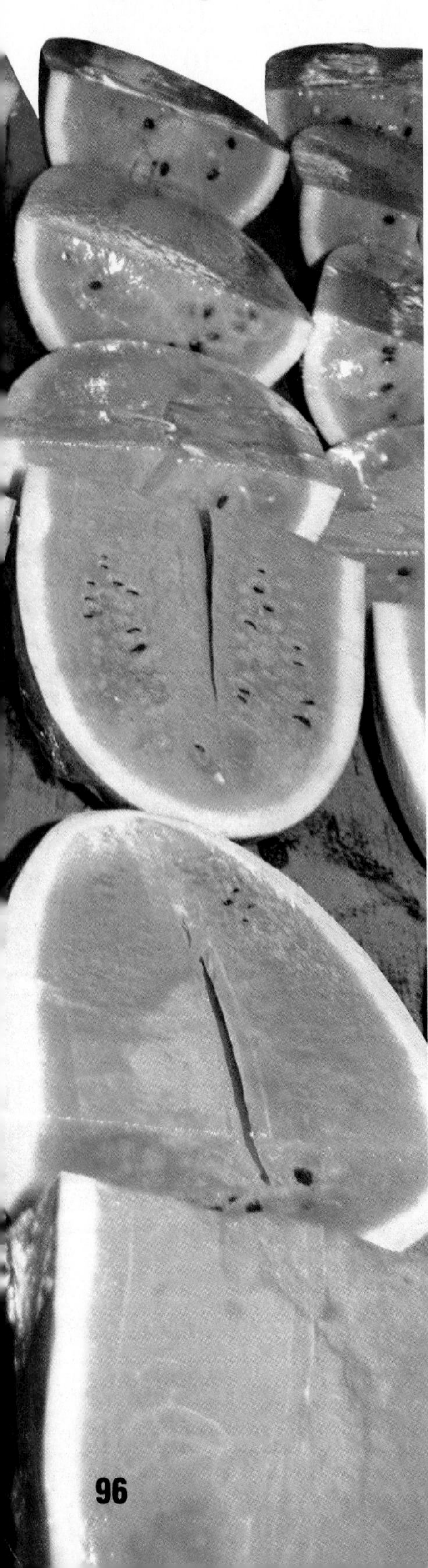

A. The largest watermelon at the state fair weighs 92 pounds. The next-largest watermelon weighs 37 pounds. What is the total number of pounds the watermelons weigh?

You can add to find how many pounds.

Add 92 + 37.

Add the ones.

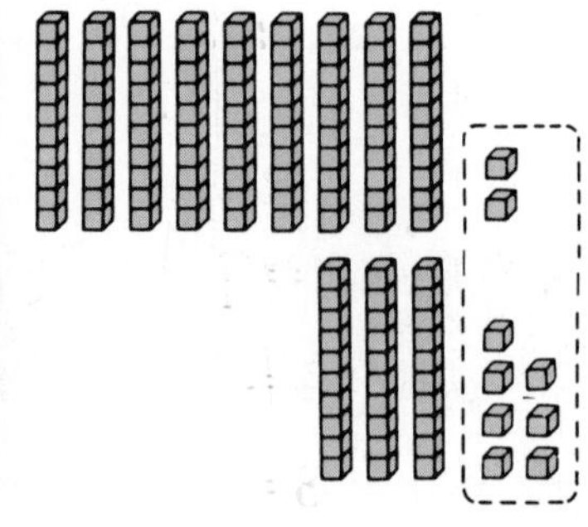

$$\begin{array}{r} 92 \\ +37 \\ \hline 9 \end{array}$$

Add the tens.
Regroup.
12 tens = 1 hundred 2 tens.

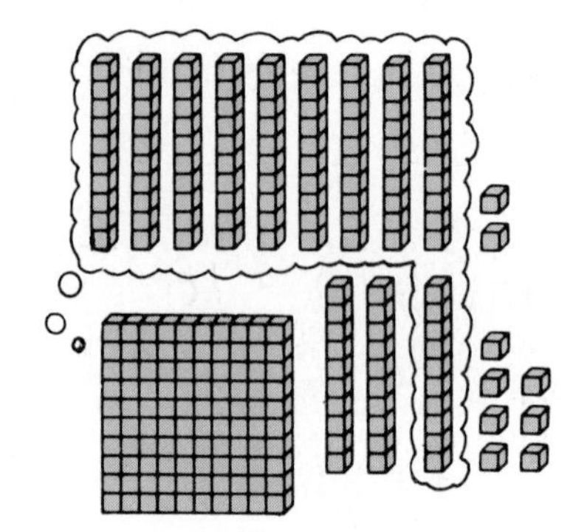

$$\begin{array}{r} 92 \\ +37 \\ \hline 129 \end{array}$$

Both watermelons weigh 129 pounds.

B. Sometimes you regroup both ones and tens.

Add 77 + 75.

Add the ones.
Regroup.
12 ones = 1 ten 2 ones

$$\begin{array}{r} 1 \\ 77 \\ +75 \\ \hline 2 \end{array}$$

Add the tens.
Regroup.
15 tens = 1 hundred 5 tens

$$\begin{array}{r} 1 \\ 77 \\ +75 \\ \hline 152 \end{array}$$

Add.

1. 45 + 71	2. 43 + 86	3. 92 + 34	4. 67 + 82	5. 37 + 70	6. 61 + 94
7. 79 + 52	8. 84 + 28	9. 96 + 47	10. 94 + 8	11. 27 + 73	12. 48 + 65
13. 25 + 76	14. 75 + 45	15. 58 + 72	16. 48 + 59	17. 65 + 46	18. 49 + 83
19. 31 + 72	20. 92 + 9	21. 71 + 32	22. 82 + 59	23. 15 + 99	24. 83 + 88

25. 76 + 33 = ____ 26. 97 + 6 = ____ 27. 31 + 79 = ____

28. 47 + 57 = ____ 29. 68 + 34 = ____ 30. 78 + 54 = ____

31. 91 + 38 = ____ 32. 85 + 78 = ____ 33. 61 + 82 = ____

Solve.

34. Lili planted a 7-foot tree. Since then, it has grown 15 feet taller. How tall is the tree now?

35. Sid grew a 39-pound melon and a 48-pound melon. Ken grew two 45-pound melons. Whose melons weigh more?

FOCUS: MENTAL MATH

It is easy to add two numbers that end in 5.
The numbers always have a sum that ends in 0.

45 + 35

Add 45 + 35.
Add the ones. 5 + 5 = 10
Then add the tens. 40 + 30 = 70 So, 70 + 10 = 80.

Add mentally.

1. 25 + 35 = ____ 2. 45 + 15 = ____ 3. 55 + 65 = ____

PROBLEM SOLVING
Identifying Extra Information

A problem may have more information than you need to answer the question.

> In 1928, a man crossed the Atlantic Ocean in a canoe that had a sail. The trip took 58 days. He rested for 7 days. Then, he returned on a steamship that traveled for 14 days. For how many days was the man at sea?

If you think a problem has extra information, follow these steps:

___ **1.** Study the question.
For how many days was the man at sea?

___ **2.** List the information in the problem.
a. The year was 1928.
b. The canoe trip took 58 days.
c. The man rested for 7 days.
d. The steamship trip took 14 days.

___ **3.** Cross out the information that will not help you answer the question. (Cross out choices *a* and *c*.)

___ **4.** Use the information that is left to solve the problem.

Solve: **58 days in the canoe**
+ 14 days in the steamship
72 days at sea

The man was at sea for 72 days.

Choose the fact that is not needed in order to solve the problem. Write the letter of the correct answer.

1. Two brothers took turns flying the first airplane. Wilbur flew it for 24 seconds and then for 59 seconds. Orville flew the plane for a total of 36 seconds. For how many seconds did Wilbur fly?

   a. Wilbur flew for 24 seconds.
   b. Orville flew for 36 seconds.
   c. Wilbur flew again for 59 seconds.

2. The first person to fly around the world alone was Mr. Post. Flying to his first stop took 1 day. In the next 7 days, he made 9 stops to complete his trip. How many days did the trip take?

   a. Flying to his first stop took 1 day.
   b. It took 7 more days to complete the trip.
   c. He made 9 stops in the next 7 days.

Solve. Use only the needed facts.

3. Amelia Earhart was the second person to fly across the Atlantic Ocean alone. Her flight took 14 hours. The first flight took 9 more hours. How long did the first flight take?

4. The first steamship to cross the Atlantic Ocean used steam power for 3 days and then used only its sails for 26 days. The ship had 32 rooms for passengers. How many days did the trip take?

5. In 1911, a 9-year-old girl and an 11-year-old girl rode horses from New York to San Francisco. They rode 58 miles per day. How far did they ride in 2 days?

★6. Columbus sailed with 3 ships. The *Santa Maria* had a crew of 40 sailors. The *Pinta* had 26 sailors, and the *Nina* had 24 sailors. How many sailors traveled with Columbus?

Column Addition

A. The biggest pizza in the world was baked in Glens Falls, New York. The cooks used 78 pounds of cheese, 45 pounds of green peppers, and 21 pounds of onions as a topping. How many pounds of topping did the cooks use for this pizza?

You can add to find how many pounds they used.

Add 78 + 45 + 21.

Add the ones.
Regroup.
14 ones = 1 ten 4 ones

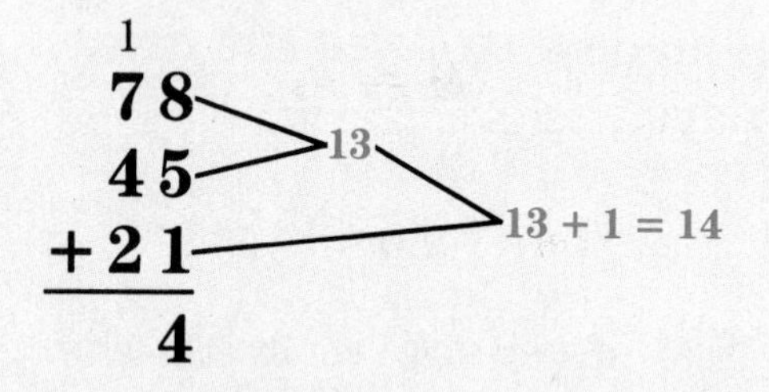

Add the tens.
Regroup.
14 tens = 1 hundred 4 tens

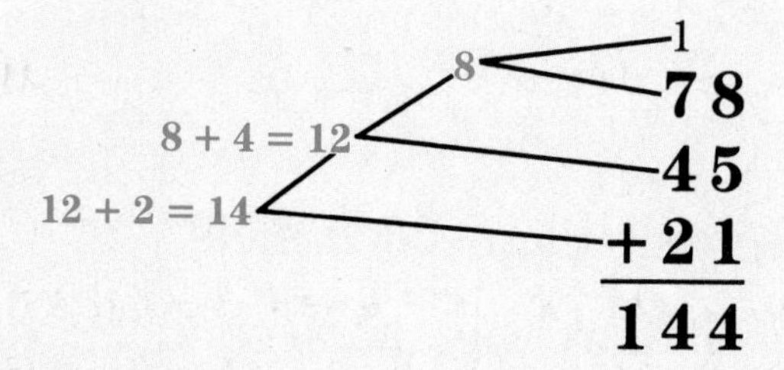

The cooks used 144 pounds of topping.

B. You can check by adding up.

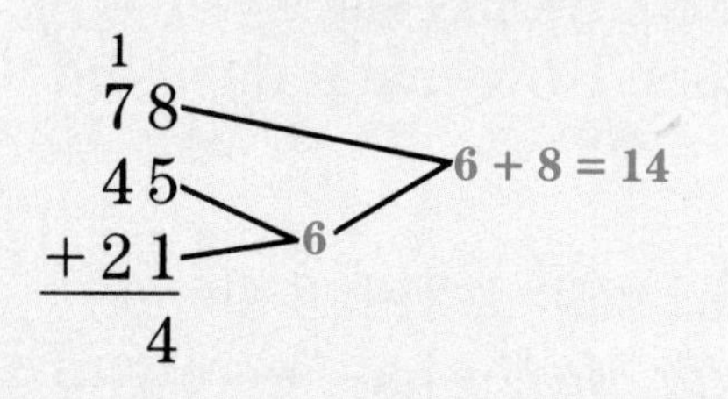

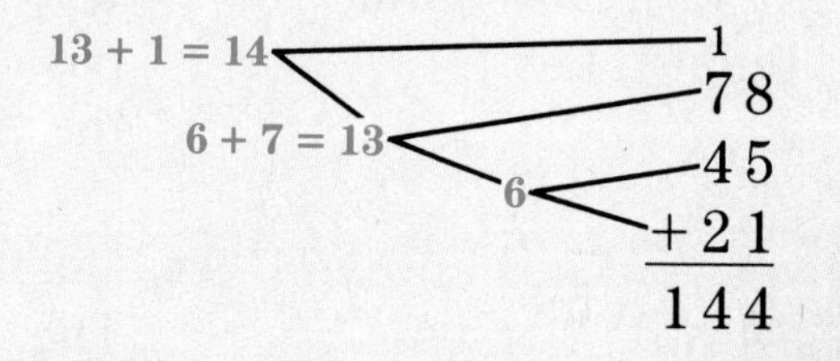

Other examples:

1	2		1
		73	42
42	77	2	34
14	55	51	28
+45	+89	+61	+13
101	221	187	117

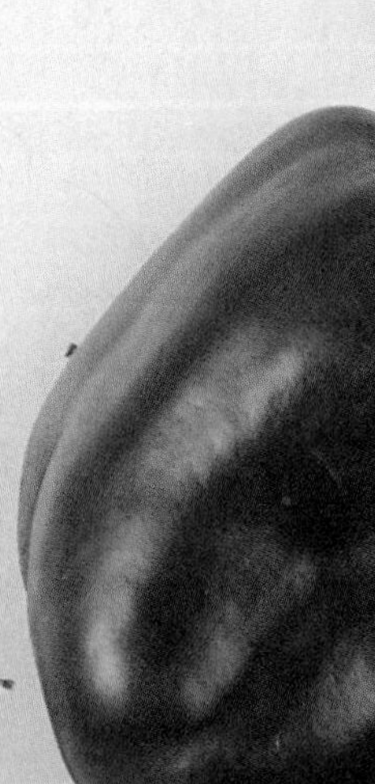

Add.

1.	2.	3.	4.	5.
14	22	48	27	62
5	16	79	10	35
+ 13	+ 31	+ 14	+ 23	+ 3

6.	7.	8.	9.	10.
11	25	33	20	43
72	67	49	74	53
34	98	88	41	7
+ 12	+ 31	+ 40	+ 13	+ 19

11. 11 + 22 + 33 = ____ **12.** 72 + 23 + 4 = ____ **13.** 10 + 38 + 21 = ____

14. 35 + 18 + 29 = ____ **15.** 59 + 15 + 21 = ____ **16.** 61 + 62 + 9 = ____

17. 81 + 7 + 13 = ____ **18.** 70 + 11 + 24 = ____ **19.** 62 + 43 + 24 = ____

20. 29 + 15 + 16 + 27 = ____ **21.** 21 + 8 + 52 + 59 = ____

Solve.

22. Clinton School sent 39 third graders, 27 fourth graders, and 8 teachers to see the world's biggest pizza. Each person ate a piece of the pizza. How many pieces of the pizza did the people from Clinton School eat?

23. A group of 72 cooks tried to bake the world's largest pizza. While the pie baked, 4 people watched it. Then 55 other people sliced it. How many people in all helped to make the pizza?

★24. Use the following information to write and solve your own word problem: 28 pounds of cheese, 9 pounds of onions, 7 pounds of peppers, 10 pounds of sausage, 15 pounds of tomato sauce.

Mental Math: Adding Tens and Hundreds

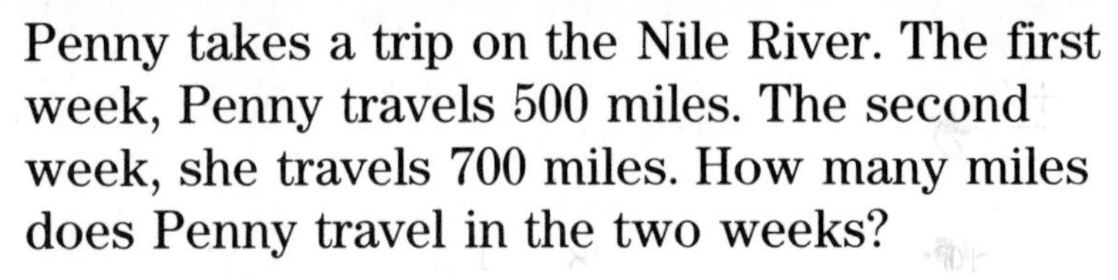

Penny takes a trip on the Nile River. The first week, Penny travels 500 miles. The second week, she travels 700 miles. How many miles does Penny travel in the two weeks?

You have to add only the number of hundreds to find how many miles she travels.

Add 500 + 700.

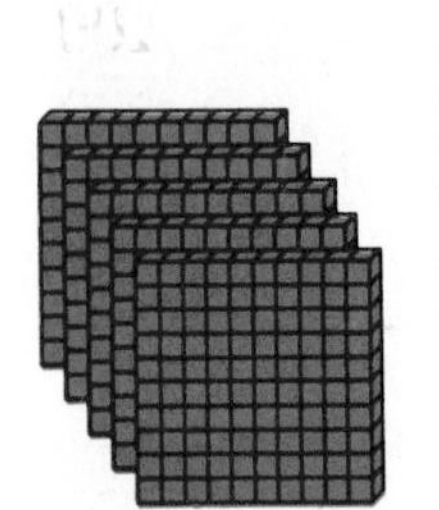 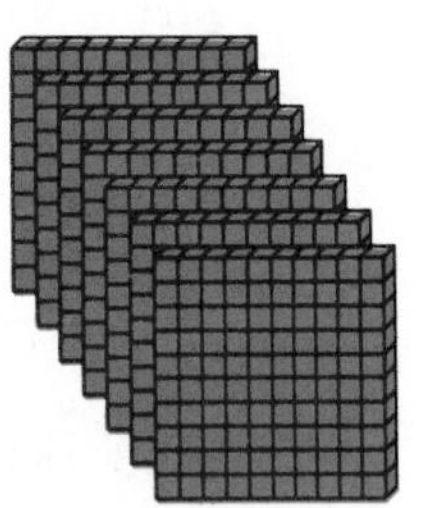 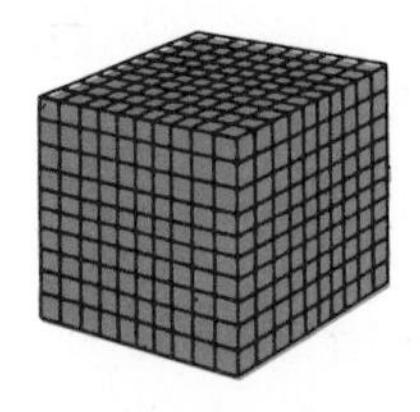 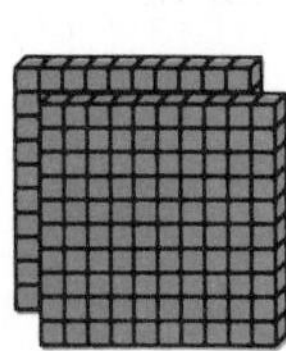

5 hundreds	+	7 hundreds	=	12 hundreds
500	+	700	=	1,200

In the two weeks, Penny travels 1,200 miles.

Other examples:

20	⟶	2 tens		
60	⟶	6 tens		
+ 30	⟶	+ 3 tens		
110		11 tens		

200	⟶	2 hundreds
+ 600	⟶	+ 6 hundreds
800		8 hundreds

Checkpoint Write the letter of the correct answer.

Add.

1. 50 + 60 =

a. 11 **b.** 110
c. 1,010 **d.** 1,100

2. 10 + 50 + 40 =

a. 10 **b.** 90
c. 100 **d.** 1,000

3. 600 + 400 + 300 =

a. 130 **b.** 300
c. 1,300 **d.** 13,000

Add.

1. 30 + 60
2. 50 + 20
3. 70 + 20
4. 80 + 30
5. 10 + 40
6. 60 + 10
7. 20 + 50 + 30
8. 40 + 30 + 50
9. 10 + 70 + 60
10. 30 + 80 + 10
11. 90 + 40 + 60
12. 60 + 10 + 90
13. 300 + 500
14. 500 + 700
15. 800 + 100
16. 100 + 400
17. 600 + 200
18. 600 + 500 + 200
19. 200 + 400 + 100
20. 700 + 300 + 400
21. 500 + 800 + 400
22. 400 + 100 + 600
23. 40 + 50 = ■
24. 30 + 20 = ■
25. 60 + 70 = ■
26. 10 + 70 = ■
27. 40 + 70 = ■
28. 80 + 20 = ■
29. 20 + 70 + 10 = ■
30. 80 + 50 + 30 = ■
31. 10 + 60 + 40 = ■
32. 90 + 30 + 50 = ■
33. 800 + 400 = ■
34. 900 + 200 = ■
35. 700 + 400 = ■
36. 400 + 100 = ■
37. 500 + 600 + 800 = ■
38. 600 + 400 + 700 = ■
39. 200 + 500 + 100 = ■
40. 700 + 100 + 800 = ■

Solve.

41. In Egypt, the Nile flows 400 miles from the Aswan Dam to Luxor. It flows 200 miles from Luxor to Cairo. How far down the river is it from the Aswan Dam to Cairo?

★42. Penny travels 1,200 miles on her trip down the Nile. She would have to go another 2,900 miles to travel the entire length of the river. How long is the Nile?

Estimating Sums

A. Sometimes you need an exact answer to make a decision. Other times all you need is an **estimate.**

There were 273 dogs, 116 cats, and 228 birds at the pet club's fair. The club members were hoping to show at least 500 pets. Did they reach their goal?

You can estimate to find whether they reached their goal.

When you estimate, look at the front digits.

Estimate: 273 + 116 + 228.

Add the front digits.	Write 0's for the other digits.
273	273
116	116
+228	+228
5	about 500

Yes, the club members reached their goal.

B. You can estimate money in the same way.

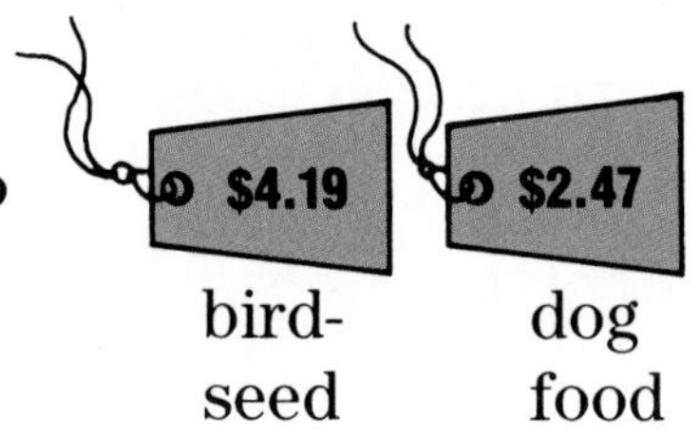

Tyrone has \$5.00. Does he have enough money to buy a box of birdseed and a bag of dog food?

Estimate: \$4.19 + \$2.47.

Line up the cents points.	Add the dollars.	Write 0's for the cents.
$4.19	$4.19	$4.19
+ 2.47	+ 2.47	+ 2.47
	6	about $6.00

$5.00 < $6.00

No, he does not have enough money for both.

Estimate. Write the letter of the correct answer.

1. $\begin{array}{r} 119 \\ +226 \\ \hline \end{array}$

a. 30 **b.** 300 **c.** 3,000

2. $\begin{array}{r} 432 \\ +219 \\ \hline \end{array}$

a. 400 **b.** 500 **c.** 600

3. $\begin{array}{r} 558 \\ +328 \\ \hline \end{array}$

a. 500 **b.** 800 **c.** 1,000

4. $\begin{array}{r} 237 \\ 315 \\ +128 \\ \hline \end{array}$

a. 300 **b.** 600 **c.** 1,000

5. $\begin{array}{r} 519 \\ 227 \\ +318 \\ \hline \end{array}$

a. 700 **b.** 900 **c.** 1,000

Estimate.

6. $\begin{array}{r} 215 \\ 107 \\ +346 \\ \hline \end{array}$

7. $\begin{array}{r} 1,258 \\ +8,364 \\ \hline \end{array}$

8. $\begin{array}{r} 5,153 \\ +1,476 \\ \hline \end{array}$

9. $\begin{array}{r} 2,176 \\ 1,059 \\ +3,108 \\ \hline \end{array}$

10. $\begin{array}{r} 4,127 \\ 2,154 \\ +1,235 \\ \hline \end{array}$

Use the price list. Estimate to decide if you have enough money.

11. Is $5.00 enough for *A* and *E*?

12. Is $5.00 enough for *D* and *E*?

13. Is $10.00 enough for *A* and *D*?

14. Is $25.00 enough for *B* and *C*?

A $1.17

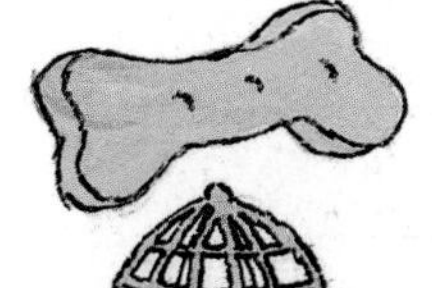

B $16.56

C $24.95

D $4.75

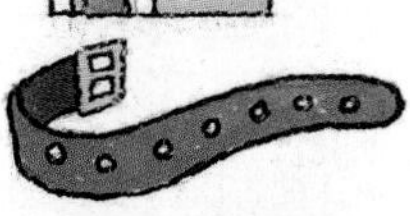

E $2.37

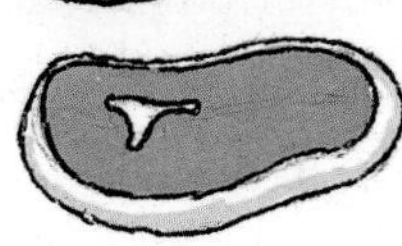

Add.

1. 28 + 51 = ____ **2.** 34 + 14 = ____ **3.** 22 + 55 = ____

4. 47 + 39 = ____ **5.** 66 + 29 = ____ **6.** 89 + 78 = ____

7. 17 + 21 + 40 = ____ **8.** 33 + 52 + 36 = ____ **9.** 30 + 15 + 53 = ____

10. 50 + 70 = ____ **11.** 400 + 400 = ____ **12.** 600 + 300 = ____

PROBLEM SOLVING
Estimation

Sometimes, you can solve a problem by estimating. At other times, it doesn't make sense to estimate.

A. Martha's club is planning a picnic. The members have $97.00 to spend on lunch. Martha orders 31 tuna sandwiches for $51.15. Then, she orders 23 egg sandwiches for $31.74. Does the club have enough money to pay for the sandwiches?

You can estimate to find if they have enough money.

Estimate $51.15 + $31.74.

Add the front digits.

```
  $51.15
+  31.74
  ------
   8
```

Write 0's for the other digits.

```
        $51.15
      +  31.74
        ------
about   $80.00
```

Yes, they have enough money.

B. The club is buying theater tickets. There are 31 people who want to see the play. Then, 24 more people decide to see the play. How many tickets should the club buy?

Should you estimate to find how many tickets the club should buy? why or why not?

You need to find an exact sum. Otherwise, you cannot be sure that there will be a ticket for each club member who wants to see the play.

Decide whether you should estimate. Write the letter of the correct answer.

1. Watermelon for all club members costs \$23.50. Juice costs \$11.25. The club has \$43.25 to spend. Do members have enough money?

   a. can estimate
   b. cannot estimate

2. A boat seats 28 people. First, 13 people board the boat. Then, 8 people board. Then, 9 more people want to board. Can they all sit in the boat?

   a. can estimate
   b. cannot estimate

Use the charts to solve. Find an estimate or an exact answer as needed.

BUS FARE TO THE PICNIC

Adults	\$0.75
Students	\$0.55
Children under 6	\$0.45

3. Linda is a student. She is taking her 5-year-old sister to the picnic. Linda gives \$2.00 to pay for their bus fare. Will she receive change?

4. Marla paid the bus fare for her brother and herself. They are both students. How much money did she give the driver?

5. A picnicker buys 2 packages of paper plates. The large package has 36 plates. The other package has 32. Are there at least 55 plates?

6. Cups are sold in packages of 24. Each person at the picnic needs 1 cup. There are 55 people at the picnic. Are 2 packages of cups enough?

NUMBER OF PEANUTS FOUND IN THE PEANUT HUNT

Team 1		Team 2		Team 3		Team 4	
Ari	25	Rusty	23	Andrea	16	Jack	12
Marla	23	Jon	14	Alvin	45	Andy	25
Martha	28	Mike	35	Marc	41	Lisa	16

7. Which team came in first? Which team came in last?

★8. Which team came in second?

Adding 3-Digit Numbers: Regrouping Once

A. Astronauts on the spacecraft *Apollo 9* spent 241 hours in space. Astronauts on *Apollo 10* spent 192 hours in space. What is the total number of hours that the astronauts spent in space?

You can add to find the number of hours.

Add 241 + 192.

Add the ones.	Add the tens. Regroup. 13 tens = 1 hundred 3 tens	Add the hundreds.
241 +192 3	1 241 +192 33	1 241 +192 433

The astronauts spent 433 hours in space.

B. You can add money the same way you add whole numbers.

Add \$3.34 + \$1.28.

Line up the cents points.	Add as you would add whole numbers.
\$3.34 + 1.28	1 \$3.34 + 1.28 \$4.62

Remember to write the dollar sign and the cents point.

Other examples:

1 270 432 +714 1,416	1 \$5.25 0.71 + 2.13 \$8.09	1 223 15 146 +311 695

Add.

1. 356 + 537
2. 129 + 252
3. 636 + 55
4. \$1.73 + 7.18
5. \$4.38 + 1.45
6. \$3.87 + 4.42
7. 647 + 271
8. 632 + 81
9. \$8.36 + 0.92
10. \$6.25 + 1.91
11. 421 + 302 + 534
12. 588 + 102 + 201
13. 714 + 502 + 231
14. \$2.50 + 1.67 + 4.12
15. \$6.43 + 7.52 + 1.04
16. 241 + 312 + 127 + 318
17. 361 + 124 + 81 + 173
18. 202 + 113 + 227 + 134
19. \$3.71 + 1.20 + 0.83 + 2.44
20. \$0.46 + 1.47 + 0.80 + 2.25

21. 922 + 601 = ____
22. 244 + 361 = ____
23. \$5.15 + \$0.66 = ____
24. 323 + 450 + 621 = ____
25. 565 + 201 + 29 = ____

Solve.

26. *Apollo 7* orbited Earth 163 times. *Apollo 9* orbited Earth 151 times. How many orbits did both make?

27. Some spacecraft orbited the moon 108 times. Others made 139 orbits. How many orbits were made in all?

CALCULATOR

Complete the pattern. Use a calculator to help you.

1. 3, 36, ____, 102, 135, ____, 201, 234, ____, 300
2. 21, 38, ____, 72, ____, 106, 123, ____, 157, ____

Adding 3-Digit Numbers: Regrouping More Than Once

The largest marching band of all time assembled in 1973. A group of 998 lined up by 10:00 A.M. Then 978 more arrived by 11:00 A.M. How many marchers were there in the band?

You can add to find how many marchers.

Add 998 + 978.

Add the ones.
Regroup. 16 ones = 1 ten 6 ones

$$\begin{array}{r} {\scriptstyle 1} \\ 998 \\ +978 \\ \hline 6 \end{array}$$

Add the tens.
Regroup. 17 tens = 1 hundred 7 tens

$$\begin{array}{r} {\scriptstyle 1\,1} \\ 998 \\ +978 \\ \hline 76 \end{array}$$

Add the hundreds.
Regroup. 19 hundreds = 1 thousand 9 hundreds

$$\begin{array}{r} {\scriptstyle 1\,1} \\ 998 \\ +978 \\ \hline 1{,}976 \end{array}$$

There were 1,976 marchers in the band.

Other examples:

$$\begin{array}{r} {\scriptstyle 1\,1} \\ 267 \\ +184 \\ \hline 451 \end{array} \qquad \begin{array}{r} {\scriptstyle 1\,1} \\ 582 \\ +\ \ 79 \\ \hline 661 \end{array} \qquad \begin{array}{r} {\scriptstyle 1\,2} \\ 917 \\ 658 \\ 45 \\ +524 \\ \hline 2{,}144 \end{array} \qquad \begin{array}{r} {\scriptstyle 1\ \ 1} \\ \$4.33 \\ 0.75 \\ +\ \ 3.12 \\ \hline \$8.20 \end{array}$$

Checkpoint Write the letter of the correct answer.

Add.

1. $\begin{array}{r} 285 \\ +719 \\ \hline \end{array}$

a. 904
b. 994
c. 1,004
d. 9,914

2. $\begin{array}{r} 571 \\ 98 \\ +163 \\ \hline \end{array}$

a. 632
b. 832
c. 822
d. 62,214

3. $1.95 + $0.68 = ■

a. $2.63
b. $2.53
c. $1.63
d. $115.13

Add.

1. $752 + 239$
2. $565 + 27$
3. $432 + 219$
4. $285 + 306$
5. $646 + 28$
6. $372 + 945$
7. $298 + 103$
8. $725 + 196$
9. $\$5.84 + 4.92$
10. $\$5.56 + 1.86$
11. $481 + 245 + 67$
12. $604 + 132 + 155$
13. $237 + 102 + 369 + 276$
14. $\$5.86 + 1.04 + 0.57$
15. $\$8.23 + 1.46 + 0.98 + 2.67$

16. $307 + 584 = \blacksquare$
17. $815 + 79 = \blacksquare$
18. $127 + 365 = \blacksquare$
19. $763 + 271 = \blacksquare$
20. $629 + 380 = \blacksquare$
21. $\$9.52 + \$4.63 = \blacksquare$
22. $514 + 209 + 77 = \blacksquare$
23. $\$2.49 + \$5.25 + \$7.80 = \blacksquare$

Solve.

24. Sally jumps rope for 135 minutes one day, 91 minutes the next day, and 114 minutes the third day. How long does she jump in all?

25. Dave did 83 sit-ups on Monday, 107 on Tuesday, and 121 on Wednesday. How many sit-ups did Dave do?

FOCUS: MENTAL MATH

Try to add 324 and 198 in your head.
Think: 198 is close to 200. Add 324 and 200.
Then subtract the 2 that you added to 198.

$324 + 200 = 524; 524 - 2 = 522$

Add mentally.

1. $284 + 599 = \blacksquare$
2. $143 + 299 = \blacksquare$
3. $499 + 396 = \blacksquare$
4. $198 + 637 = \blacksquare$

★5. $163 + 301 = \blacksquare$

★6. $301 + 252 = \blacksquare$

More Practice, page 408

Adding 4-Digit Numbers

A group of people gathered to play the world's largest game of musical chairs. There were 2,407 people ready to play. Then, another 2,896 arrived. How many people played the game?

You can add to find how many people played.

Add 2,407 + 2,896.

Add the ones. Regroup the 13 ones.	Add the tens. Regroup the 10 tens.	Add the hundreds. Regroup the 13 hundreds.	Add the thousands.
1	1 1	1 1 1	1 1 1
2,407	2,407	2,407	2,407
+2,896	+2,896	+2,896	+2,896
3	03	303	5,303

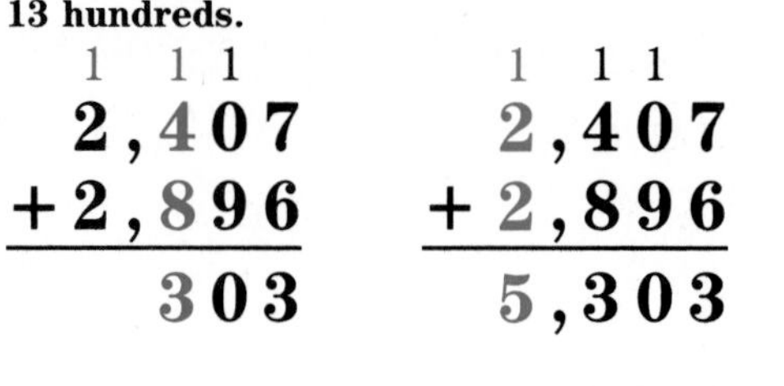

5,303 people played the game.

Other examples:

			1 1
	1 1	1	7,856
2,310	4,513	$36.20	6,415
+ 4,056	+2,849	+ 8.34	+1,213
6,366	7,362	$44.54	15,484

Checkpoint Write the letter of the correct answer.

Add.

1. 4,758 + 1,693

a. 8,341 **b.** 6,451 **c.** 5,341 **d.** 6,441

2. 3,931 + 1,536 + 5,988

a. 2,455 **b.** 11,355 **c.** 11,445 **d.** 11,455

3. $45.47 + $28.95 = ▒

a. $73.42 **b.** $74.42 **c.** $75.42 **d.** $76.42

Add.

1. $\begin{array}{r} 3,154 \\ +2,362 \\ \hline \end{array}$

2. $\begin{array}{r} 4,285 \\ +3,194 \\ \hline \end{array}$

3. $\begin{array}{r} 7,853 \\ +1,080 \\ \hline \end{array}$

4. $\begin{array}{r} 6,482 \\ +3,057 \\ \hline \end{array}$

5. $\begin{array}{r} 5,387 \\ +\ \ 709 \\ \hline \end{array}$

6. $\begin{array}{r} 6,507 \\ +\ \ 536 \\ \hline \end{array}$

7. $\begin{array}{r} 2,332 \\ +5,758 \\ \hline \end{array}$

8. $\begin{array}{r} 6,469 \\ +1,923 \\ \hline \end{array}$

9. $\begin{array}{r} 5,586 \\ +1,609 \\ \hline \end{array}$

10. $\begin{array}{r} \$29.24 \\ +\ \ 41.83 \\ \hline \end{array}$

11. $\begin{array}{r} 3,472 \\ +3,469 \\ \hline \end{array}$

12. $\begin{array}{r} \$65.98 \\ +\ \ 43.75 \\ \hline \end{array}$

13. $\begin{array}{r} \$70.39 \\ +\ \ 25.78 \\ \hline \end{array}$

14. $\begin{array}{r} 5,237 \\ +\ \ 782 \\ \hline \end{array}$

15. $\begin{array}{r} \$98.22 \\ +\ \ 46.75 \\ \hline \end{array}$

16. $\begin{array}{r} 1,739 \\ +4,625 \\ \hline \end{array}$

17. $\begin{array}{r} 3,579 \\ 4,138 \\ +1,253 \\ \hline \end{array}$

18. $\begin{array}{r} 8,364 \\ 5,907 \\ +4,716 \\ \hline \end{array}$

★19. $\begin{array}{r} 4,538 \\ 1,409 \\ 1,711 \\ +2,626 \\ \hline \end{array}$

★20. $\begin{array}{r} \$22.70 \\ 8.25 \\ 36.42 \\ +\ \ 59.81 \\ \hline \end{array}$

21. 1,497 + 2,361 = ____

22. 5,842 + 6,590 = ____

23. $71.53 + $2.89 = ____

24. 7,188 + 1,452 = ____

Solve.

25. Ed spent $27.96 to go to the game. He spent $10.07 for food and $10.54 for T-shirts. How much did he spend?

26. Seats cost $12.75 for adults, and $6.25 for children. How much do seats cost for Ty and his parents?

ANOTHER LOOK

Use the fewest coins to show the amount.

1. $0.59

____ quarters ____ dimes

____ nickels ____ pennies

2. $1.15

____ quarters ____ dimes

____ nickels ____ pennies

More Practice, page 408

PROBLEM SOLVING

Using Outside Sources Including the Infobank

Sometimes the information you need to solve a problem is not given in the problem. You can often go to newspapers, magazines, or books to find the information you need.

The computer was invented 286 years after the first adding machine was invented. When was the computer invented?

You could answer this question if you knew when the adding machine was invented. In the back of this book, there are several charts and pictures that give information. These are part of the Infobank. Look on page 401 for the chart that will tell you when the adding machine was invented. When you have this information, answer the question.

adding machine		years		computer
1642	+	286	=	1928

The computer was invented in 1928.

Would you need to look for more information to answer each question? Write *yes* or *no*.

1. Which fabulous first invention took place before the year A.D. 1000?

2. Crossword puzzles first appeared in 1875. Children's magazines were first published in 1751. How many years separate the two?

Solve each problem. Use the Infobank on pages 401–404 if you need other facts.

3. A turkey travels 55 miles in an hour. Is that farther than a pig can travel in an hour?

4. Mr. Steele is shopping at Billy's Discount Store. He has $4.50. Can he buy both light bulbs and batteries?

5. Jack planted a garden. He grew 34 yellow flowers, 12 white flowers, and 22 purple flowers. How many flowers did he grow?

6. The Greentown Band has 12 French-horn players. How many fewer French-horn players does the New York Philharmonic have?

7. The *Atocha* is a sunken treasure ship. Did divers find more silver bars or more gold items on it?

8. A fox can travel 42 miles in an hour. Which animal can travel 3 more miles in an hour than a fox can?

9. The zipper was invented in 1891. What other fabulous first invention took place in the 1800's?

10. A zebra can travel 9 more miles in an hour than a greyhound can. How far can a zebra travel in an hour?

11. The Declaration of Independence was signed in 1776. The telephone was invented 100 years later. What year was that?

12. On Wilson School's Career Day, 8 students signed up to talk to the artist. How many of those 8 students were in the fourth grade?

READING MATH

Read this riddle.

Why did the cat walk to the store?
He didn't know how to drive.

What is the first word you read? What is the fifth word? What are the last two words?

How do you read a book? Your eyes follow a path.

- You start at the top of the page.
- You read from left to right across each line.
- You go back to the left and move down one line.
- Then you read from the left to the right again.

As we read math our eyes move in many different directions. How would you solve this problem? What number would you write first? second? third?

$$\begin{array}{r} 37 \\ +48 \\ \hline \end{array} \qquad \begin{array}{r} 37 \\ +48 \\ \hline 5 \end{array} \qquad \begin{array}{r} {}^{1} \\ 37 \\ +48 \\ \hline 5 \end{array} \qquad \begin{array}{r} {}^{1} \\ 37 \\ +48 \\ \hline 85 \end{array}$$

To add these numbers you start on the right. Then you work to the left. You can add up or you can add down.

Copy the problem below on a separate sheet of paper. Solve. Notice the direction in which you read the numbers. Notice the direction in which you write them.

$$\begin{array}{r} 5{,}736 \\ +\ 1{,}489 \\ \hline \end{array}$$

GROUP PROJECT

People Who Count

The problem: Think about how many people you know. How many people do you think you and your classmates know in all?

Before beginning, guess how many people you and your classmates know. Then copy the chart. Fill in the names of everybody you know. Find the total.

Hints

- Think of all the people you meet every day (at the store, around your neighborhood).
- Remember your teacher and your classmates.
- Try not to count the same person twice.

PEOPLE I KNOW

Family	Friends	People in the community	Neighbors	Others

Totals

Now have everybody in the class add the numbers in their separate categories. Who has the most family? friends? neighbors? Add all your totals. Was your guess close to the final number? Did the final number surprise you?

CHAPTER TEST

Add. (pages 94, 96 and 100)

1. $\begin{array}{r} 29 \\ +\ 14 \\ \hline \end{array}$

2. $\begin{array}{r} 15 \\ +\ 24 \\ \hline \end{array}$

3. $\begin{array}{r} 67 \\ +\ 35 \\ \hline \end{array}$

4. $\begin{array}{r} 49 \\ +\ 60 \\ \hline \end{array}$

5. $\begin{array}{r} 99 \\ +\ 89 \\ \hline \end{array}$

6. $\begin{array}{r} 63 \\ +\ 87 \\ \hline \end{array}$

7. 14 + 15 + 60 = ___

8. 34 + 18 + 91 + 15 = ___

Solve. (page 98)

9. Last year, Marty went to 9 hockey games, 23 movies, 15 baseball games, and 4 rock concerts. How many sports events did Marty go to last year?

10. Sue bought 3 plaid shirts for $12.50 each and 2 red shirts for $15.00 each. She bought 4 pairs of socks on sale. How many shirts did Sue buy?

Add. (pages 102, 108, 110, and 112)

11. $\begin{array}{r} \$1.89 \\ +\ \ 3.07 \\ \hline \end{array}$

12. $\begin{array}{r} 796 \\ +\ 358 \\ \hline \end{array}$

13. $\begin{array}{r} 862 \\ +\ 138 \\ \hline \end{array}$

14. $\begin{array}{r} \$5.05 \\ +\ \ 3.22 \\ \hline \end{array}$

15. $\begin{array}{r} \$19.36 \\ +\ \ 11.07 \\ \hline \end{array}$

16. $\begin{array}{r} \$74.37 \\ +\ \ 3.48 \\ \hline \end{array}$

17. 200 + 500 + 300 = ___

18. 359 + 426 + 214 = ___

19. 250 + 65 + 196 + 114 = ___

20. $15.76 + $19.20 + $8.79 = ___

21. $4.00 + $3.50 + $1.75 = ___

22. 4,653 + 466 + 1,193 = ___

Estimate the sum. (page 104)

23. 433 + 907 = ▬

24. 2,789 + 7,332 = ▬

25. $1.75 + $3.45 = ▬

26. 290 + 346 + 778 = ▬

27. 7,405 + 4,120 + 5,528 = ▬

Add. (page 112)

28. 4,221 + 3,352

29. 6,253 + 4,124

30. 8,266 + 2,980

31. 1,978 + 4,056

Estimate to solve. (page 106)

32. Steve spent $10.00 on a radio. He bought earphones for $7.95 and batteries for $3.78. About how much money did Steve spend?

33. A table in the store costs $50.50. A lamp costs $8.95, and 2 chairs cost $20.00 each. About how much does the furniture cost?

BONUS

Add.

1. 81,467 + 17,655

2. 30,099 + 49,878

3. 41,371 + 25,997

4. 93,293 + 78,697

5. 11,153 + 98,968

6. 78,465 + 41,807

RETEACHING

A. Add 12 + 15 + 59.

Add the ones.
Regroup.

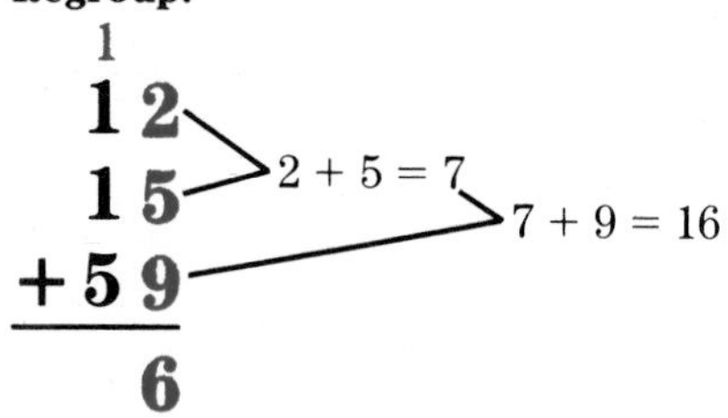

Add the tens.

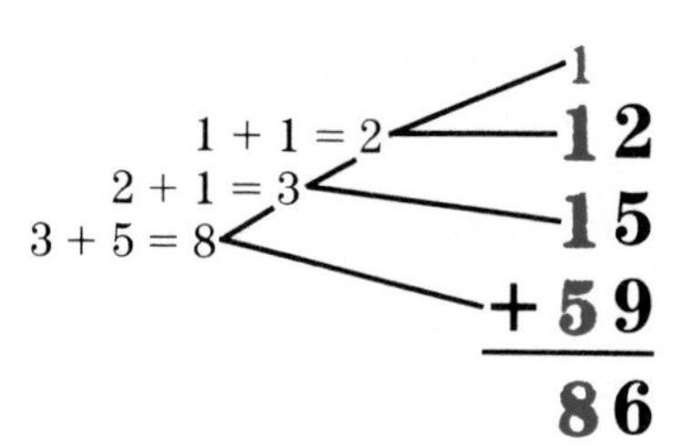

B. You can check your answer by adding up.

```
  1
  12 ── 14 + 2 = 16
  15 ── 5 + 9 = 14
+ 59
  6
```

```
7 + 1 = 8 ── 1
6 + 1 = 7 ── 12
5 + 1 = 6 ── 15
            +59
             86
```

Add.

1. 31 27 + 19	**2.** 53 7 + 48	**3.** 16 34 + 11	**4.** 28 43 + 14	**5.** 52 9 + 70
6. 472 38 + 286	**7.** 829 412 + 253	**8.** 773 151 + 63	**9.** 555 314 + 672	**10.** 260 747 + 414
11. 3,565 1,823 + 5,702	**12.** 6,703 3,816 + 7,312	**13.** 1,717 7,459 + 6,852	**14.** 2,348 4,692 + 6,370	

ENRICHMENT

Input/Output

This machine operates like a simple computer. It performs an operation on the input to produce the output.

1,364 ⟵ input
ADD 3,798 ⟵ operation
5,162 ⟵ output

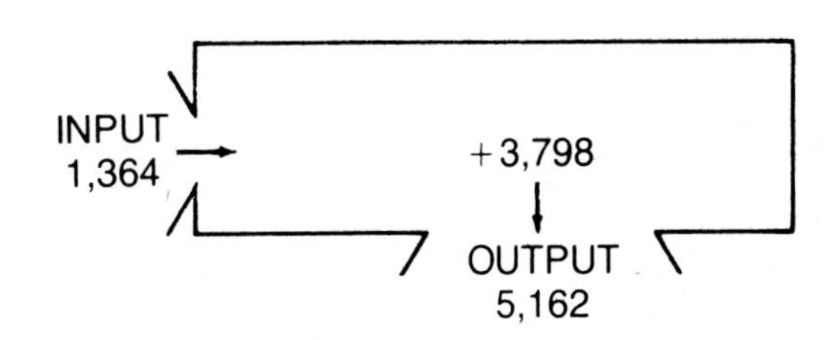

Solve.

1.
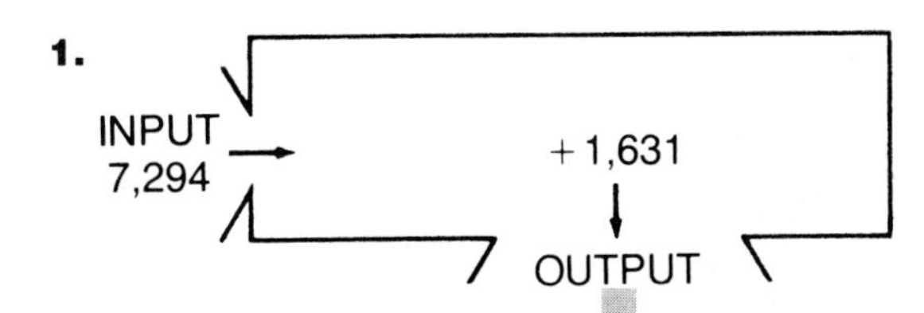

2.

3.
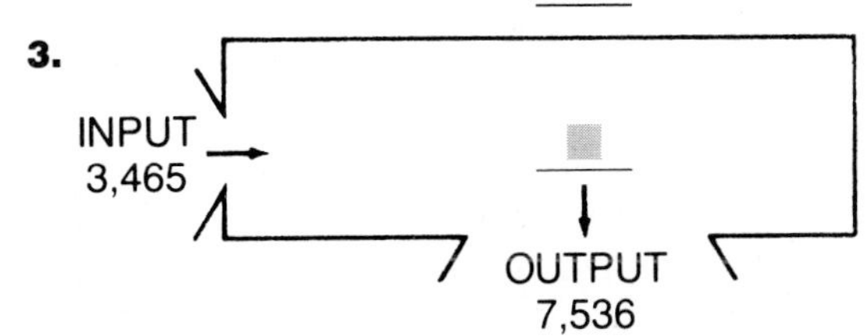

4.
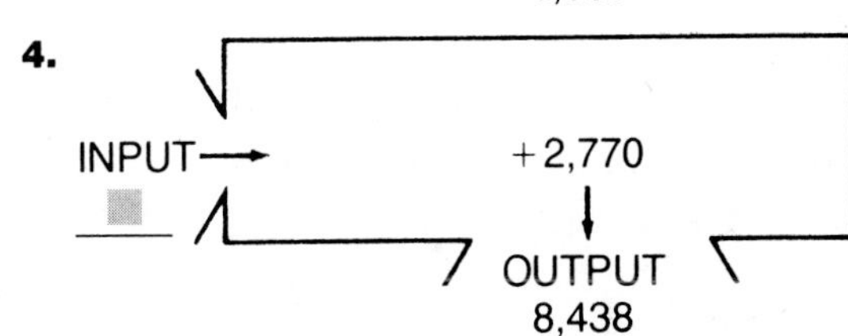

5.
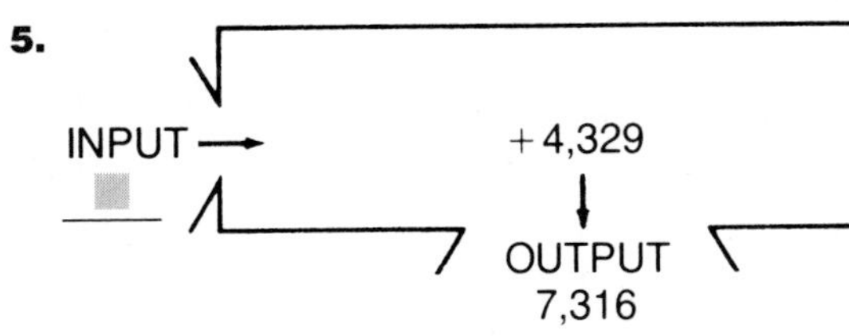

6.
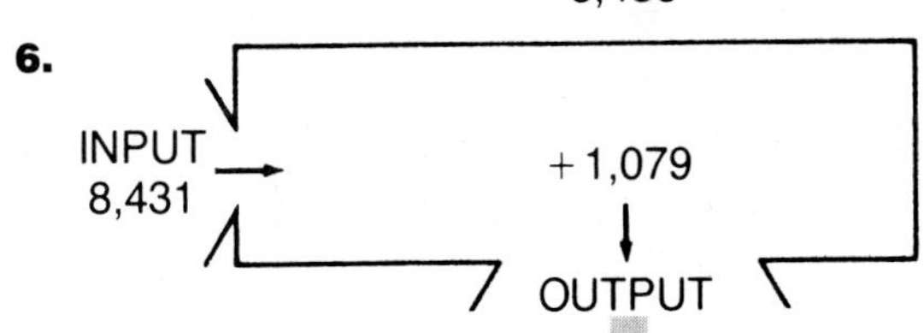

★7.
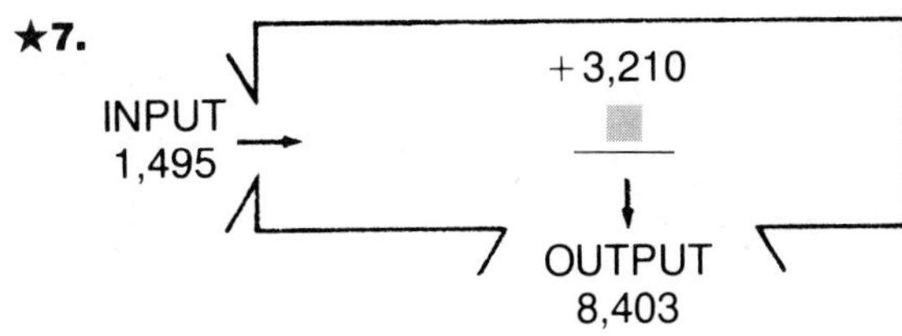

★8.
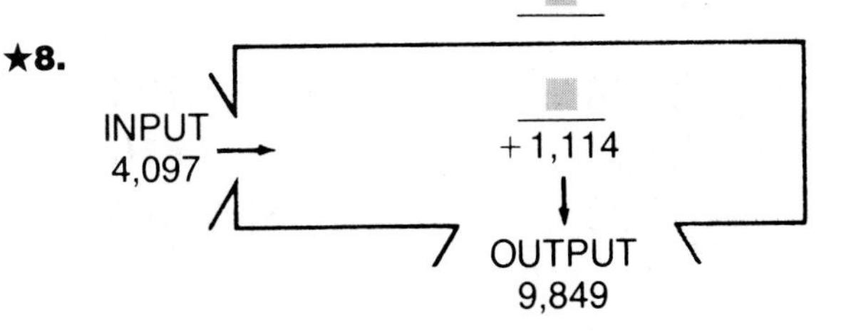

9. Which machine has the greatest output?

10. Which machine has the greatest input?

TECHNOLOGY

BASIC is a computer language that many computers use. To make your computer understand you, you need to know a few commands in BASIC. They are easy. The first one is called the PRINT command. When you are at the computer, you can type these letters.

You will see this on the screen.

PRINT “HELLO”

At the end of the line, press the RETURN or the ENTER key. As soon as you press the key,

HELLO

will appear on the screen.

If you are using a computer, try it. Congratulations! You just gave the computer the instruction to print HELLO, and it did.

Here's another. If you type

PRINT “LUNCHTIME”

and press RETURN, the computer prints this.

LUNCHTIME

When you use the PRINT statement with quotation marks, the computer prints whatever is between the quotation marks.

You can also use the PRINT statement to do math for you, like a calculator.

Type this with no quotation marks.	The computer prints this.
PRINT 7 + 1	8
Type this PRINT statement.	The computer prints this.
PRINT 4 − 2	2

Write what you think each PRINT statement will make the computer print.

1. PRINT "MY DOG HAS FLEAS."
2. PRINT "4 PLUS 4 IS 8"
3. PRINT "4 + 4"
4. PRINT 4 + 4
5. PRINT 9 − 2
6. PRINT "9 − 2"
7. PRINT "SEVEN"
8. PRINT "7"

Complete the PRINT statement so that it will make the computer print what is shown. For Exercises 9–12, use quotation marks only if indicated.

	Command	Printout
9.	PRINT "▒"	HI!
10.	PRINT "▒"	12
11.	PRINT ▒	12
12.	PRINT "▒"	6 + 6
13.	▒	5
14.	▒	6 PLUS 4 IS 10
15.	▒	SEVENTEEN

CUMULATIVE REVIEW

Write the letter of the correct answer.

1. Round 146 to the nearest hundred.

   a. 100 b. 150
   c. 200 d. not given

2. $\begin{array}{r} \blacksquare \\ +\ 4 \\ \hline 6 \end{array}$

   a. 2
   b. 3
   c. 10
   d. not given

3. What is the missing number?
   15, 20, ▇, 30, 35

   a. 5 b. 25
   c. 30 d. not given

4. What ordinal word comes after fifty-first?

   a. fifteenth b. fiftieth
   c. fifty-second d. not given

5. What is the time?

   a. 1:09 b. 1:45
   c. 9:01 d. not given

6. What is the amount?

   a. 40¢ b. 45¢ c. 55¢ d. not given

7.

MAY						
S	M	T	W	T	F	S
	1	2	3	4	5	6
7	8	9	10	11	12	13
14	15	16	17	18	19	20
21	22	23	24	25	26	27
28	29	30	31			

   What is the date of the third Saturday?

   a. May 6 b. May 20
   c. May 21 d. not given

8. It is 2:40. What time will it be in 20 minutes?

   a. 2:60 b. 3:00
   c. 4:40 d. not given

9. Dan has $0.39. Bruce gives him $0.85. How much money does Dan have?

   a. 14¢ b. 25¢
   c. 54¢ d. not given

10.

Story Time	12:50
Music	1:10
Science	1:30

   It is time to start Story Time. How many minutes will it be until Science starts?

   a. 40 minutes b. 70 minutes
   c. 90 minutes d. not given

5 SUBTRACTION OF WHOLE NUMBERS

Did you know that hundreds of kinds of animals are in danger? Some people help to protect these animals by keeping close records and by tracking the animals. Are there any endangered species in your state? Are these animals increasing or decreasing in number?

Regrouping Tens

There are 3 tens and 2 ones.
You can regroup 1 ten as 10 ones.

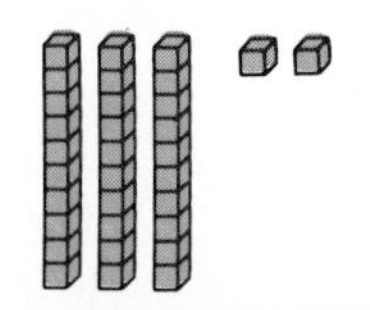

Tens	Ones
3	2

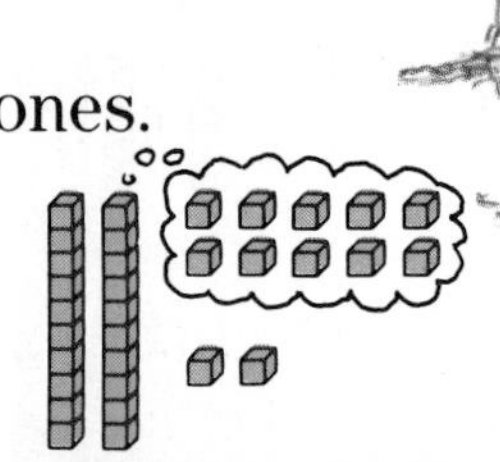

Tens	Ones
2	12

2 tens and 12 ones is the same amount as
3 tens and 2 ones.

Regroup 1 ten into 10 ones. Write the new number of tens and ones.

1.

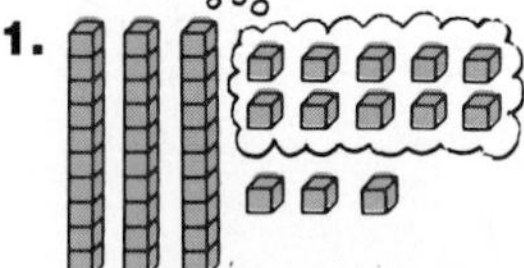

Tens	Ones
3	3

2.

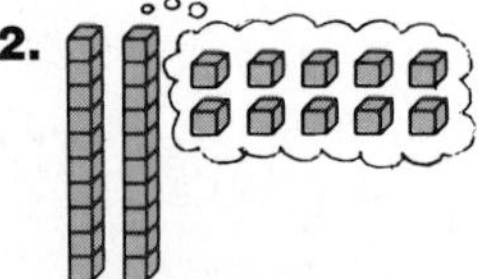

Tens	Ones
2	0

3.

Tens	Ones
1	4

4.

Tens	Ones
3	6

5.

Tens	Ones
6	1

6.

Tens	Ones
2	4

7.

Tens	Ones
5	5

8.

Tens	Ones
4	6

Regroup 1 ten into 10 ones. Write the new number of tens and ones.

9.

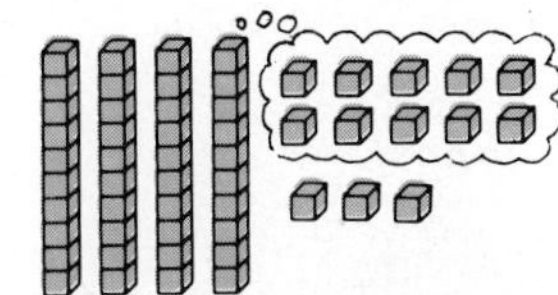

Tens	Ones
4	3

10.

Tens	Ones
1	0

11.

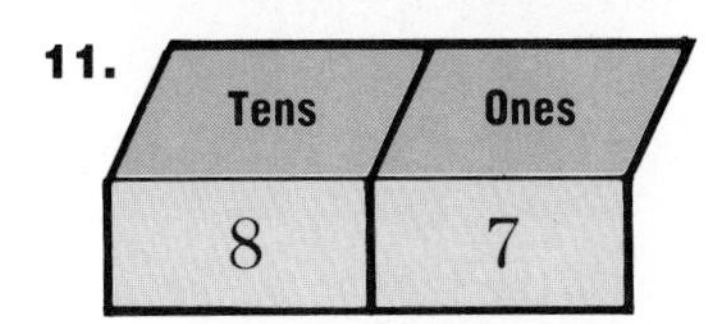

12.

Tens	Ones
2	2

13.

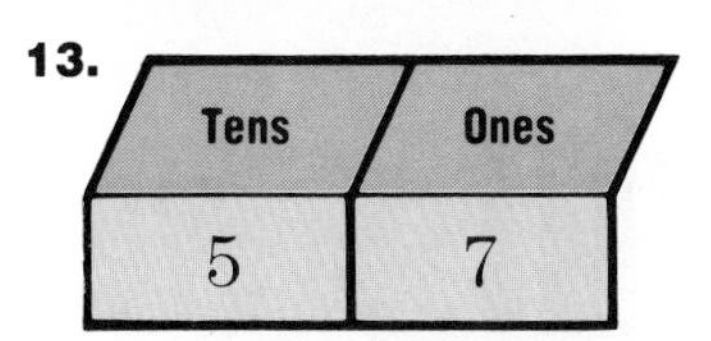

14.

Tens	Ones
1	2

15.

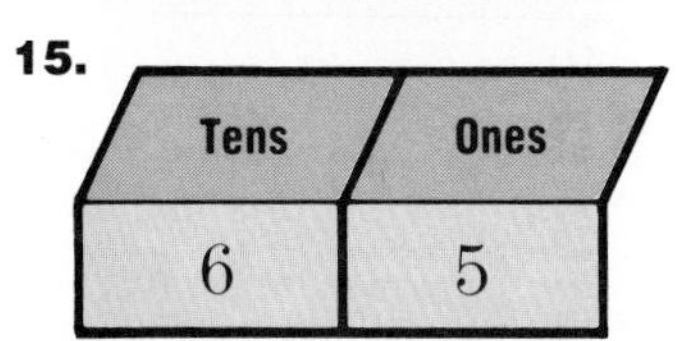

16.

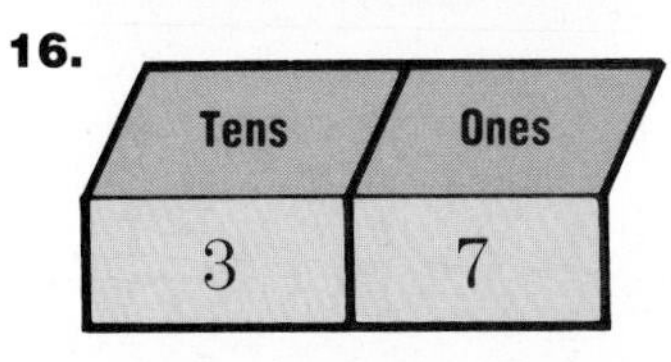

17.

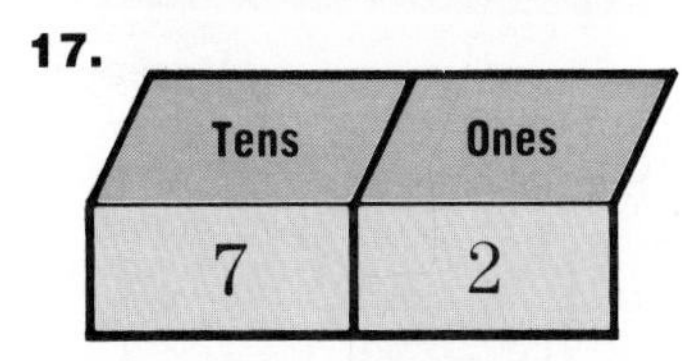

18.

Tens	Ones
5	2

FOCUS: MENTAL MATH

It is easy to add multiples of 10.

$40 + 50 =$ ____ $400 + 300 =$ ____

$4 + 5 = 9$ So, $40 + 50 = 90$. $4 + 3 = 7$ So, $400 + 300 = 700$.

Add mentally.

1. $20 + 30 =$ ____
2. $70 + 10 =$ ____
3. $50 + 30 =$ ____
4. $800 + 100 =$ ____
5. $900 + 700 =$ ____

★6. $6{,}000 + 7{,}000 =$ ____

Subtracting 2-Digit Numbers: Regrouping Tens

A. In one hour, an eagle can fly 60 miles. A roadrunner can run 19 miles in the same time. How many miles farther can an eagle travel in one hour?

You can subtract to find how much farther an eagle can travel in one hour.

Subtract 60 − 19.

Not enough ones.

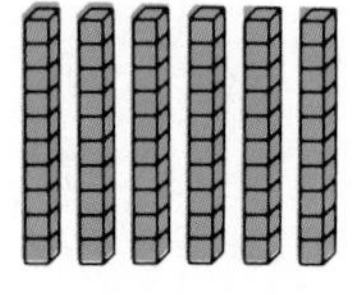

$$\begin{array}{r} 60 \\ -19 \\ \hline \end{array}$$

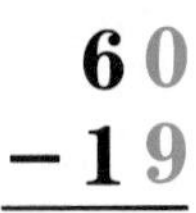

Regroup.
1 ten = 10 ones

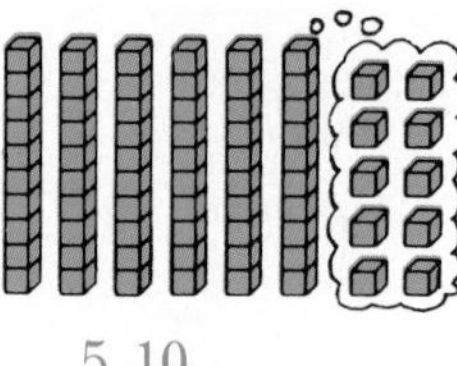

$$\begin{array}{r} \scriptstyle 5\ 10 \\ \not{6}\not{0} \\ -19 \\ \hline \end{array}$$

Subtract the ones.

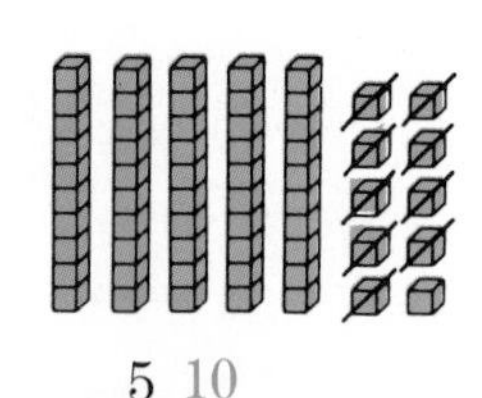

$$\begin{array}{r} \scriptstyle 5\ 10 \\ \not{6}\not{0} \\ -19 \\ \hline 1 \end{array}$$

Subtract the tens.

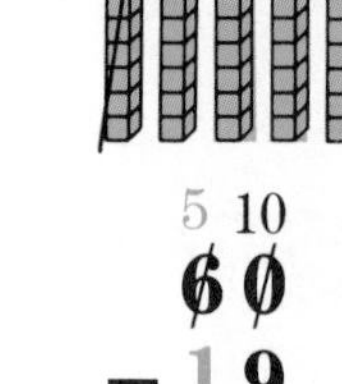

$$\begin{array}{r} \scriptstyle 5\ 10 \\ \not{6}\not{0} \\ -19 \\ \hline 41 \end{array}$$

An eagle can travel 41 miles farther in one hour.

B. You can check subtraction by adding.

Check.

$$\begin{array}{r} 60 \\ -\ 19 \\ \hline 41 \end{array} \qquad \begin{array}{r} 41 \\ +\ 19 \\ \hline 60 \end{array}$$

Checkpoint Write the letter of the correct answer.

Subtract.

1. $\begin{array}{r} 40 \\ -\ 16 \\ \hline \end{array}$

a. 24 **b.** 34 **c.** 44 **d.** 56

2. $\begin{array}{r} 62 \\ -\ \ 8 \\ \hline \end{array}$

a. 64 **b.** 70 **c.** 54 **d.** 66

3. 85 − 79 = ____

a. 6 **b.** 14 **c.** 16 **d.** 164

4. 34 − 6 = ____

a. 32 **b.** 28 **c.** 40 **d.** 38

Subtract.

1. $99 - 70$	2. $87 - 77$	3. $45 - 15$	4. $78 - 33$	5. $65 - 10$	6. $96 - 31$
7. $12 - 7$	8. $28 - 19$	9. $84 - 19$	10. $67 - 28$	11. $18 - 9$	12. $71 - 38$
13. $23 - 14$	14. $75 - 68$	15. $91 - 82$	16. $60 - 8$	17. $44 - 27$	18. $53 - 28$
19. $39 - 17$	20. $42 - 5$	21. $67 - 32$	22. $50 - 13$	23. $30 - 17$	24. $79 - 59$
25. $68 - 39$	26. $23 - 18$	27. $87 - 64$	28. $62 - 29$	29. $35 - 2$	30. $76 - 48$

31. $43 - 22 =$ ___	32. $66 - 28 =$ ___	33. $72 - 5 =$ ___
34. $40 - 17 =$ ___	35. $71 - 52 =$ ___	36. $57 - 28 =$ ___
37. $60 - 33 =$ ___	38. $81 - 24 =$ ___	39. $42 - 29 =$ ___
40. $92 - 63 =$ ___	41. $76 - 45 =$ ___	42. $63 - 59 =$ ___
43. $87 - 76 =$ ___	44. $57 - 35 =$ ___	45. $24 - 9 =$ ___
46. $39 - 30 =$ ___	47. $74 - 28 =$ ___	48. $62 - 37 =$ ___

Solve. Use the Infobank on page 401.

49. How many more miles can a jackrabbit travel in one hour than a greyhound?

50. If a turkey and a pelican raced for one hour, which one would win?

51. If a pig and a chicken run for one hour, how many more miles does the pig travel than the chicken?

★52. Which animal is fastest? How many more miles could it go in one hour than the slowest animal?

More Practice, page 409

PROBLEM SOLVING
Making an Organized List

A list can help you solve some problems.

Roy has a picture frame that holds 2 pictures. He has 4 pictures. They are pictures of an owl, a wolf, a tiger, and a bear. List all the different combinations of 2 pictures that Roy can put in the frame.

Choose one of the animals to begin your list. List all the combinations that can be made.

Next, list all the combinations that can be made by using the picture of another animal. You've already paired the owl with the wolf; so, don't use that combination again.

Now, list all combinations that use the tiger's picture. Don't repeat pairs you've already listed.

The only animal left is the bear. But you've already listed all the combinations that can be made with the bear's picture.

Count the number of combinations.

Roy can choose from 6 different combinations of 2 pictures.

owl
owl–wolf
owl–tiger
owl–bear

wolf
wolf–tiger
wolf–bear

tiger
tiger—bear

Copy and complete the list. Then solve.

1. Amy has 6 roses. The roses are either red or white. How many different combinations of colors could she have?

Red roses		White roses
0	+	6
1	+	5
2	+	4
3	+	3
■	+	■
■	+	■
■	+	■

2. Amy has an odd number of red roses in the group of 6. Does she also have an odd number of white roses?

Solve.

3. Molly has 1 nickel, 1 dime, and 2 quarters. She buys popcorn and pays the cashier the exact price. She uses 2 coins. What are all the possible prices of the popcorn? Copy and complete the list.

5¢ + 10¢ = 15¢
5¢ + 25¢ = 30¢
10¢ + ■ = ■
■ + ■ = ■

4. The zoo gives each animal in the Jungle House a 2-digit code number. Only 5, 6, 7, and 8 are used. One number can be used for both digits, as in 55. How many code numbers can be made? Copy and complete the list.

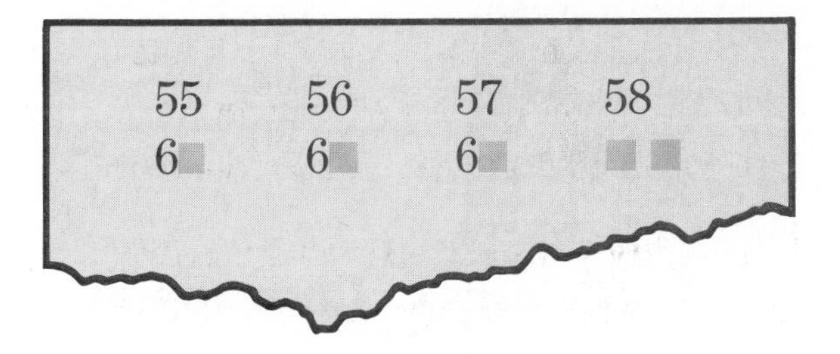

5. The zoo uses even numbered codes for birds. How many 2-digit even numbers can be made from 5, 6, 7, and 8? Make your own list.

Mental Math: Subtracting Tens and Hundreds

There are two young elephants in the zoo. Daisy weighs 700 pounds and Lulu weighs 1,200 pounds. How much more does Lulu weigh?

An easy way to find the answer is to think about subtracting hundreds.

Subtract 1200 − 700.

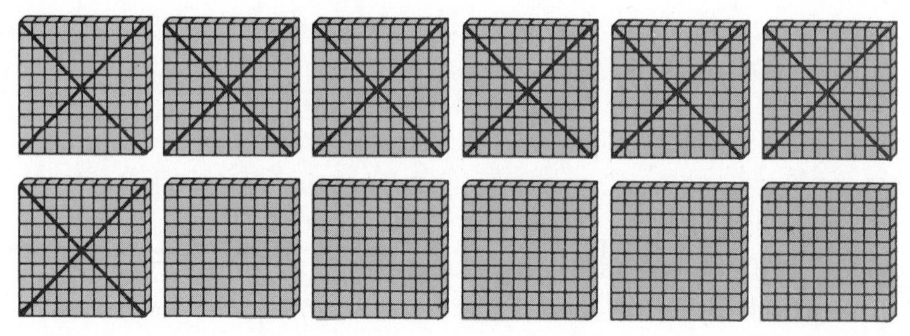

12 hundreds − 7 hundreds = 5 hundreds
1200 − 700 = 500

Lulu weighs 500 pounds more.

Another example:

170	⟶	17 tens
− 90	⟶	− 9 tens
80		8 tens

Checkpoint Write the letter of the correct answer.

Subtract.

1. $70 - 50$
   - a. 2
   - b. 20 tens
   - c. 20
   - d. 120

2. $900 - 400$
   - a. 500 hundreds
   - b. 500
   - c. 5
   - d. 1,300

3. $150 - 60$
   - a. 90 tens
   - b. 90
   - c. 190
   - d. 210

4. $1{,}400 - 700$
   - a. 700
   - b. 1,700
   - c. 2,100
   - d. 7,000

Subtract.

1. $\begin{array}{r} 80 \\ -50 \\ \hline \end{array}$	2. $\begin{array}{r} 90 \\ -10 \\ \hline \end{array}$	3. $\begin{array}{r} 60 \\ -30 \\ \hline \end{array}$	4. $\begin{array}{r} 70 \\ -40 \\ \hline \end{array}$	5. $\begin{array}{r} 40 \\ -30 \\ \hline \end{array}$
6. $\begin{array}{r} 170 \\ -\ \ 80 \\ \hline \end{array}$	7. $\begin{array}{r} 180 \\ -\ \ 90 \\ \hline \end{array}$	8. $\begin{array}{r} 120 \\ -\ \ 60 \\ \hline \end{array}$	9. $\begin{array}{r} 160 \\ -\ \ 90 \\ \hline \end{array}$	10. $\begin{array}{r} 120 \\ -\ \ 90 \\ \hline \end{array}$
11. $\begin{array}{r} 800 \\ -200 \\ \hline \end{array}$	12. $\begin{array}{r} 200 \\ -100 \\ \hline \end{array}$	13. $\begin{array}{r} 800 \\ -600 \\ \hline \end{array}$	14. $\begin{array}{r} 1{,}200 \\ -\ \ 800 \\ \hline \end{array}$	15. $\begin{array}{r} 1{,}400 \\ -\ \ 900 \\ \hline \end{array}$

16. $160 - 70 =$ ___ 17. $90 - 40 =$ ___ 18. $130 - 60 =$ ___

19. $900 - 400 =$ ___ 20. $1{,}500 - 600 =$ ___ 21. $1{,}400 - 800 =$ ___

Solve.

22. An elephant can eat 180 pounds of hay every day. The zookeeper brings 90 pounds of hay. How much more hay is needed?

23. An elephant drank 30 gallons of water in the morning and 90 gallons in the afternoon. How many gallons of water did he drink?

24. A zookeeper brings 1,200 pounds of fruit for a group of elephants. They eat 900 pounds of fruit. How much fruit is left?

★25. One elephant weighs 6,100 pounds. Another weighs 5,800 pounds. How much more does the larger elephant weigh?

FOCUS: MENTAL MATH

When both numbers end in the same digit, the answer to a subtraction problem ends in 0.

$86 - 56 = 30$

Think: $6 - 6 = 0$.

Subtract mentally.

1. $45 - 25 =$ ___ 2. $68 - 38 =$ ___ 3. $42 - 12 =$ ___

4. $94 - 24 =$ ___ 5. $63 - 53 =$ ___ 6. $32 - 22 =$ ___

Estimating Differences

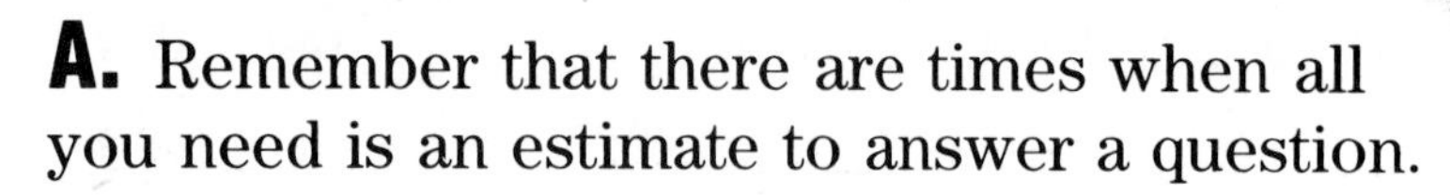

A. Remember that there are times when all you need is an estimate to answer a question.

The largest clam weighed 576 pounds. The largest octopus weighed 118 pounds. About how much more did the clam weigh?

You can estimate to find about how much more the clam weighed.

When you estimate, look at the front digits.

Estimate: 576 − 118.

Subtract the front digits.	Write 0's for the other digits.
$\begin{array}{r} 576 \\ -118 \\ \hline 4 \end{array}$	$\begin{array}{r} 576 \\ -118 \\ \hline \text{about } 400 \end{array}$

The clam weighed about 400 pounds more.

B. You can estimate money in the same way.

Marianne has \$7.25 to spend at the zoo. Admission is \$3.40. About how much money will she have left to spend on lunch?

Estimate: \$7.25 − \$3.40.

Line up the cents points.	Subtract the dollars.	Write 0's for the cents.
$\begin{array}{r} \$7.25 \\ -\ 3.40 \\ \hline \end{array}$	$\begin{array}{r} \$7.25 \\ -\ 3.40 \\ \hline 4 \end{array}$	$\begin{array}{r} \$7.25 \\ -\ 3.40 \\ \hline \text{about } \$4.00 \end{array}$

She will have about \$4.00 to spend on lunch.

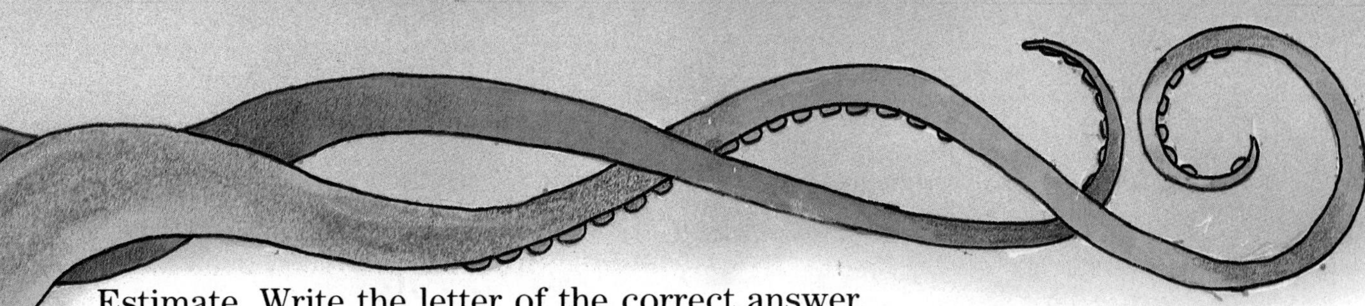

Estimate. Write the letter of the correct answer.

1. 528 − 219
- a. 100
- b. 200
- c. 300

2. 757 − 139
- a. 200
- b. 600
- c. 800

3. $8.73 − 2.68
- a. $6.00
- b. $8.00
- c. $10.00

4. 529 − 435
- a. 10
- b. 100
- c. 1,000

5. $8.84 − 5.95
- a. $3.00
- b. $6.00
- c. $10.00

6. 417 − 128
- a. 300
- b. 400
- c. 500

7. 6,275 − 1,089
- a. 5,000
- b. 7,000
- c. 9,000

8. 8,917 − 1,876
- a. 700
- b. 800
- c. 7,000

Estimate.

9. 929 − 331

10. 7,319 − 1,043

11. 1,006 − 500

12. 9,102 − 2,499

Solve.

13. There are 255 monkeys in the zoo. About 160 of them are male. About how many monkeys are female?

14. The zoo has 485 different cages. Over 300 of them have been cleaned. About how many cages need to be cleaned?

15. There are 27 kinds of birds in an outdoor cage, and 33 kinds of birds in an indoor cage. How many kinds of birds are there in both cages?

16. Pedro has $2.35. A ride on the train costs $2.00. If Pedro buys a train ticket, will he have enough money to buy a fruit drink that costs $1.00?

PROBLEM SOLVING
Estimation

Some problems can be solved by estimating. Sometimes you need to find the exact answer.

Rich found a baby raccoon and wanted to keep it. His mother said he could if he bought a cage for it. Rich makes a list of his earnings. If the cage costs $6.25, will Rich make enough money to buy the cage this week?

MONEY EARNED IN 1 WEEK	
Mowing lawns	$2.00
Helping Mr. Burke deliver papers	$1.50
Allowance	$2.25
Washing the car	$1.75

Rich decides to estimate his earnings.

Estimate $2.00 + $1.50 + $2.25 + $1.75.

Add the front digits.

```
 $2.00
  1.50
  2.25
+ 1.75
------
  6
```

Write 0's for the other digits.

```
       $2.00
        1.50
        2.25
      + 1.75
      ------
about  $6.00
```

The estimate is $6.00. But that is very close to the price for the cage. Rich can't be sure he will make enough money. To solve this problem, he needs to find the exact sum.

```
Exact sum:  $2.00
             1.50
             2.25
           + 1.75
           ------
            $7.50
```

Rich will make $7.50 this week. He will make enough money to buy the cage this week.

Decide whether you can use an estimate to solve each problem. Write *estimate* or *exact answer.*

1. Rich needs a quart of milk and a baby bottle for his raccoon Saucy. Look at the prices. Is $1.50 enough money for both?

2. Rich wants to buy a toy for Saucy to chew on. The toy costs $5.79. Another toy costs $4.58. Rich has $9.95. Does he have enough to buy both toys?

Solve. Find an estimate or an exact answer as needed.

3. Rich needed some wood to build a den for Saucy. He figured he needed about 144 inches of wood. In the barn, he found 2 boards. They were both 84 inches long. Does Rich have enough wood?

4. To build the den, Rich needs 48 nails. In one jar, Rich finds 19 nails. If he finds 30 more nails, will he have enough?

5. One day Saucy found that she could open the refrigerator door by herself. She ate a box of muffins, a bag of walnuts, and a bunch of grapes. Look at the prices. Will Rich be able to replace the food eaten if he has $2.25?

6. Rich wants to order *Wildlife* magazine. He saves $2.35 one week, $1.35 the next week, and $1.15 the third week. If the magazine costs $5.50, does he have enough?

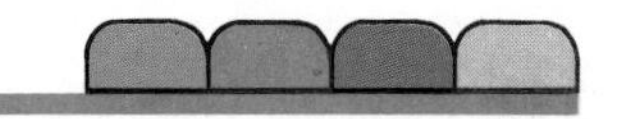

Subtracting 3-Digit Numbers: Regrouping Once

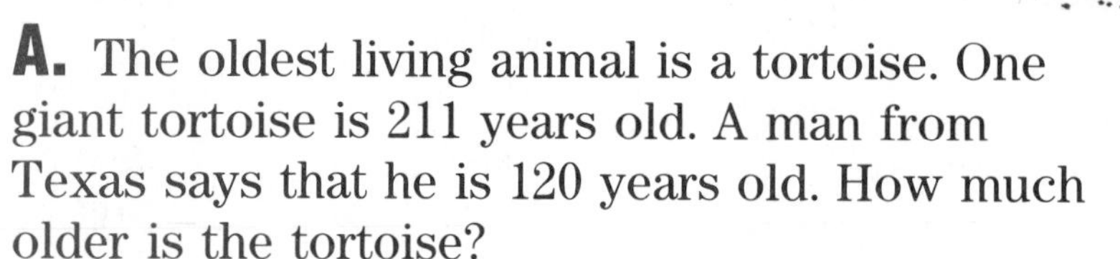

A. The oldest living animal is a tortoise. One giant tortoise is 211 years old. A man from Texas says that he is 120 years old. How much older is the tortoise?

You can subtract to find how much older the tortoise is.

Subtract 211 − 120.

Subtract the ones.	**Regroup. Subtract the tens.**	**Subtract the hundreds.**
$\begin{array}{r} 211 \\ -120 \\ \hline 1 \end{array}$	$\begin{array}{r} \scriptstyle 1\ 11 \\ \not2\not11 \\ -120 \\ \hline 91 \end{array}$	$\begin{array}{r} \scriptstyle 1\ 11 \\ \not2\not11 \\ -120 \\ \hline 91 \end{array}$

Sometimes there is no digit in the hundreds place of the difference.

The tortoise is 91 years older.

Other examples:

$\begin{array}{r} \scriptstyle 5\,16 \\ 5\not6\not6 \\ -339 \\ \hline 227 \end{array}$ $\quad$ $\begin{array}{r} \scriptstyle 3\,12 \\ \not4\not27 \\ -\ \ 52 \\ \hline 375 \end{array}$ $\quad$ $\begin{array}{r} \scriptstyle 8\,10 \\ 6\not9\not0 \\ -352 \\ \hline 338 \end{array}$

B. You can subtract money in the same way.

Subtract \$6.47 − \$3.18.

Line up the cents points.	Subtract as you would whole numbers.	Write the dollar sign and the cents point.
$\begin{array}{r} \$6.47 \\ -\ \ 3.18 \\ \hline \end{array}$	$\begin{array}{r} \scriptstyle 3\,17 \\ \$6.\not4\not7 \\ -\ \ 3.18 \\ \hline 3\ 29 \end{array}$	$\begin{array}{r} \scriptstyle 3\,17 \\ \$6.\not4\not7 \\ -\ \ 3.18 \\ \hline \$3.29 \end{array}$

Subtract.

1. $585 - 301$
2. $937 - 20$
3. $778 - 666$
4. $867 - 501$
5. $868 - 24$
6. $423 - 304$
7. $740 - 528$
8. $426 - 17$
9. $980 - 359$
10. $678 - 129$
11. $746 - 683$
12. $364 - 90$
13. $522 - 382$
14. $638 - 267$
15. $419 - 93$
16. $\$9.82 - 7.64$
17. $\$5.61 - 1.81$
18. $\$3.36 - 2.15$
19. $\$7.28 - 4.37$
20. $\$4.51 - 3.49$
21. $345 - 124$
22. $\$8.36 - 6.42$
23. $590 - 289$
24. $\$2.08 - 1.91$
25. $617 - 83$
26. $189 - 176 =$ ____
27. $446 - 234 =$ ____
28. $782 - 81 =$ ____
29. $690 - 85 =$ ____
30. $837 - 229 =$ ____
31. $462 - 138 =$ ____
32. $735 - 572 =$ ____
33. $321 - 240 =$ ____
34. $676 - 583 =$ ____
35. $\$5.67 - \$4.73 =$ ____
36. $\$7.76 - \$6.52 =$ ____
37. $\$2.61 - \$1.39 =$ ____
38. $873 - 667 =$ ____
39. $485 - 92 =$ ____
40. $\$9.30 - \$7.18 =$ ____

MIDCHAPTER REVIEW

Subtract.

1. $79 - 70$
2. $37 - 17$
3. $87 - 29$
4. $61 - 48$
5. $35 - 7$
6. $80 - 10$
7. $160 - 60$
8. $600 - 90$
9. $900 - 200$
10. $1,100 - 800$

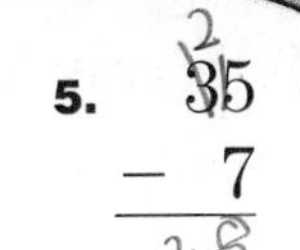

Subtracting 3-Digit Numbers: Regrouping Twice

In 1981, there were 172 whooping cranes left in the world. By 1985, there were 361 whooping cranes. How many more whooping cranes were there by 1985?

You can subtract to find how many more.

Subtract 361 − 172.

Regroup. Subtract the ones.	Regroup. Subtract the tens.	Subtract the hundreds.
5 11	15 2 5 11	15 2 5 11
361	361	361
−172	−172	−172
9	89	189

There were 189 more whooping cranes in 1985 than in 1981.

Other examples:

11 4 1 10	13 3 3 16	12 4 2 14	12 5 2 10
520	446	$5.34	$6.30
−143	− 88	− 1.55	− 1.75
377	358	$3.79	$4.55

Checkpoint Write the letter of the correct answer.

Subtract.

1. 386 − 199	2. 877 − 379	3. $5.40 − 2.91	4. 365 − 78 = ■
a. 97	a. 498	a. $2.49	a. 197
b. 187	b. 502	b. $3.49	b. 297
c. 197	c. 598	c. $3.51	c. 287
d. 287	d. 1,256	d. $8.31	d. 443

Subtract.

1. 941 − 355
2. 321 − 123
3. 584 − 385
4. 734 − 348
5. 766 − 267

6. 418 − 397
7. 634 − 256
8. 250 − 97
9. 933 − 844
10. 843 − 44

11. $5.16 − 2.18
12. $3.92 − 1.99
13. $8.15 − 7.16
14. $7.60 − 5.68
15. $8.11 − 4.99

16. 460 − 69 = ■
17. 918 − 99 = ■
18. 314 − 87 = ■
19. 210 − 197 = ■
20. 655 − 257 = ■
21. 534 − 35 = ■
22. $8.27 − $0.59 = ■
23. $6.56 − $3.88 = ■
24. $7.22 − $0.89 = ■

Solve.

25. A black bear weighs 256 pounds. A sun bear weighs 69 pounds. How much heavier is the black bear?

26. Some trees grow 135 feet tall. Woodpeckers live in the treetops. Chimpanzees live 68 feet above ground. How much higher do the woodpeckers live?

27. The tiny hummingbird eats 49 small meals per day. How many meals does it eat in 2 days?

28. Sylvia saw a blue whale that was 100 feet long. Then she saw a pygmy whale 18 feet long. How much longer was the blue whale?

FOCUS: ESTIMATION

Is $10.00 enough to make each purchase? Write *yes* or *no*.

1. poster—$3.85
   T-shirt—$8.76

2. button—$0.95
   hat—$5.76
   pennant—$4.57

3. book—$3.65
   record—$3.75
   magazine—$1.98

Subtracting with Zeros

A. A bat flew 202 miles. It flew 179 miles before it was stopped by a storm. How many miles did it fly after the storm?

You can subtract to find the number of miles.

Subtract 202 − 179.

Regroup the hundreds.	Regroup the tens.	Subtract.
$\begin{array}{r} \scriptstyle 1\;10\;\; \\ \not{2}\,\not{0}\,2 \\ -179 \\ \hline \end{array}$	$\begin{array}{r} \scriptstyle 9\;\;\; \\ \scriptstyle 1\;\not{10}\,12 \\ \not{2}\,\not{0}\,\not{2} \\ -179 \\ \hline \end{array}$	$\begin{array}{r} \scriptstyle 9\;\;\; \\ \scriptstyle 1\;\not{10}\,12 \\ \not{2}\,\not{0}\,\not{2} \\ -179 \\ \hline 23 \end{array}$

It flew 23 miles after the storm.

B. Sometimes there is a zero in both the ones and the tens places.

Subtract 400 − 237.

Regroup the hundreds.	Regroup the tens.	Subtract.
$\begin{array}{r} \scriptstyle 3\;10\;\; \\ \not{4}\,\not{0}\,0 \\ -237 \\ \hline \end{array}$	$\begin{array}{r} \scriptstyle 9\;\;\; \\ \scriptstyle 3\;\not{10}\,10 \\ \not{4}\,\not{0}\,\not{0} \\ -237 \\ \hline \end{array}$	$\begin{array}{r} \scriptstyle 9\;\;\; \\ \scriptstyle 3\;\not{10}\,10 \\ \not{4}\,\not{0}\,\not{0} \\ -237 \\ \hline 163 \end{array}$

Checkpoint Write the letter of the correct answer.

Subtract.

1. $\begin{array}{r} 803 \\ -\,408 \\ \hline \end{array}$

a. 392 **b.** 395
c. 305 **d.** 1,211

2. $\begin{array}{r} \$3.00 \\ -\;\;1.27 \\ \hline \end{array}$

a. $0.83 **b.** 173
c. $1.73 **d.** $2.83

3. 500 − 76 = ■

a. 24 **b.** 524
c. 424 **d.** 576

Subtract.

1. 603 − 590	2. 601 − 503	3. 407 − 54	4. \$3.08 − 2.99	5. \$9.05 − 3.19
6. 400 − 206	7. 600 − 333	8. 200 − 184	9. \$9.00 − 7.89	10. \$8.00 − 0.23
11. 401 − 39	12. \$6.00 − 0.55	13. 300 − 29	14. 500 − 78	15. \$2.09 − 0.78
16. 705 − 607	17. 409 − 247	18. \$1.00 − 0.89	19. 600 − 407	20. \$8.01 − 7.09

21. 702 − 490 = ____ 22. 903 − 890 = ____ 23. \$8.05 − \$5.20 = ____

24. 700 − 569 = ____ 25. 400 − 375 = ____ 26. \$5.00 − \$2.59 = ____

27. \$3.01 − \$0.98 = ____ 28. 340 − 212 = ____ 29. 200 − 77 = ____

30. 480 − 322 = ____ 31. \$6.02 − \$1.98 = ____ 32. \$8.05 − \$0.43 = ____

Solve.

33. There are 900 kinds of bats. Of these, 97 kinds live only in West Africa. How many kinds of bats are found in the rest of the world?

★34. One book about bats costs \$3.09. Another book costs \$2.77. If Tracy has \$6.00, how much change will she receive if she buys both books?

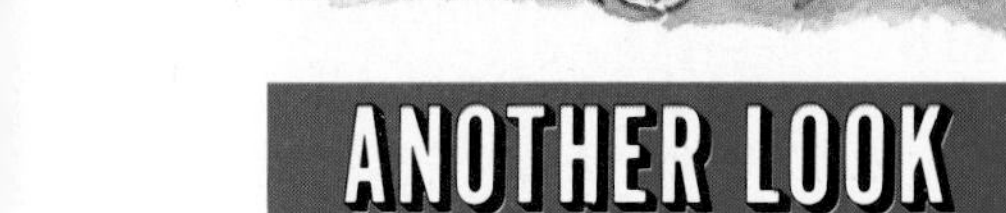

Write the numbers in order from the least to the greatest.

1. 40, 56, 27, 45, 30
2. 76, 67, 72, 38, 74
3. 12, 17, 23, 11, 27

PROBLEM SOLVING

Writing a Number Sentence

Writing a number sentence can help you solve a word problem. A number sentence can show you how to use the numbers you know to find the number you need.

Mammals are animals that breathe air and give milk to their young. Lions, deer, and whales are all mammals. There are 135 groups of mammals. Of these, 17 groups, such as whales, seals, and otters, live in the water all or most of the time. How many mammal groups live on land?

1. List what you know and what you need to find.

know There are 135 mammal groups. 17 groups live in the water.

find How many mammal groups live on land?

2. Think about what your list tells you about whether to + or −. Write a number sentence about this problem. Use ■ to stand for the number you need to find.

17	+	■	=	135
water groups		land groups		animal groups

3. Solve. Write the answer.

$135 - 17 = ■$
$135 - 17 = 118$

118 mammal groups live on land.

Write the letter of the correct number sentence.

1. Some mammals, such as the kangaroo, have pouches. There are 256 kinds of pouched mammals. Of these, 173 kinds live only in or near Australia. The rest live in North or South America. How many kinds live in North or South America?

   a. $256 + 173 = \blacksquare$
   b. $256 - 173 = \blacksquare$

2. Many dogs are raised to do certain jobs. The American Kennel Club lists 33 kinds of working or herding dogs. They list 95 kinds of dogs that are raised as pets or for other uses. How many kinds of dogs are listed?

   a. $95 + 33 = \blacksquare$
   b. $95 - 33 = \blacksquare$

Write a number sentence if you can, and solve.

3. All mammals do not have the same number of teeth. A panda has 50 teeth. A skunk has 34 teeth. How many more teeth does a panda have than a skunk?

4. The Iditarod is a dogsled race held in Alaska. One racer owns 98 sled dogs. This year, 14 of them will race. How many of his dogs will not race?

5. A mouse lives for about 3 years. A lion can live for 24 years. A hippo can live for 16 more years than a lion can. How old can a hippo live to be?

6. One wolf pack had 8 wolves. A group of lions had 12 lions. A cheetah family had a mother and 8 young. Which group was the smallest?

★7. One polar bear spent 127 days near the sea. Then it moved west for 95 days. It then returned to the sea for 185 days. How many days did it spend near the sea?

★8. A pod is a group of whales. One pod had 47 whales. More joined from other pods. Then 12 babies were born. Now the pod has 87 whales. How many joined from other pods?

Subtracting 4-Digit Numbers

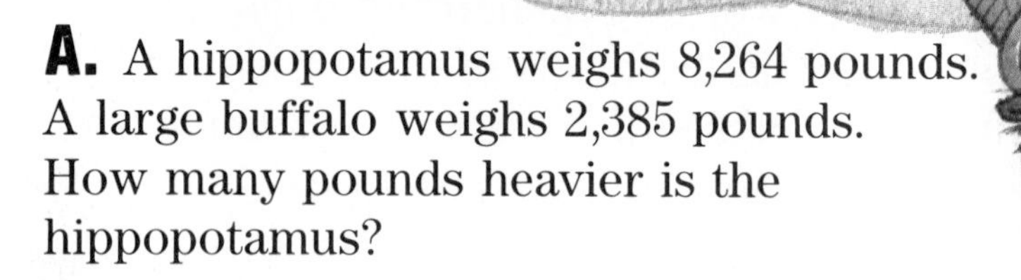

A. A hippopotamus weighs 8,264 pounds. A large buffalo weighs 2,385 pounds. How many pounds heavier is the hippopotamus?

You can subtract to find how much heavier the hippopotamus is.

Subtract 8,264 − 2,385.

Regroup. Subtract the ones.	Regroup. Subtract the tens.	Regroup. Subtract the hundreds.	Subtract the thousands.
5 14	15 1 5 14	11 15 7 1 5 14	11 15 7 1 5 14
8,264	8,264	8,264	8,264
−2,385	−2,385	−2,385	−2,385
9	79	879	5,879

The hippopotamus is 5,879 pounds heavier.

Other examples:

	9 7 10 10	9 5 10 15	15 4 5 13
6,459	7,800	4,605	5,632
−3,247	−4,319	−3,297	−2,781
3,212	3,481	1,308	2,851

B. You can subtract money in the same way.

Line up the cents points.	Subtract as you would whole numbers.	Write the dollar sign and the cents point.
	7 10	7 10
$28.00	$28.00	$28.00
− 10.90	− 10.90	− 10.90
	17 10	$17.10

Subtract.

1. 7,689 − 5,872
2. 6,594 − 728
3. 4,853 − 571
4. 7,853 − 2,622
5. $65.82 − 17.64
6. $46.00 − 2.78
7. $43.57 − 28.73
8. $34.92 − 13.76
9. 9,503 − 7,876
10. 6,057 − 3,849
11. 4,002 − 3,765
12. 4,803 − 2,895
13. 8,700 − 4,589
14. 6,200 − 2,364
15. 9,800 − 3,972
16. 7,200 − 3,547
17. 7,634 − 2,892
18. $58.28 − 37.17
19. 4,600 − 3,469
20. $62.37 − 18.46

21. 6,737 − 4,625 = ____
22. 6,981 − 4,760 = ____
23. 5,239 − 4,328 = ____
24. $87.45 − $32.81 = ____
25. $78.00 − $69.27 = ____
26. $37.18 − $5.29 = ____
27. 8,300 − 5,624 = ____
28. $54.61 − $35.73 = ____
29. 8,945 − 6,588 = ____
30. 6,765 − 5,899 = ____
31. 3,503 − 876 = ____
32. $48.37 − $26.33 = ____
33. 5,831 − 2,795 = ____
34. $46.93 − $18.70 = ____

FOCUS: REASONING

You can sort these shapes in different ways.

a. 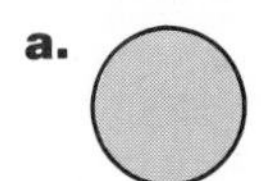 b. c. d. e. f. g. h.

One way to sort them is by color: *a,e,h* *b,c* *d,f,g*
Find two other ways to sort the shapes.

More Practice, page 409

LOGICAL REASONING

Sometimes you can solve a problem by ruling out possible answers.

Ann, Pat, and Bob are friends. The tallest of the three lives next to Ann. Pat is the shortest. Bob lives next to Pat. Who is the tallest?

Make a table to help answer the question. Write *no* next to each person who is *not* the tallest.

	Tallest	
Ann	no	The tallest person lives *next to* Ann. So, Ann is not the tallest.
Pat	no	Pat is the *shortest.* So, Pat is not the tallest.
Bob		

So, Bob is the tallest.

Solve. Copy and use the tables.

Mark, Joe, and Meg live on the same block. One of their houses is red, one is blue, and one is green. Mark lives in the green house. The house on the left is blue. Joe lives in the middle house.

	Joe's house	Meg's house	Mark's house
green			
blue			
red			

1. What color is Joe's house?

2. What color is Meg's house?

Andy, Brett and Carla are the only members of a stamp club. Andy is the treasurer's best friend. Brett is the vice president.

3. Who is the president of the club?

	President
Andy	
Brett	
Carla	

GROUP PROJECT

Zippy Zoo

The problem: You are the zookeeper of the Central City Zoo. The animals have gotten together. They want to get paid by the hour. You have to decide how much to pay each animal.

Key Facts

- The elephant helps around the zoo by carrying heavy objects.
- The lion sleeps about 20 hours per day.
- The panda attracts many visitors to the zoo because she is so rare.
- The snake lies on a rock in the sun and doesn't move much.
- The seals are always playing. Many people visit just to see them.
- The polar bear just had a cub. The cub is learning to swim. People like to watch the cub.
- Many people visit the monkeys.
- The two snow leopards are popular because they are beautiful. But they are shy. Sometimes they hide behind rocks and are hard to see.
- The bats and the owls sleep during the day.

CHAPTER TEST

Subtract. (pages 128, 138, 140, and 142)

1. $\begin{array}{r} 98 \\ -\ 65 \\ \hline \end{array}$

2. $\begin{array}{r} 86 \\ -\ 57 \\ \hline \end{array}$

3. $\begin{array}{r} 75 \\ -\ 69 \\ \hline \end{array}$

4. $\begin{array}{r} 309 \\ -\ 207 \\ \hline \end{array}$

5. $\begin{array}{r} 563 \\ -\ 184 \\ \hline \end{array}$

6. $\begin{array}{r} \$7.31 \\ -\ 5.06 \\ \hline \end{array}$

7. $\begin{array}{r} 600 \\ -\ 378 \\ \hline \end{array}$

8. $\begin{array}{r} 88 \\ -\ 59 \\ \hline \end{array}$

9. $\begin{array}{r} 943 \\ -\ 462 \\ \hline \end{array}$

10. \$8.56 − \$4.88 = ■

11. \$8.03 − \$6.19 = ■

12. 619 − 327 = ■

13. 942 − 378 = ■

Write a number sentence. Solve. (page 144)

14. There are 485 birds in the wild-bird park. In the winter, 98 geese fly south. How many birds are left in the park?

15. Jeffrey counted 55 fireflies in the air one night. Mary counted 82. How many more fireflies did Mary see that night?

Subtract. (page 146)

16. $\begin{array}{r} 4{,}892 \\ -\ 1{,}491 \\ \hline \end{array}$

17. $\begin{array}{r} 9{,}345 \\ -\ 6{,}486 \\ \hline \end{array}$

18. $\begin{array}{r} 8{,}076 \\ -\ 7{,}876 \\ \hline \end{array}$

19. $\begin{array}{r} \$67.42 \\ -\ 35.90 \\ \hline \end{array}$

20. $\begin{array}{r} 5{,}009 \\ -\ 1{,}317 \\ \hline \end{array}$

21. $\begin{array}{r} \$99.14 \\ -\ 37.75 \\ \hline \end{array}$

Subtract. (page 146)

22. \$25.62 − \$18.14 = ■

23. \$54.00 − \$34.85 = ■

24. 3,988 − 2,059 = ■

25. 7,490 − 5,070 = ■

Estimate the difference. (page 134)

26. 943 − 462 = ■

27. 619 − 327 = ■

28. \$8.56 − \$4.81 = ■

29. 3,709 − 1,235 = ■

30. 6,854 − 4,929 = ■

31. \$9.15 − \$5.40 = ■

Solve. Find an estimate or an exact answer as needed. (page 136)

32. Emma wants to buy some movie posters. A poster costs \$4.95. Emma has \$10.00. Does she have enough to buy 2 posters?

33. Billy buys his mother perfume for her birthday. The perfume costs \$16.50. Billy has \$20.00. Does he have enough to also buy a toy for \$5.39?

BONUS

Subtract.

1. 64,590 − 31,475

2. 76,835 − 68,742

3. 75,531 − 36,900

4. 49,858 − 48,956

5. 62,533 − 50,987

6. 97,056 − 89,508

7. 81,094 − 29,949

8. 28,500 − 19,571

RETEACHING

Subtract 402 − 347.

Regroup hundreds.

$$\begin{array}{r} \scriptstyle 3\ 10 \\ \not{4}\not{0}2 \\ -347 \\ \hline \end{array}$$

Regroup tens.

$$\begin{array}{r} \scriptstyle 9 \\ \scriptstyle 3\ \not{10}\ 12 \\ \not{4}\not{0}\not{2} \\ -347 \\ \hline \end{array}$$

Subtract.

$$\begin{array}{r} \scriptstyle 9 \\ \scriptstyle 3\ \not{10}\ 12 \\ \not{4}\not{0}\not{2} \\ -347 \\ \hline 55 \end{array}$$

Complete.

1. $302 - 41$ **2.** $706 - 608$ **3.** $207 - 89$ **4.** $\$6.05 - 3.99$ **5.** $\$3.02 - 1.25$

6. $400 - 203$ **7.** $900 - 512$ **8.** $100 - 34$ **9.** $\$5.00 - 4.68$ **10.** $\$8.00 - 0.56$

11. $309 - 32$ **12.** $700 - 413$ **13.** $400 - 155$ **14.** $\$6.03 - 0.79$ **15.** $\$9.05 - 2.27$

16. $602 - 26$ **17.** $500 - 409$ **18.** $901 - 102$ **19.** $\$3.07 - 0.89$ **20.** $\$8.01 - 7.79$

21. $\$6.04 - 1.43$ **22.** $306 - 18$ **23.** $503 - 305$ **24.** $\$1.08 - 0.39$ **25.** $\$7.07 - 0.28$

26. 405 − 298 = ____ **27.** 606 − 59 = ____ **28.** 102 − 63 = ____

29. 207 − 135 = ____ **30.** 801 − 739 = ____ **31.** \$5.00 − \$4.18 = ____

32. \$8.00 − \$1.22 = ____ **33.** 700 − 352 = ____ **34.** \$8.03 − \$2.77 = ____

ENRICHMENT

Addition and Subtraction

Add and subtract from left to right.

CUMULATIVE REVIEW

Write the letter of the correct answer.

1. What is the missing number?

$7 + \_\_\_ = 16$
$\_\_\_ + 7 = 16$
$16 - \_\_\_ = 7$
$16 - 7 = \_\_\_$

a. 6
b. 8
c. 9
d. not given

2. Order the numbers from the greatest to the least.
460, 645, 599

a. 460, 599, 645 **b.** 599, 460, 645
c. 645, 599, 460 **d.** not given

3. What is the amount?

a. \$5.15 **b.** \$6.15
c. \$15.15 **d.** not given

4. Round 76 to the nearest 10.

a. 70 **b.** 75 **c.** 100 **d.** not given

5. Estimate:
$348 + 309 = \_\_\_$

a. 600 **b.** 657
c. 900 **d.** 348,309

6.
$$\begin{array}{r} 189 \\ 320 \\ +\,249 \\ \hline \end{array}$$

a. 648
b. 759
c. 61,418
d. not given

7. $\$10.43 + \$7.07 = \_\_\_$

a. \$17.40 **b.** \$17.50
c. \$18.50 **d.** not given

8.
$$\begin{array}{r} 1{,}023 \\ +\,4{,}675 \\ \hline \end{array}$$

a. 348
b. 5,698
c. 5,968
d. not given

9. Lynn sells 169 tickets. Ruth sells 208 tickets. About how many tickets do they sell?

a. 200 **b.** 300
c. 377 **d.** 477

10. How many students are there in grade 3?

GRADE 3 STUDENTS

One 🯅 equals 2 students.
Room 10 🯅🯅🯅🯅🯅🯅🯅🯅🯅🯅
Room 11 🯅🯅🯅🯅🯅🯅🯅🯅🯅🯅🯅🯅
Room 12 🯅🯅🯅🯅🯅🯅🯅🯅🯅🯅🯅🯅🯅🯅

a. 32 students **b.** 54 students
c. 64 students **d.** not given

6 MULTIPLICATION FACTS

Do you make models? Do you collect stickers, magazines, or coins? Do you take photos? About how much time do you spend on your hobby? What do you think would be a good way to estimate the time you spend?

The Meaning of Multiplication

Jane makes bookmarks. She puts dried flowers on them. She puts 2 flowers on each bookmark. How many flowers does she need to make 3 bookmarks?

You can add to find how many flowers.

$2 + 2 + 2 = 6$
3 twos = 6

You can also multiply to find how many.

$3 \times 2 = 6$

Read this fact: Three times two equals six.
Jane needs 6 flowers.

Complete each pair of number sentences.

1. 

$3 + 3 =$ 6

2 threes = 6

2. 

$2 + 2 + 2 + 2 =$ 8

4 twos = 8

3. 

$6 + 6 =$ 12

2 sixes = 12

4. 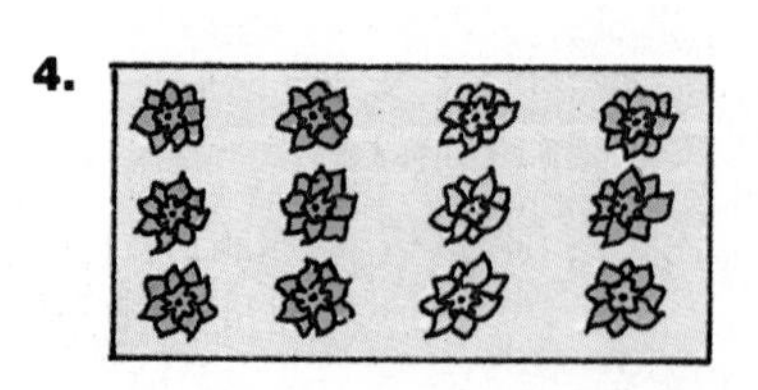

$3 + 3 + 3 + 3 =$ 12

4 threes = 12

Complete each pair of number sentences.

5.

$4 + 4 + 4 + 4 =$ ____

$4 \times 4 =$ ____

6.

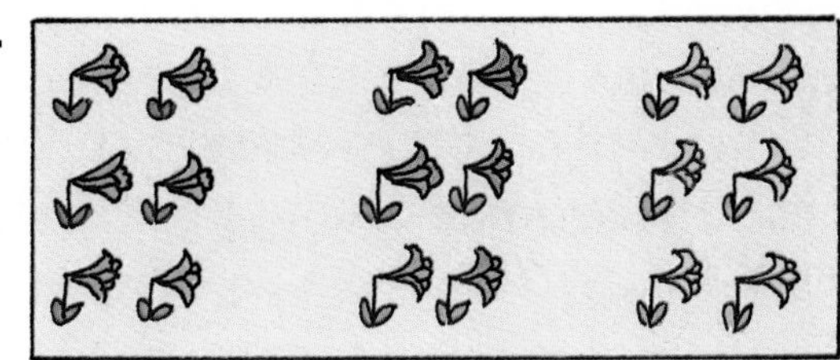

$6 + 6 + 6 =$ ____

$3 \times 6 =$ ____

7.

$5 + 5 + 5 =$ ____

$3 \times 5 =$ ____

8.

$2 + 2 + 2 + 2 + 2 + 2 + 2 =$ ____

$7 \times 2 =$ ____

Write an addition sentence and a multiplication sentence for each group.

9.

____ + ____ = ____

____ × ____ = ____

10.

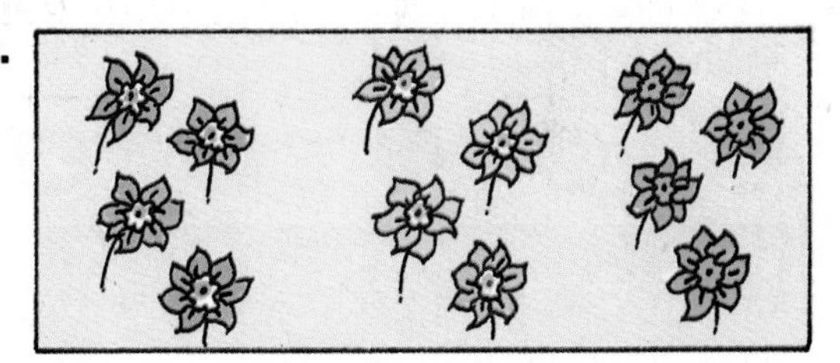

____ + ____ + ____ = ____

____ × ____ = ____

11.

12.

Multiplying Twos

Jim makes finger puppets from tennis balls. He uses a button for each eye. Today, Jim plans to make 4 puppets. Each puppet needs 2 eyes. How many buttons does Jim need?

You can multiply to find how many buttons.

$$2 + 2 + 2 + 2 = 8$$

$$\underset{\text{factor}}{4} \times \underset{\text{factor}}{2} = \underset{\text{product}}{8}$$

$$\begin{array}{r} 2 \\ \times 4 \\ \hline 8 \end{array} \begin{array}{l} \longleftarrow \text{factor} \\ \longleftarrow \text{factor} \\ \longleftarrow \text{product} \end{array}$$

Jim needs 8 buttons.

Count the groups of two to find the product.

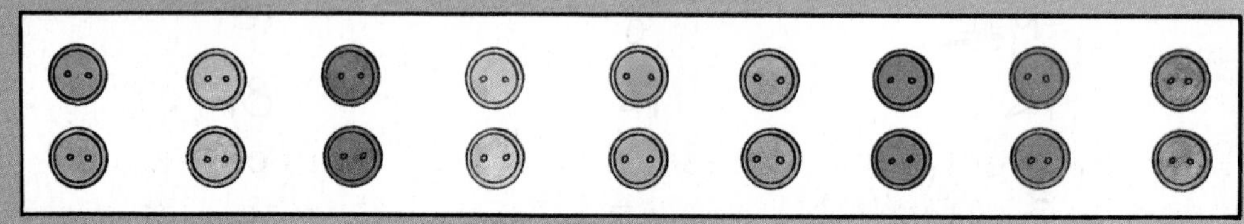

1. $1 \times 2 =$ ____
2. $2 \times 2 =$ ____
3. $3 \times 2 =$ ____
4. $5 \times 2 =$ ____
5. $6 \times 2 =$ ____
6. $7 \times 2 =$ ____
7. $8 \times 2 =$ ____
8. $9 \times 2 =$ ____

Complete each pair of number sentences.

9.

$2 + 2 + 2 =$ ____

$3 \times 2 =$ ____

10.

$2 + 2 + 2 + 2 + 2 + 2 =$ ____

$6 \times 2 =$ ____

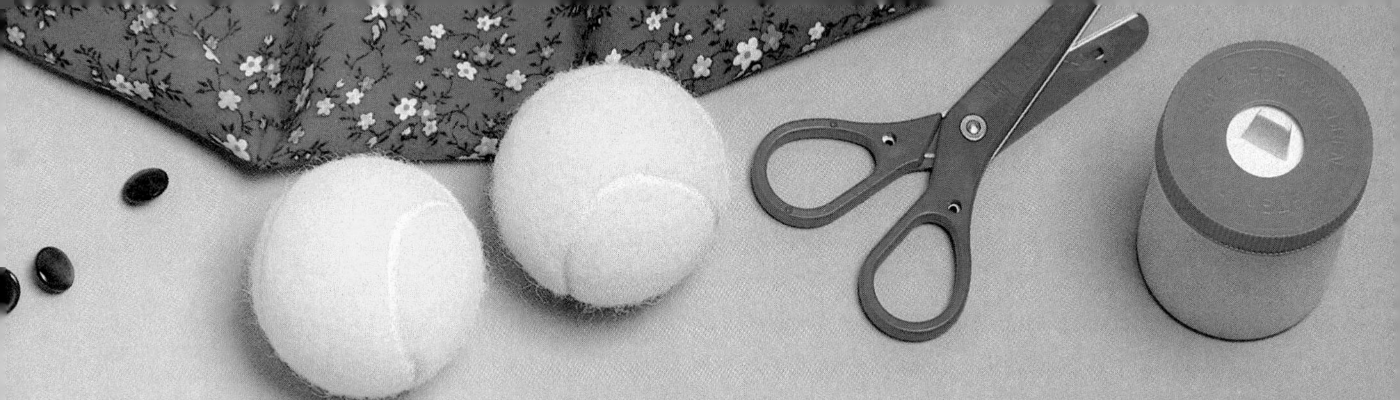

Multiply.

11. 2×3 **12.** 2×4 **13.** 2×6 **14.** 2×7 **15.** 2×8 **16.** 2×5 **17.** 2×9

18. 2×7 **19.** 2×1 **20.** 2×6 **21.** 2×2 **22.** 2×9 **23.** 2×5 **24.** 2×3

25. $8 \times 2 =$ ___ **26.** $4 \times 2 =$ ___ **27.** $7 \times 2 =$ ___ **28.** $9 \times 2 =$ ___

29. $3 \times 2 =$ ___ **30.** $6 \times 2 =$ ___ **31.** $2 \times 2 =$ ___ **32.** $5 \times 2 =$ ___

Solve.

33. Jim, Tara, and Nan perform a puppet show. Each child uses 2 puppets. How many puppets do they use in all?

34. Tara made 18 finger puppets. She sold 10 of them before the show. How many puppets did Tara have left?

35. Jim and his friends perform puppet shows on 5 nights. They perform 2 shows each night. How many shows do they perform?

★36. Kelly sells 6 pairs of tickets two hours before the show. She sells 3 tickets five minutes before the show. How many tickets does she sell in all?

ANOTHER LOOK

Add.

1. $327 + 482$ **2.** $618 + 723$ **3.** $465 + 287$ **4.** $609 + 718$ **5.** $561 + 259$

Multiplying Threes

Sue likes to make potato prints. On one day, she prints stars on notepaper. She makes 4 groups with 3 stars in each group. How many stars does Sue print on the paper?

You can multiply to find how many stars.

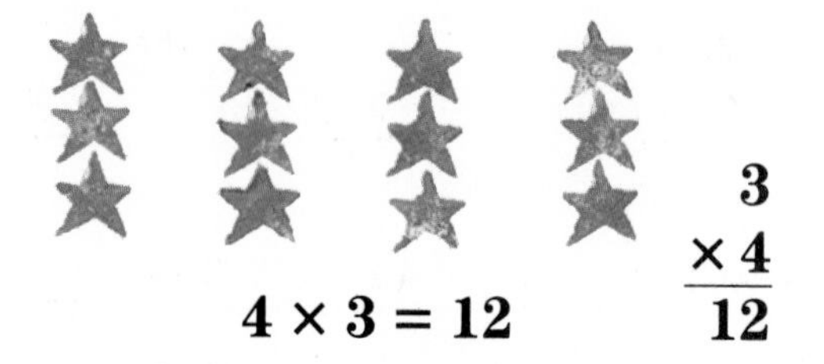

$4 \times 3 = 12$

$$\begin{array}{r} 3 \\ \times\, 4 \\ \hline 12 \end{array}$$

Sue prints 12 stars on the paper.

Count the groups of three to find the product.

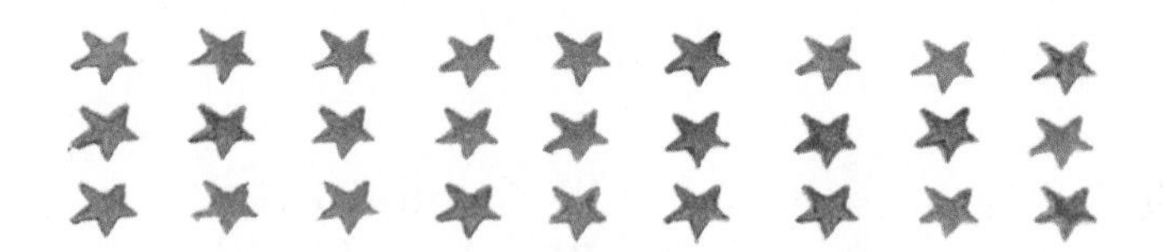

1. $1 \times 3 =$ 3 **2.** $2 \times 3 =$ 6 **3.** $3 \times 3 =$ 9 **4.** $5 \times 3 =$ 15

5. $6 \times 3 =$ 18 **6.** $7 \times 3 =$ 21 **7.** $8 \times 3 =$ 24 **8.** $9 \times 3 =$ 27

Complete each pair of number sentences.

9.

$3 + 3 + 3 + 3 + 3 + 3 =$ 18

$6 \times 3 =$ 18

10.

$3 + 3 + 3 =$ 9

$3 \times 3 =$ 9

Multiply.

11. $\begin{array}{r} 3 \\ \times\, 4 \\ \hline 12 \end{array}$ **12.** $\begin{array}{r} 3 \\ \times\, 7 \\ \hline 21 \end{array}$ **13.** $\begin{array}{r} 3 \\ \times\, 2 \\ \hline 6 \end{array}$ **14.** $\begin{array}{r} 3 \\ \times\, 5 \\ \hline 15 \end{array}$ **15.** $\begin{array}{r} 3 \\ \times\, 9 \\ \hline 27 \end{array}$ **16.** $\begin{array}{r} 3 \\ \times\, 3 \\ \hline 9 \end{array}$ **17.** $\begin{array}{r} 3 \\ \times\, 6 \\ \hline 18 \end{array}$

Multiply.

18. 3×1 = 3	**19.** 3×7 = 21	**20.** 3×6 = 18	**21.** 3×9 = 27	**22.** 3×2 = 6	**23.** 3×5 = 15	**24.** 3×8 = 24
25. 3×4 = 12	**26.** 2×1 = 2	**27.** 2×8 = 16	**28.** 3×7 = 21	**29.** 3×3 = 9	**30.** 3×2 = 6	**31.** 2×3 = 6
32. 2×2 = 4	**33.** 3×9 = 27	**34.** 2×6 = 12	**35.** 2×5 = 10	**36.** 2×9 = 18	**37.** 3×7 = 21	**38.** 3×8 = 24

39. $8 \times 3 =$ 24	**40.** $4 \times 3 =$ 12	**41.** $7 \times 3 =$ 21	**42.** $6 \times 3 =$ 18
43. $4 \times 3 =$ 12	**44.** $5 \times 3 =$ 15	**45.** $9 \times 3 =$ 27	**46.** $1 \times 3 =$ 3
47. $2 \times 3 =$ 6	**48.** $6 \times 3 =$ 18	**49.** $9 \times 2 =$ 18	**50.** $3 \times 3 =$ 9
51. $7 \times 2 =$ 14	**52.** $4 \times 3 =$ 12	**53.** $6 \times 2 =$ 12	**54.** $8 \times 3 =$ 24

Solve.

55. Sue printed flowers on paper cups for a party. She printed 6 rows of 3 flowers on each cup. How many flowers did she print on each cup?

★**56.** Sue pays 50¢ for a box of envelopes. The box has 9 different colors of envelopes. There are 3 of each color. Do the envelopes cost more or less than 1¢ each?

CALCULATOR

You can use your calculator to solve.

1. Start with 0. Add 2 until you have 18. How many times did you add 2? Write a multiplication sentence for this. $2 \times 9 = 18$

2. Start with 0. Add 3 until you have 18. How many times did you add 3? Write a multiplication sentence for this. $3 \times 6 = 18$

PROBLEM SOLVING
Choosing the Operation

When a problem has two or more groups to be joined together, you can add to find the total. Sometimes you can multiply.

A. Juan and his friends make their own band instruments. Juan makes 3 drums. Each drum needs 2 skins. How many drum skins does Juan need?

You know			You want to find	You can
how many groups.	how many in each group.	the number in each group is the same.	how many in all.	ADD or MULTIPLY.

You can add.
$2 + 2 + 2 = 6$ drum skins

You can multiply.
$3 \times 2 = 6$ drum skins

Juan needs 6 drum skins.

B. Juan punches holes in the skins to make the drums. He punches 36 holes in the big drum skin. He punches 24 holes in the middle drum skin and 18 holes in the small drum skin. How many holes does Juan punch?

You know		You want to find	You can
how many in each group.	the groups are not the same size.	how many in all.	ADD.

You can add.
$36 + 24 + 18 = 78$ holes in all

You cannot multiply.

Juan punches 78 holes.

Decide whether you can multiply to solve each problem. Write the letter of the correct answer.

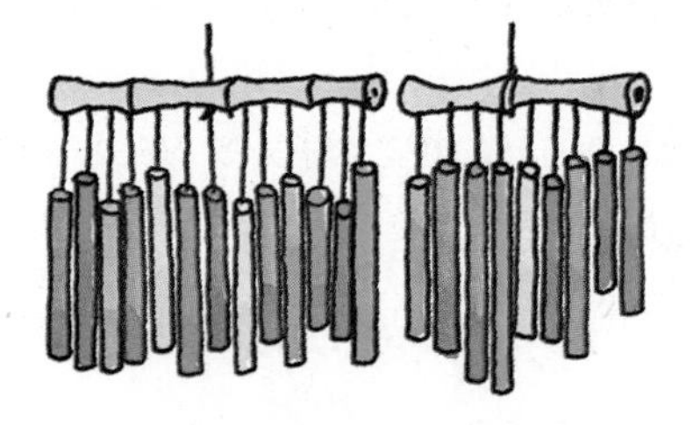

1. Ray made 2 sets of chimes. He used 13 pieces of pipe in the first set of chimes. He used 9 pieces of pipe in the second set. How many pieces of pipe did Ray use?

   a. can multiply b. cannot multiply

2. Claudia made a tin-can tambourine. First, she punched 8 holes in a tin can. Then, she tied a string with 3 bottle caps through each hole. How many bottle caps did she need?

   a. can multiply b. cannot multiply

Solve.

3. Leroy made 3 rattles. He took 3 yogurt cups and put 9 dried beans into one of them. He put 6 beans into another. He put 12 beans into the third. How many dried beans did he use?

   a can multiply b. cannotmultiply 3+9+12

4. Ray and Claudia both played 2 different kinds of instruments. Each of the other 3 band members played 1 kind of instrument. How many kinds of instruments were played in the band?

   8

5. The band practiced for 1 day per week at Claudia's house and for 2 days per week at Roosevelt's house. For how many days per week did they practice at the two houses?

   1+2=3

6. Each of the 5 band members learned 2 new songs. Then they taught their songs to the other members. How many new songs did the band learn?

   2x5=10

7. Bob can make 3 whistles from one long reed. If he has 4 reeds, how many whistles can he make?

8. Ray wrote 4 new songs. Claudia wrote 8 new songs. How many new songs did they write altogether?

   12

Multiplying Fours

Greg's dad gave him a box for his rock collection. Greg put the rocks in 5 groups. There were 4 rocks in each group. How many rocks were there in the box?

You can multiply to find how many rocks were in the box.

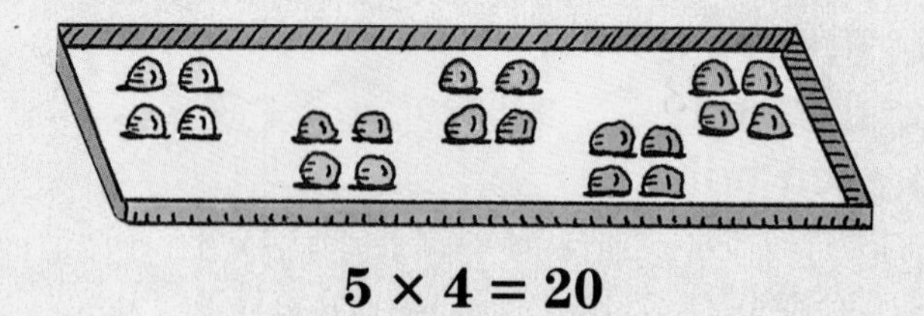

$5 \times 4 = 20$

$$\begin{array}{r} 4 \\ \times 5 \\ \hline 20 \end{array}$$

There were 20 rocks in the box.

Count the groups of four to find the product.

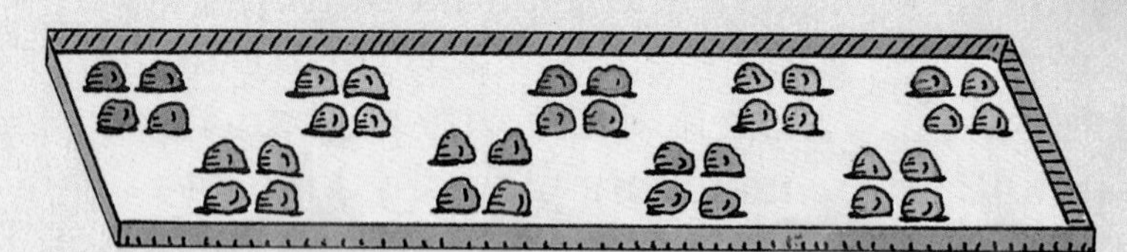

1. $1 \times 4 =$ 4
2. $2 \times 4 =$ 8
3. $3 \times 4 =$ 12
4. $4 \times 4 =$ 16
5. $6 \times 4 =$ 24
6. $7 \times 4 =$ 28
7. $8 \times 4 =$ 32
8. $9 \times 4 =$ 36

Complete each pair of number sentences.

9.

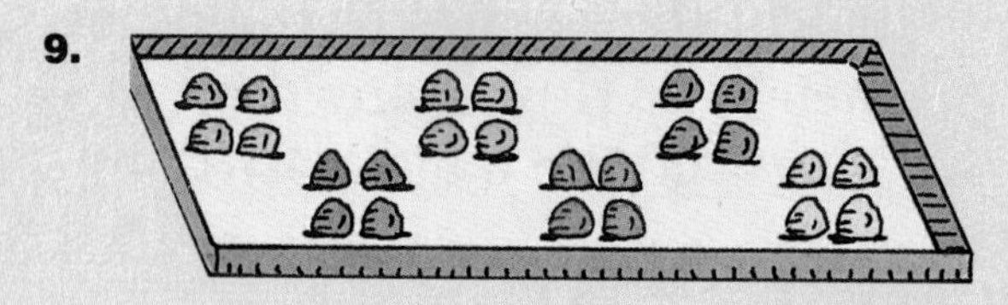

$4 + 4 + 4 + 4 + 4 + 4 =$ 24

$6 \times 4 =$ 24

10.

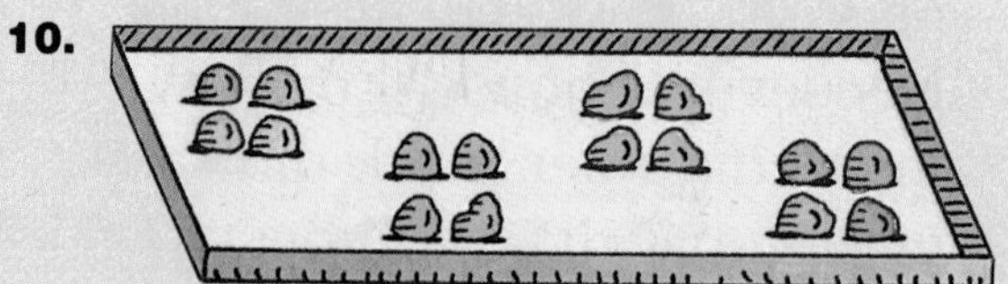

$4 + 4 + 4 + 4 =$ 16

$4 \times 4 =$ 16

Multiply.

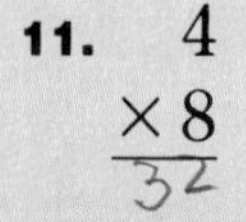

11. $\begin{array}{r} 4 \\ \times 8 \\ \hline 32 \end{array}$
12. $\begin{array}{r} 4 \\ \times 5 \\ \hline 20 \end{array}$
13. $\begin{array}{r} 4 \\ \times 7 \\ \hline 28 \end{array}$
14. $\begin{array}{r} 4 \\ \times 3 \\ \hline 12 \end{array}$
15. $\begin{array}{r} 4 \\ \times 2 \\ \hline 8 \end{array}$
16. $\begin{array}{r} 4 \\ \times 4 \\ \hline 16 \end{array}$
17. $\begin{array}{r} 4 \\ \times 6 \\ \hline 24 \end{array}$

Multiply.

18. $\begin{array}{r} 4 \\ \times 4 \\ \hline \end{array}$ 19. $\begin{array}{r} 4 \\ \times 8 \\ \hline \end{array}$ 20. $\begin{array}{r} 4 \\ \times 6 \\ \hline \end{array}$ 21. $\begin{array}{r} 4 \\ \times 5 \\ \hline \end{array}$ 22. $\begin{array}{r} 4 \\ \times 9 \\ \hline \end{array}$ 23. $\begin{array}{r} 4 \\ \times 5 \\ \hline \end{array}$ 24. $\begin{array}{r} 4 \\ \times 7 \\ \hline \end{array}$

25. $\begin{array}{r} 2 \\ \times 9 \\ \hline \end{array}$ 26. $\begin{array}{r} 3 \\ \times 8 \\ \hline \end{array}$ 27. $\begin{array}{r} 2 \\ \times 6 \\ \hline \end{array}$ 28. $\begin{array}{r} 3 \\ \times 7 \\ \hline \end{array}$ 29. $\begin{array}{r} 3 \\ \times 4 \\ \hline \end{array}$ 30. $\begin{array}{r} 2 \\ \times 5 \\ \hline \end{array}$ 31. $\begin{array}{r} 3 \\ \times 1 \\ \hline \end{array}$

32. $2 \times 4 =$ ____ 33. $9 \times 4 =$ ____ 34. $7 \times 4 =$ ____ 35. $1 \times 4 =$ ____

36. $7 \times 3 =$ ____ 37. $5 \times 2 =$ ____ 38. $6 \times 2 =$ ____ 39. $4 \times 3 =$ ____

Solve.

40. Kate built dollhouse furniture for the hobby show. She made 5 sets of chairs. Each set had 4 chairs. How many chairs did Kate build?

41. Nat made 4 model rockets and 7 model planes for the hobby show. How many models did he make altogether?

42. On one page of Bob's stamp collection, there are 4 rows of stamps with 4 stamps in each row. How many stamps are there on the page?

★43. Both Fred and Roy have rock collections. Each has 4 special rocks. Each gives 1 rock away. How many rocks are left altogether?

FOCUS: MENTAL MATH

You can skip-count to help you find multiplication facts. 4 skips of 4 = 16 or $4 \times 4 = 16$. Find the product by skip counting.

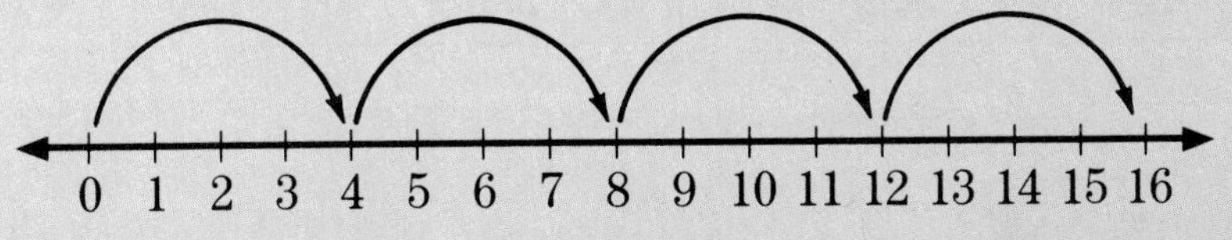

1. $2 \times 4 =$ ____ 2. $4 \times 4 =$ ____ 3. $1 \times 4 =$ ____ 4. $3 \times 4 =$ ____

Multiplying Fives

Ken's hobby is baking. He baked pretzels for his class. He put 6 groups of pretzels on a tray. There were 5 pretzels in each group. How many pretzels did he bake?

You can multiply to find how many pretzels.

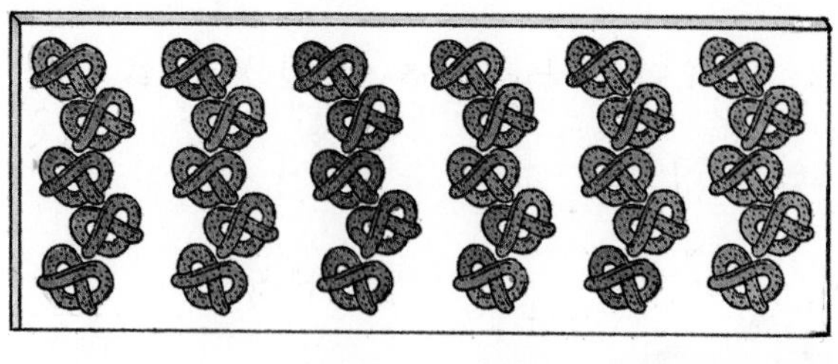

$6 \times 5 = 30$

$$\begin{array}{r} 5 \\ \times\, 6 \\ \hline 30 \end{array}$$

Ken baked 30 pretzels.

Count the groups of five to find the product.

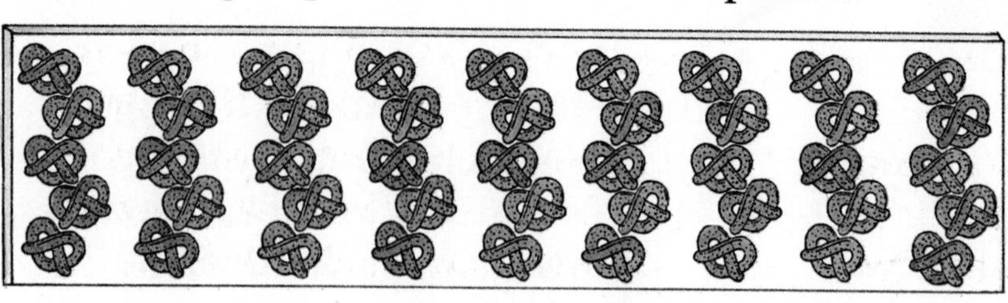

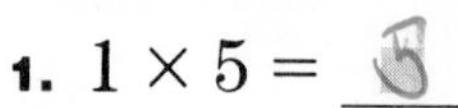

1. $1 \times 5 =$ ____
2. $2 \times 5 =$ ____
3. $3 \times 5 =$ ____
4. $4 \times 5 =$ ____
5. $5 \times 5 =$ ____
6. $7 \times 5 =$ ____
7. $8 \times 5 =$ ____
8. $9 \times 5 =$ ____

Complete each pair of number sentences.

9.

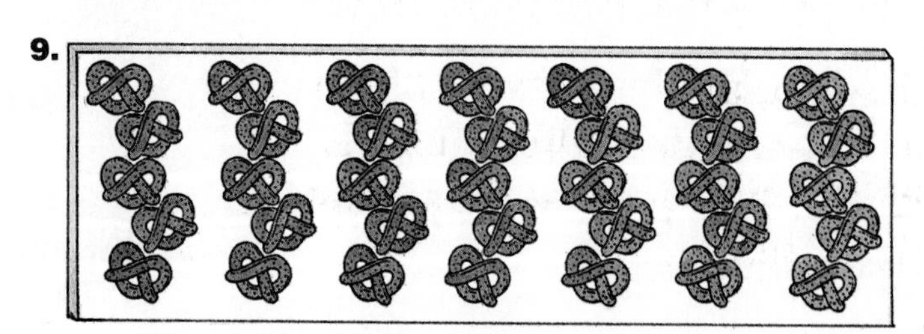

$5 + 5 + 5 + 5 + 5 + 5 + 5 =$ ____

$7 \times 5 =$ ____

10.

$5 + 5 + 5 + 5 + 5 =$ ____

$5 \times 5 =$ ____

Multiply.

11. 5×4	12. 5×6	13. 5×3	14. 5×7	15. 5×9	16. 5×2	17. 5×1
18. 3×8	19. 4×6	20. 4×4	21. 2×2	22. 3×9	23. 5×5	24. 3×7
25. 5×3	26. 3×6	27. 5×8	28. 4×9	29. 2×6	30. 2×9	31. 4×7

32. $9 \times 5 =$ ___	33. $6 \times 5 =$ ___	34. $4 \times 5 =$ ___	35. $8 \times 5 =$ ___
36. $7 \times 4 =$ ___	37. $3 \times 5 =$ ___	38. $5 \times 2 =$ ___	39. $5 \times 1 =$ ___
40. $7 \times 5 =$ ___	41. $2 \times 5 =$ ___	42. $3 \times 3 =$ ___	43. $4 \times 3 =$ ___

Solve.

44. Nora takes pictures as a hobby. She can take 12 pictures on one roll of film. She has taken 7 pictures. How many more pictures can Nora take on one roll of film?

★45. Jim needs to set up chairs for 42 people to watch a puppet show. He sets up 6 rows with 5 chairs in each row. How many more chairs does he need to set up?

FOCUS: MENTAL MATH

Skip-counting by fives will help you find multiplication facts. It will also help you multiply 5 by large numbers.

Count by fives.

1. 5, 10, ___, ___, ___, ___, ___, ___, ___

2. 55, 60, ___, ___, ___, ___, ___, ___, ___

Order in Multiplication

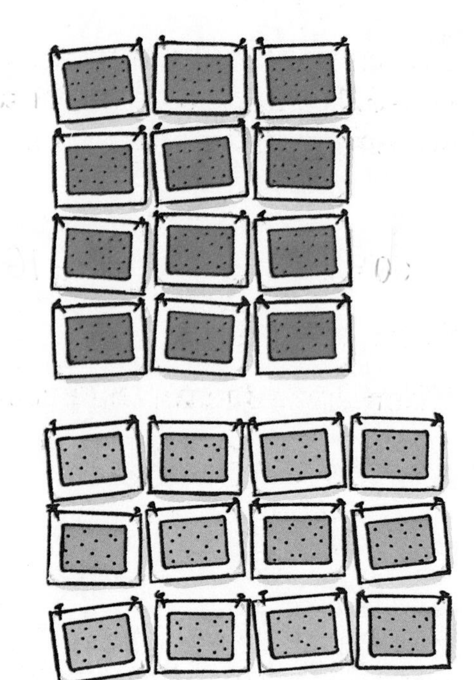

Jerry makes prints as a hobby.
He tapes some prints on the walls.

On one wall, he puts 4 rows with
3 prints in each row.

$4 \times 3 = 12$ $\quad \begin{array}{r}3\\ \times 4\\ \hline 12\end{array}$

On another wall, he puts 3 rows
with 4 prints in each row.

$3 \times 4 = 12$ $\quad \begin{array}{r}4\\ \times 3\\ \hline 12\end{array}$

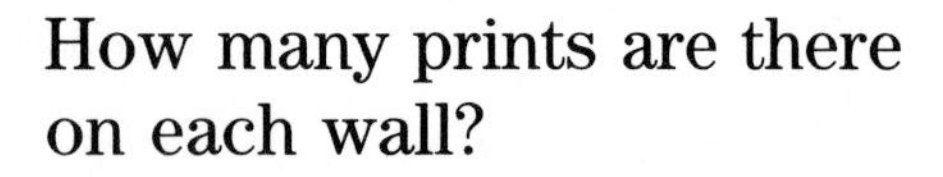

How many prints are there
on each wall?

There are 12 prints on each wall.
You can multiply numbers in any order.
The product is always the same.

Multiply.

1. $\begin{array}{r}4\\ \times 6\\ \hline 24\end{array}$ $\begin{array}{r}6\\ \times 4\\ \hline 24\end{array}$ **2.** $\begin{array}{r}2\\ \times 3\\ \hline 6\end{array}$ $\begin{array}{r}3\\ \times 2\\ \hline 6\end{array}$ **3.** $\begin{array}{r}4\\ \times 7\\ \hline 28\end{array}$ $\begin{array}{r}7\\ \times 4\\ \hline 28\end{array}$ **4.** $\begin{array}{r}3\\ \times 6\\ \hline 18\end{array}$ $\begin{array}{r}6\\ \times 3\\ \hline 18\end{array}$

5. $\begin{array}{r}2\\ \times 9\\ \hline 18\end{array}$ $\begin{array}{r}9\\ \times 2\\ \hline 18\end{array}$ **6.** $\begin{array}{r}3\\ \times 5\\ \hline 15\end{array}$ $\begin{array}{r}5\\ \times 3\\ \hline 15\end{array}$ **7.** $\begin{array}{r}4\\ \times 2\\ \hline 8\end{array}$ $\begin{array}{r}2\\ \times 4\\ \hline 8\end{array}$ **8.** $\begin{array}{r}5\\ \times 7\\ \hline 35\end{array}$ $\begin{array}{r}7\\ \times 5\\ \hline 35\end{array}$

9. $6 \times 5 = 30$ $5 \times 6 = 30$ **10.** $3 \times 4 = 12$ $4 \times 3 = 12$ **11.** $8 \times 3 = 24$ $3 \times 8 = 24$ **12.** $8 \times 2 = 16$ $2 \times 8 = 16$

13. $6 \times 4 = 24$ $4 \times 6 = 24$ **14.** $7 \times 3 = 21$ $3 \times 7 = 21$ **15.** $9 \times 5 = 45$ $5 \times 9 = 45$ **16.** $2 \times 6 = 12$ $6 \times 2 = 12$

17. $9 \times 3 = 27$ $3 \times 9 = 27$ **18.** $8 \times 5 = 40$ $5 \times 8 = 40$ **19.** $5 \times 2 = 10$ $2 \times 5 = 10$ **20.** $8 \times 4 = 32$ $4 \times 8 = 32$

Multiplying with Zero and One

A. Jerry wants to tack pictures on the bulletin board. He finds 3 boxes of tacks. All the boxes are empty. How many tacks are there in all?

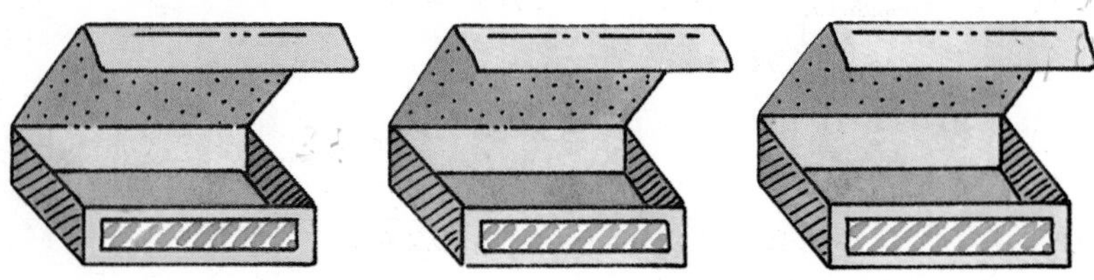

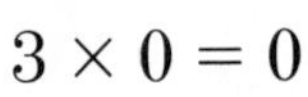

$3 \times 0 = 0$

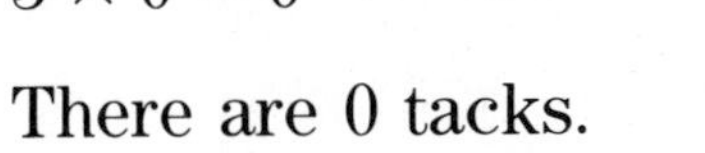

There are 0 tacks.

When 0 is a factor, the product is 0.

$3 \times 0 = 0$, or $0 \times 3 = 0$.

$$\begin{array}{r} 0 \\ \times 3 \\ \hline 0 \end{array} \qquad \begin{array}{r} 3 \\ \times 0 \\ \hline 0 \end{array}$$

B. Jerry finds 3 more boxes of tacks. Each box has 1 tack. How many tacks does Jerry have in all?

$3 \times 1 = 3$

Jerry has 3 tacks.

When 1 is a factor, the product is the other factor.

$3 \times 1 = 3$, or $1 \times 3 = 3$.

$$\begin{array}{r} 1 \\ \times 3 \\ \hline 3 \end{array} \qquad \begin{array}{r} 3 \\ \times 1 \\ \hline 3 \end{array}$$

Multiply.

1. $\begin{array}{r} 8 \\ \times 0 \\ \hline \end{array}$ **2.** $\begin{array}{r} 2 \\ \times 1 \\ \hline \end{array}$ **3.** $\begin{array}{r} 0 \\ \times 4 \\ \hline \end{array}$ **4.** $\begin{array}{r} 7 \\ \times 1 \\ \hline \end{array}$ **5.** $\begin{array}{r} 1 \\ \times 0 \\ \hline \end{array}$ **6.** $\begin{array}{r} 0 \\ \times 6 \\ \hline \end{array}$ **7.** $\begin{array}{r} 3 \\ \times 1 \\ \hline \end{array}$

8. $1 \times 4 =$ ____ **9.** $5 \times 0 =$ ____ **10.** $6 \times 1 =$ ____ **11.** $0 \times 3 =$ ____

12. $0 \times 8 =$ ____ **13.** $9 \times 0 =$ ____ **14.** $3 \times 1 =$ ____ **15.** $1 \times 2 =$ ____

16. $6 \times 2 =$ ____ **17.** $5 \times 4 =$ ____ **18.** $1 \times 5 =$ ____ **19.** $4 \times 7 =$ ____

Multiplying Sixes

Kim belongs to a camping club. She once brought 3 bags of apples to the camp. Each bag contained 6 apples. How many apples did Kim bring?

You can multiply to find how many apples.

$3 \times 6 = 18$ $\quad \begin{array}{r} 6 \\ \times 3 \\ \hline 18 \end{array}$

> Remember: you can multiply numbers in any order

Kim brought 18 apples.

Count the groups of six to find the product.

1. $1 \times 6 =$ ____ 2. $2 \times 6 =$ ____ 3. $4 \times 6 =$ ____ 4. $5 \times 6 =$ ____

5. $6 \times 6 =$ ____ 6. $7 \times 6 =$ ____ 7. $8 \times 6 =$ ____ 8. $9 \times 6 =$ ____

Multiply.

9. $\begin{array}{r} 5 \\ \times 6 \\ \hline \end{array}$ 10. $\begin{array}{r} 6 \\ \times 2 \\ \hline \end{array}$ 11. $\begin{array}{r} 3 \\ \times 6 \\ \hline \end{array}$ 12. $\begin{array}{r} 6 \\ \times 9 \\ \hline \end{array}$ 13. $\begin{array}{r} 6 \\ \times 7 \\ \hline \end{array}$ 14. $\begin{array}{r} 1 \\ \times 6 \\ \hline \end{array}$ 15. $\begin{array}{r} 4 \\ \times 6 \\ \hline \end{array}$

16. $\begin{array}{r} 6 \\ \times 8 \\ \hline \end{array}$ 17. $\begin{array}{r} 6 \\ \times 9 \\ \hline \end{array}$ 18. $\begin{array}{r} 6 \\ \times 4 \\ \hline \end{array}$ 19. $\begin{array}{r} 6 \\ \times 3 \\ \hline \end{array}$ 20. $\begin{array}{r} 6 \\ \times 8 \\ \hline \end{array}$ 21. $\begin{array}{r} 6 \\ \times 7 \\ \hline \end{array}$ 22. $\begin{array}{r} 6 \\ \times 6 \\ \hline \end{array}$

Multiply.

23. 6×6
24. 3×6
25. 6×8
26. 6×7
27. 6×2
28. 4×6
29. 9×6
30. 6×5
31. 6×6
32. 3×0
33. 2×7
34. 6×8
35. 4×6
36. 0×5
37. 3×3
38. 1×6
39. 4×7
40. 4×4
41. 1×8
42. 2×6
43. 5×8

44. $2 \times 6 =$ ___
45. $9 \times 6 =$ ___
46. $4 \times 6 =$ ___
47. $5 \times 6 =$ ___
48. $3 \times 5 =$ ___
49. $3 \times 3 =$ ___
50. $9 \times 5 =$ ___
51. $6 \times 2 =$ ___
52. $3 \times 6 =$ ___
53. $7 \times 6 =$ ___
54. $7 \times 0 =$ ___
55. $9 \times 1 =$ ___

Solve.

56. Each of 6 campers carried 3 water bottles. How many water bottles did the group carry?

★57. There were 6 boats that held 4 campers each. Another 4 boats held 2 campers each. Could the boats hold 48 campers at one time?

MIDCHAPTER REVIEW

Multiply.

1. 2×8
2. 3×6
3. 5×7
4. 4×3
5. 2×6
6. 0×9
7. 3×9

8. $9 \times 4 =$ ___
9. $5 \times 5 =$ ___
10. $8 \times 1 =$ ___
11. $2 \times 5 =$ ___
12. $4 \times 6 =$ ___
13. $7 \times 6 =$ ___
14. $6 \times 5 =$ ___
15. $6 \times 6 =$ ___

More Practice, page 410

PROBLEM SOLVING

Solving Two-Step Problems/Making a Plan

A problem may need more than one step in order to be solved. Making a plan can help you solve such a problem.

> Renee's crafts club used to have 4 members. Last week, 3 more joined. Today, each member is making 2 tie-dyed T-shirts. How many shirts are the club members making today?

Needed Data: the total number of members

Plan

Step 1: Find the total number of members.
Step 2: Find the total number of shirts.

Step 1: Add to find the total number of members.

$$\begin{array}{rl} 4 & \text{(club members)} \\ +3 & \text{(new members)} \\ \hline 7 & \text{(total number of members)} \end{array}$$

Step 2: Multiply to find the total number of shirts.

$$\begin{array}{rl} 7 & \text{(number of members)} \\ \times 2 & \text{(shirts made by each)} \\ \hline 14 & \text{(total number of shirts)} \end{array}$$

The club members are making 14 shirts today.

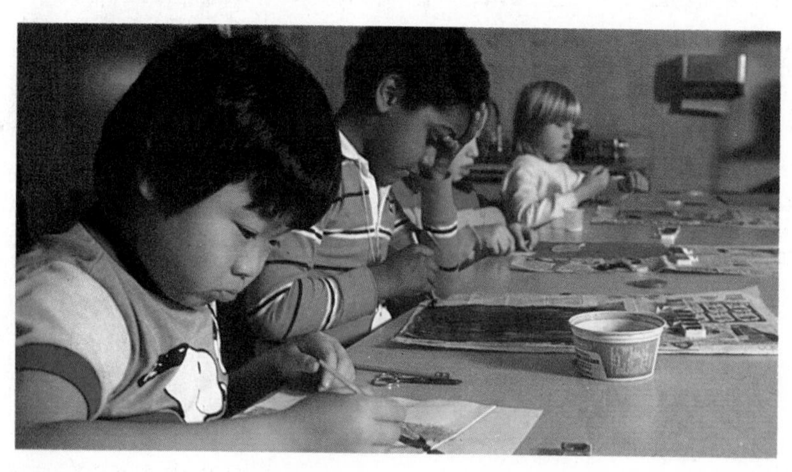

Complete the plan. Write the missing step.

1. The club members make 5 belts. Each belt is made by braiding 3 leather strips. The club has 16 leather strips on hand. How many strips will be left?

   Step 1: Find the total number of strips used for 5 belts.

   Step 2:

2. Lee has a book that contains 36 patterns for sweaters. Fay has another book that has 16 patterns. Club members have made 47 of the sweaters. How many are left to make?

   Step 1:

   Step 2: Find the number of sweaters that are left to make.

Make a plan for each problem. Solve.

3. Each of 6 club members makes 2 clay bookends. The firing oven holds only 8 bookends at a time. After the oven fires the first batch, how many bookends must still be fired?

4. Lee buys 17 used crafts magazines. Erica buys 12. Later, they find that their club already has copies of 6 of these magazines. How many of these magazines did they not already have?

5. The club members are making a quilt. The quilt will have 6 rows of 6 squares each. They have already made 34 squares. How many squares must they still make?

6. Fay, Erica, and Mark each make a leather bookbag. Ty and Renee each make a cloth bag. Each bag needs 2 zippers. How many zippers are needed for all the bags?

★7. There are 2 cards of buttons. Ty plans to use 7 from a card of 12 buttons. Renee will use 6 from a card of 8. How many buttons will be left?

★8. Ty strings 9 wooden beads. Fay strings twice as many. Erica strings 8 more than Fay and Ty combined. How many beads do the 3 club members string?

Multiplying Sevens

Mike and Ann like to work in their garden. Last week, they planted 6 groups of tulip bulbs. They put 7 bulbs in each group. How many bulbs did they plant?

You can multiply to find how many bulbs.

$6 \times 7 = 42$, or $7 \times 6 = 42$.

$$\begin{array}{r} 7 \\ \times 6 \\ \hline 42 \end{array} \quad \textbf{OR} \quad \begin{array}{r} 6 \\ \times 7 \\ \hline 42 \end{array}$$

They planted 42 bulbs.

Count the groups of seven to find the product.

1. $1 \times 7 =$ ____ **2.** $2 \times 7 =$ ____ **3.** $3 \times 7 =$ ____ **4.** $4 \times 7 =$ ____

5. $5 \times 7 =$ ____ **6.** $7 \times 7 =$ ____ **7.** $8 \times 7 =$ ____ **8.** $9 \times 7 =$ ____

Copy the chart. Multiply to complete.

9.

×	1	2	3	4	5	6	7	8	9
7									

Multiply.

10. 7×8 **11.** 7×7 **12.** 7×4 **13.** 7×5 **14.** 7×9 **15.** 7×3 **16.** 7×6

17. 5×4 **18.** 6×5 **19.** 4×3 **20.** 2×3 **21.** 3×8 **22.** 2×7 **23.** 9×7

24. 9×4 **25.** 4×7 **26.** 5×5 **27.** 6×7 **28.** 4×2 **29.** 8×7 **30.** 4×8

31. 5×7 **32.** 6×6 **33.** 3×7 **34.** 9×3 **35.** 4×4 **36.** 1×7 **37.** 7×7

38. $1 \times 7 =$ ____ **39.** $6 \times 7 =$ ____ **40.** $2 \times 7 =$ ____ **41.** $4 \times 7 =$ ____

42. $0 \times 9 =$ ____ **43.** $8 \times 1 =$ ____ **44.** $8 \times 4 =$ ____ **45.** $7 \times 4 =$ ____

46. $7 \times 5 =$ ____ **47.** $7 \times 3 =$ ____ **48.** $7 \times 7 =$ ____ **49.** $2 \times 2 =$ ____

Solve.

50. Members of a garden club work in a city park. One day, 7 club members planted rosebushes in the park. Each person planted 7 bushes. How many rosebushes did they plant?

★51. Club members dug 7 rows in a garden. They planted 7 bushes in each row. They dug 3 more rows and planted 4 trees in each row. How many more bushes than trees did they plant?

ANOTHER LOOK

Subtract.

1. $742 - 589 =$ ____ **2.** $428 - 165 =$ ____ **3.** $984 - 299 =$ ____

4. $332 - 255 =$ ____ **5.** $659 - 372 =$ ____ **6.** $407 - 229 =$ ____

7. $813 - 625 =$ ____ **8.** $562 - 407 =$ ____ **9.** $348 - 197 =$ ____

More Practice, page 410

Multiplying Eights and Nines

Ty and Tanya make stuffed animals. They keep the animals on 9 shelves. There are 8 animals on each shelf. How many animals do Ty and Tanya have?

You can multiply to find how many animals.

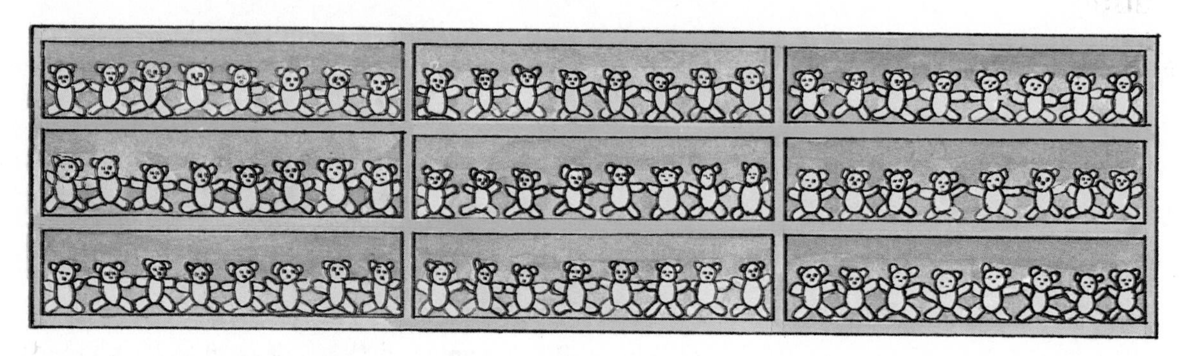

$$9 \times 8 = 72 \qquad \begin{array}{r} 8 \\ \times 9 \\ \hline 72 \end{array}$$

Ty and Tanya have 72 animals.

Count the bears to find these products.

1. $1 \times 8 =$ ____
2. $2 \times 8 =$ ____
3. $3 \times 8 =$ ____
4. $4 \times 8 =$ ____
5. $5 \times 8 =$ ____
6. $6 \times 8 =$ ____
7. $7 \times 8 =$ ____
8. $8 \times 8 =$ ____

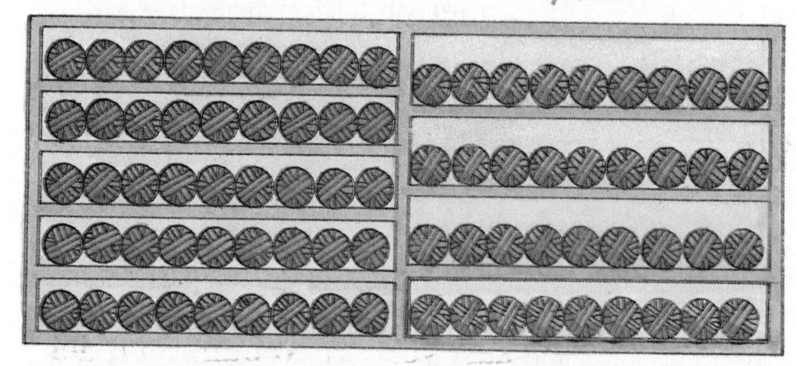

Count the balls of yarn to find these products.

9. $1 \times 9 =$ ____
10. $2 \times 9 =$ ____
11. $3 \times 9 =$ ____
12. $4 \times 9 =$ ____
13. $5 \times 9 =$ ____
14. $6 \times 9 =$ ____
15. $7 \times 9 =$ ____
16. $9 \times 9 =$ ____

Copy each chart. Multiply to complete.

17.

×	1	2	3	4	5	6	7	8	9
8									

18.

×	1	2	3	4	5	6	7	8	9
9									

Multiply.

19. $\begin{array}{r} 3 \\ \times 9 \\ \hline \end{array}$ **20.** $\begin{array}{r} 5 \\ \times 4 \\ \hline \end{array}$ **21.** $\begin{array}{r} 8 \\ \times 2 \\ \hline \end{array}$ **22.** $\begin{array}{r} 6 \\ \times 7 \\ \hline \end{array}$ **23.** $\begin{array}{r} 1 \\ \times 9 \\ \hline \end{array}$ **24.** $\begin{array}{r} 7 \\ \times 9 \\ \hline \end{array}$ **25.** $\begin{array}{r} 6 \\ \times 5 \\ \hline \end{array}$

26. $\begin{array}{r} 4 \\ \times 7 \\ \hline \end{array}$ **27.** $\begin{array}{r} 7 \\ \times 2 \\ \hline \end{array}$ **28.** $\begin{array}{r} 9 \\ \times 3 \\ \hline \end{array}$ **29.** $\begin{array}{r} 5 \\ \times 7 \\ \hline \end{array}$ **30.** $\begin{array}{r} 3 \\ \times 6 \\ \hline \end{array}$ **31.** $\begin{array}{r} 2 \\ \times 6 \\ \hline \end{array}$ **32.** $\begin{array}{r} 8 \\ \times 1 \\ \hline \end{array}$

33. $\begin{array}{r} 8 \\ \times 4 \\ \hline \end{array}$ **34.** $\begin{array}{r} 7 \\ \times 6 \\ \hline \end{array}$ **35.** $\begin{array}{r} 5 \\ \times 9 \\ \hline \end{array}$ **36.** $\begin{array}{r} 6 \\ \times 9 \\ \hline \end{array}$ **37.** $\begin{array}{r} 4 \\ \times 3 \\ \hline \end{array}$ **38.** $\begin{array}{r} 5 \\ \times 8 \\ \hline \end{array}$ **39.** $\begin{array}{r} 9 \\ \times 0 \\ \hline \end{array}$

40. $8 \times 4 =$ ____ **41.** $3 \times 5 =$ ____ **42.** $7 \times 9 =$ ____ **43.** $5 \times 2 =$ ____

44. $6 \times 0 =$ ____ **45.** $4 \times 3 =$ ____ **46.** $8 \times 6 =$ ____ **47.** $1 \times 9 =$ ____

Solve.

48. Amelia's hobby is making quilts. One of her quilts has a sunburst design. There are 9 rows of suns with 8 suns in each row. Write a number sentence that shows how many suns are on the quilt.

49. Ivan makes his own writing paper and then boxes it for gifts. He makes 4 different designs. He puts 9 sheets of each design into one box. How many sheets of paper does he put into the box?

50. Rosa makes animals out of shells. She uses 8 shells to make each animal. How many shells does Rosa need to make 6 animals?

51. Last year, Paco built 5 model rockets. He built 2 more last month. His cousin Carmen has built 9 model rockets. Who has built more rockets?

More Practice, page 410

PROBLEM SOLVING
Identifying Needed Information

Some problems do not have all the information you need to answer the question.

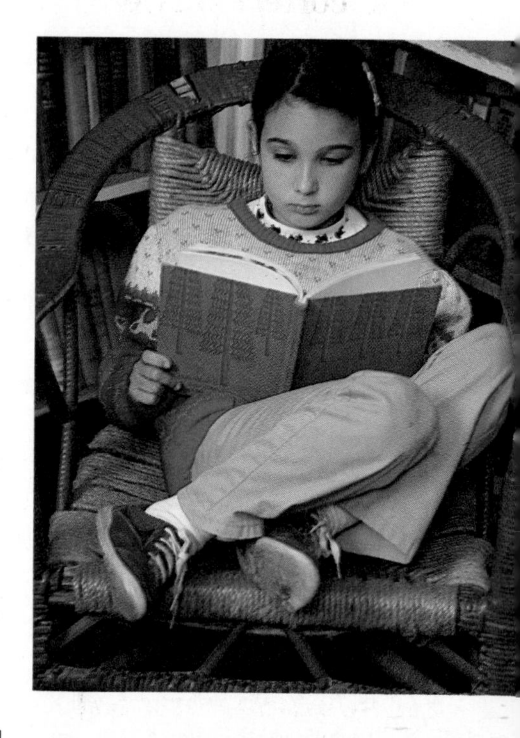

Brenda's hobby is collecting old books. This week, she bought several mystery books from the 1940's. Each book has 2 stories in it. How many stories is that in all?

First decide what information you need to solve the problem. Then make a checklist to help you decide how to find the information.

What information is needed?
the number of books Brenda bought

Checklist	Yes	No
Do I already know the information?		✓
Can I find the information		
in a magazine article?		✓
in a reference book?		✓
from another person?		✓
Am I really stuck?	✓	

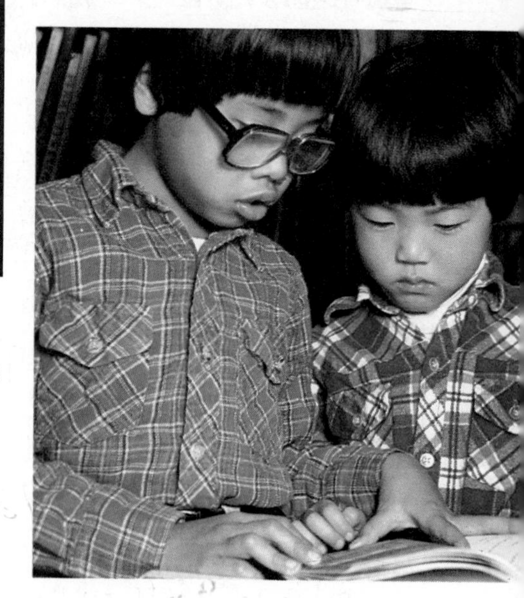

There is no way to find this information. When you are given a problem like this, you can write as your answer *There is not enough information.*

Choose the information you would need in order to solve each problem. Write the letter of the correct answer.

1. Brenda sees a copy of *Young Detectives* that costs $2.25. She has already bought the same book at a lower price. How much has she saved?
   - a. when she bought the book
   - b. the lower price
   - c. the number of pages

2. Brenda has read each of the Mystery Ace books 3 times. How many times has Brenda read the *Mystery Ace* books?
   - a. the number of books
   - b. the length of each book
   - c. the author of the books

Write what information, if any, you would need to solve each problem. Solve any problem that can be solved.

3. The first book that starred The Treasure Hunter is now worth $157. The first Nurse Kay Adams book is worth more. How much more is it worth?

4. Brenda has the first Jenny Star book. It was printed in 1954. She also has the first Science Twins book. Which book is older?

5. Pete bought 350 see-through book covers. When he used all the covers, he still had 129 books with no covers. How many books does he have?

6. Brenda has 89 Doctor Meg books. She plans to trade 35 Pilot Tales books for 13 more Doctor Meg books. How many Doctor Meg books will she have then?

★7. Pete has enough Cowboys books to read a different one every day in the month of June and have 9 left. How many books does he have?

★8. Movie Detective is a series of mystery books. One book is issued each month. Pete buys every book issued from 1977 and 1978. How many books does he buy?

Three Factors

A. Rhea plans to knit 2 sweaters for each of her 4 large dogs. She will need 3 balls of yarn for each sweater. How many balls of yarn will she need?

You can multiply to find how many balls of yarn.

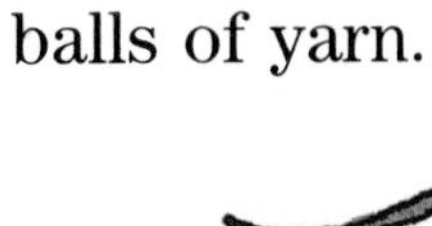

dogs → **4** x sweaters → **2** x balls of yarn → **3** =

8 x **3** = **24** balls of yarn

↑ sweaters ↑ balls of yarn per sweater

Rhea will need 24 balls of yarn.

B. You can change the grouping of the factors. The product is always the same.

$(4 \times 2) \times 3 = 24$	$4 \times (2 \times 3) = 24$
$8 \times 3 = 24$	$4 \times 6 = 24$

Checkpoint Write the letter of the correct answer.

Multiply.

1. $(3 \times 3) \times 1 =$
   - a. 6
   - b. 7
   - c. 9
   - d. 10

2. $2 \times (3 \times 1) =$
   - a. 5
   - b. 6
   - c. 7
   - d. 8

3. $3 \times (3 \times 2) =$
   - a. 8
   - b. 9
   - c. 15
   - d. 18

Complete.

1. $(2 \times 4) \times 1$

   $8 \times 1 =$ ___

2. $2 \times (4 \times 1)$

   $2 \times 4 =$ ___

3. $(3 \times 3) \times 2$

   ___ $\times 2 =$ ___

4. $3 \times (3 \times 2)$

   $3 \times$ ___ $=$ ___

Multiply.

5. $(2 \times 2) \times 5 =$ ___
6. $(5 \times 1) \times 4 =$ ___
7. $(4 \times 1) \times 4 =$ ___
8. $4 \times (2 \times 4) =$ ___
9. $(1 \times 9) \times 2 =$ ___
10. $6 \times (7 \times 0) =$ ___

Group two factors and then multiply.

11. $3 \times 1 \times 4 =$ ___
12. $2 \times 1 \times 9 =$ ___
13. $2 \times 5 \times 1 =$ ___
14. $4 \times 1 \times 8 =$ ___
15. $1 \times 2 \times 4 =$ ___
16. $4 \times 0 \times 5 =$ ___

Solve.

17. Rhea has 3 aquariums. She has a different kind of fish in each aquarium. There are 4 of each kind of fish. How many fish does Rhea have in all?

18. Rhea feeds her 4 parrots 2 times each day. She gives each parrot 2 pieces of banana at each feeding. How many pieces of banana must Rhea prepare for her parrots each day?

CHALLENGE

Find the number.

1. If you multiply this number by 4, you get 32. ___

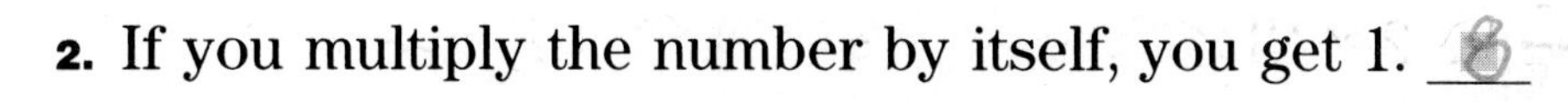

2. If you multiply the number by itself, you get 1. ___

3. If you multiply a number by it, you get the same number. ___

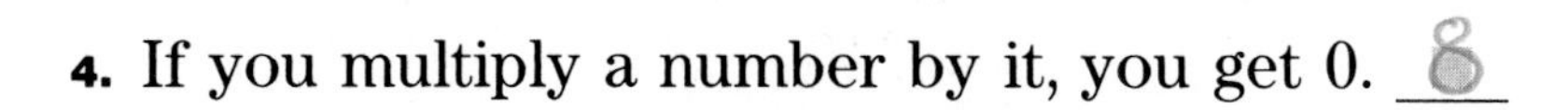

4. If you multiply a number by it, you get 0. ___

CALCULATOR

You can use your calculator to count by fours. Try this addition problem on your calculator.

$$4 + 4 + 4 + 4 = ■$$

Press these keys. Watch the display.

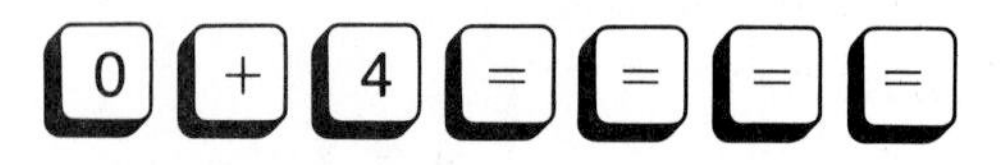

You see 4, 8, 12, and 16 because

$0 + 4 = 4$	or	$1 \times 4 = 4$
$4 + 4 = 8$	or	$2 \times 4 = 8$
$4 + 4 + 4 = 12$	or	$3 \times 4 = 12$
$4 + 4 + 4 + 4 = 16$	or	$4 \times 4 = 16$

You can count by fours to practice the multiplication facts for 4.

Use your calculator to practice multiplication facts. Be sure to clear the calculator after each exercise.

	Start	Number of =	Display	Multiplication Sentence
Example	0 + 8	7	56	$7 \times 8 = 56$
1.	0 + 8	6	48	
2.	0 + 8	5	■	
3.	0 + 9	7	■	
4.	0 + 5	■	20	
5.	■	3	24	
6.	■	■	■	$3 \times 4 = 12$

GROUP PROJECT

Lunch for Larry

The problem: Larry taught you how to play chess. To thank him, you have invited him to lunch. You go to a restaurant. Larry asks you to order for him. Look at the Key Facts and at the menu. Decide what you can order without going over your budget.

Key Facts

- You have $15.00.
- You want to order food that Larry likes.

LAKESIDE RESTAURANT

Soup of the Day	
Chicken	$1.50
Salads	
Tossed Green Salad	$1.00
Cook's Salad	$2.25
Hamburger	$2.00
Cheeseburger	$3.00
Drinks	
Milk	$0.35
Apple Drink	$0.45
Lemonade	$0.60

Main Course	
Meat-Loaf Dish with Vegetables	$2.75
Chicken Pot Pie	$2.25
Steak Dinner with Baked Potato	$4.50
Seafood Dinner	$3.75
Desserts	
Fresh Fruit	$1.25
Carrot Cake	$1.75

CHAPTER TEST

Multiply. (pages 158, 160, 162, 164, 166, 170, 174, and 176)

1. 2×6 **2.** 3×4 **3.** 3×6 **4.** 3×8 **5.** 4×8

6. 7×1 **7.** 8×8 **8.** 8×6 **9.** 9×7 **10.** 9×5

11. 6×6 **12.** 5×5 **13.** 0×5 **14.** 7×7 **15.** 8×7

16. $8 \times 2 =$ ____ **17.** $4 \times 9 =$ ____ **18.** $5 \times 3 =$ ____

19. $2 \times 9 =$ ____ **20.** $9 \times 9 =$ ____ **21.** $3 \times 6 =$ ____

Solve. (page 172)

22. Pete counts 9 ponies in the circus tent. Each pony has 2 riders. Then 3 ponies and their riders leave the tent. How many riders are there still inside the tent?

23. There are 4 jugglers in the tent. When the show begins, each juggler is juggling 6 hoops. During the show, 8 hoops were dropped. How many hoops were not dropped?

Multiply. (page 180)

24. $(3 \times 3) \times 2 =$ ____

25. $(7 \times 5) \times 0 =$ ____

26. $(2 \times 3) \times 8 =$ ____

27. $6 \times (3 \times 1) =$ ____

28. $(2 \times 2) \times 9 =$ ____

29. $(9 \times 0) \times 9 =$ ____

What information do you need to solve? (page 178)

30. Kerry buys movie tickets for each of her friends. Tickets are $3.00 each. How much money does Kerry spend?

31. Mercedes plays some jazz records for her sister. Each record has 8 songs. How many songs do they play in all?

Solve. (page 162)

32. Bob sells newspapers. He sells 62 on Friday, 38 on Saturday, and 85 on Sunday. How many papers does he sell in the three days?

33. Sylvia is training for a race. She runs 6 miles each day after school. How many miles does Sylvia run in 5 days?

BONUS

Find the pattern. Then use it to complete the chart.

1.

×	1	2	3	4	5	6	7	8	9
3									

2.

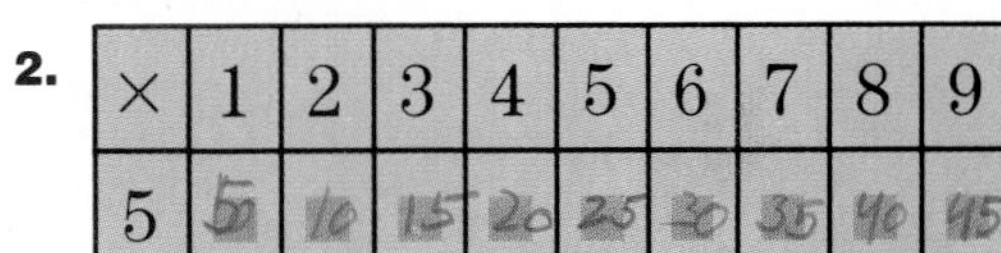

×	1	2	3	4	5	6	7	8	9
5									

3.

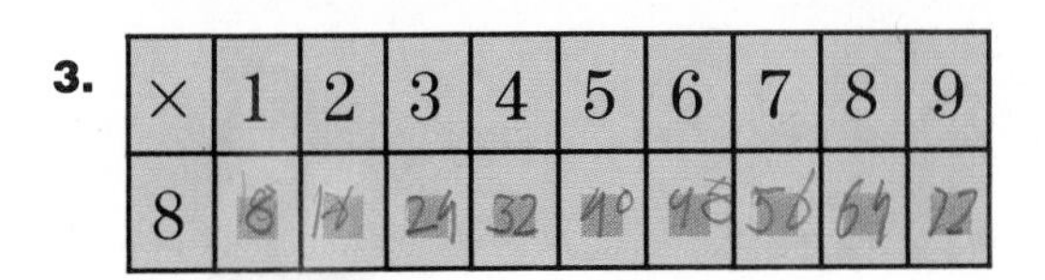

×	1	2	3	4	5	6	7	8	9
8									

RETEACHING

You can multiply numbers in any order. The product is always the same.

$2 \times 5 = 10$ $\quad \begin{array}{r} 5 \\ \times 2 \\ \hline 10 \end{array}$

$5 \times 2 = 10$ $\quad \begin{array}{r} 2 \\ \times 5 \\ \hline 10 \end{array}$

Multiply.

1. $\begin{array}{r} 3 \\ \times 6 \\ \hline \end{array}$ $\quad \begin{array}{r} 6 \\ \times 3 \\ \hline \end{array}$

2. $\begin{array}{r} 2 \\ \times 9 \\ \hline \end{array}$ $\quad \begin{array}{r} 9 \\ \times 2 \\ \hline \end{array}$

3. $\begin{array}{r} 4 \\ \times 8 \\ \hline \end{array}$ $\quad \begin{array}{r} 8 \\ \times 4 \\ \hline \end{array}$

4. $\begin{array}{r} 3 \\ \times 7 \\ \hline \end{array}$ $\quad \begin{array}{r} 7 \\ \times 3 \\ \hline \end{array}$

5. $\begin{array}{r} 5 \\ \times 3 \\ \hline \end{array}$ $\quad \begin{array}{r} 3 \\ \times 5 \\ \hline \end{array}$

6. $\begin{array}{r} 3 \\ \times 4 \\ \hline \end{array}$ $\quad \begin{array}{r} 4 \\ \times 3 \\ \hline \end{array}$

7. $\begin{array}{r} 9 \\ \times 6 \\ \hline \end{array}$ $\quad \begin{array}{r} 6 \\ \times 9 \\ \hline \end{array}$

8. $\begin{array}{r} 4 \\ \times 5 \\ \hline \end{array}$ $\quad \begin{array}{r} 5 \\ \times 4 \\ \hline \end{array}$

9. $\begin{array}{r} 7 \\ \times 5 \\ \hline \end{array}$ $\quad \begin{array}{r} 5 \\ \times 7 \\ \hline \end{array}$

10. $\begin{array}{r} 9 \\ \times 8 \\ \hline \end{array}$ $\quad \begin{array}{r} 8 \\ \times 9 \\ \hline \end{array}$

11. $8 \times 5 =$ ____ $\quad 5 \times 8 =$ ____
12. $8 \times 2 =$ ____ $\quad 2 \times 8 =$ ____
13. $7 \times 6 =$ ____ $\quad 6 \times 7 =$ ____
14. $6 \times 4 =$ ____ $\quad 4 \times 6 =$ ____
15. $9 \times 3 =$ ____ $\quad 3 \times 9 =$ ____
16. $8 \times 3 =$ ____ $\quad 3 \times 8 =$ ____
17. $6 \times 5 =$ ____ $\quad 5 \times 6 =$ ____
18. $7 \times 4 =$ ____ $\quad 4 \times 7 =$ ____
19. $7 \times 8 =$ ____ $\quad 8 \times 7 =$ ____
20. $8 \times 6 =$ ____ $\quad 6 \times 8 =$ ____
21. $9 \times 4 =$ ____ $\quad 4 \times 9 =$ ____
22. $5 \times 9 =$ ____ $\quad 9 \times 5 =$ ____

ENRICHMENT

Tree Diagrams

Dirk is dressing. He wears a blue shirt. He can add a green tie or a red tie. He can add a green sweater or an orange sweater. How many different outfits can he make?
You can use a **tree diagram** to find how many.

A tree diagram has a branch for each possible outcome.

You can also multiply to find how many different outfits.

Multiply 1 shirt $\times$ 2 ties $\times$ 2 sweaters.
$$1 \times 2 \times 2 = 4$$

Dirk can make 4 different outfits.

Solve by drawing a tree diagram. Check by multiplying.

1. Sue has 3 kinds of crackers and 5 flavors of soup. How many different meals can she make?

2. Billy has peanut butter, 2 kinds of bread, and 3 kinds of jam. How many different sandwiches can he make?

3. Ann has 3 hats, 2 tops, and 3 skirts. How many different outfits can she choose to wear?

4. Count 3 kinds of items in your classroom. How many combinations are possible?

CUMULATIVE REVIEW

Write the letter of the correct answer.

1. Write the number:
thirty thousand sixty-one

a. 3,061 **b.** 30,061
c. 34,610 **d.** not given

2. Write the number.
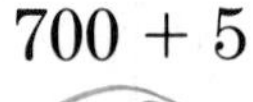
700 + 5

a. 705 **b.** 750
c. 7005 **d.** not given

3. Write the time.

a. 3:40
b. 6:05
c. 8:15
d. not given

4.
$$\begin{array}{r} 6{,}521 \\ +\,3{,}479 \\ \hline \end{array}$$

a. 9,990
b. 9,999
c. 10,000
d. not given

5.
$$\begin{array}{r} 45 \\ 23 \\ +\,69 \\ \hline \end{array}$$

a. 138
b. 127
c. 139
d. not given

6.
$$\begin{array}{r} 345 \\ -\,119 \\ \hline \end{array}$$

a. 237
b. 226
c. 464
d. not given

7. Round 232 to the nearest 10.

a. 200 **b.** 230
c. 250 **d.** not given

8.
$$\begin{array}{r} 3{,}489 \\ -\,1{,}450 \\ \hline \end{array}$$

a. 2,039
b. 2,129
c. 4,939
d. not given

9. 4,128 − 3,217 = ____

a. 911 **b.** 1,111
c. 1,811 **d.** not given

10.
$$\begin{array}{r} \$51.26 \\ -\ \ 47.35 \\ \hline \end{array}$$

a. $3.91
b. $30.01
c. $46.11
d. not given

11. Sue has $10.00. She buys a puzzle for $3.15. About how much money does Sue have left?

a. $6.00 **b.** $7.00
c. $13.15 **d.** $41.50

12. Sharon sells 85 bags of peanuts during the football game. Barbara sells 70 bags. How many more bags does Sharon sell than Barbara?

a. 10 bags **b.** 15 bags
c. 70 bags **d.** not given

Suppose the class was going to make a mural showing the four seasons. How would you decide what to picture? What would be a good way to divide the work and the wall space?

7 DIVISION FACTS

The Meaning of Division

A. The students are making costumes for the school talent show. They need 24 green buttons. There are 8 buttons on each card. How many cards of buttons do they need?

Find how many groups of 8 there are in 24.

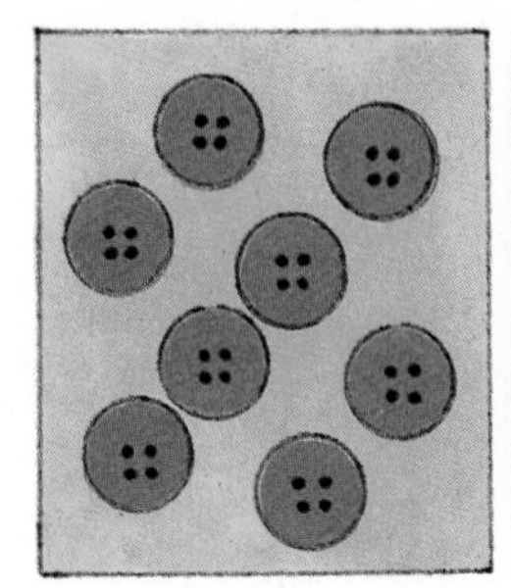 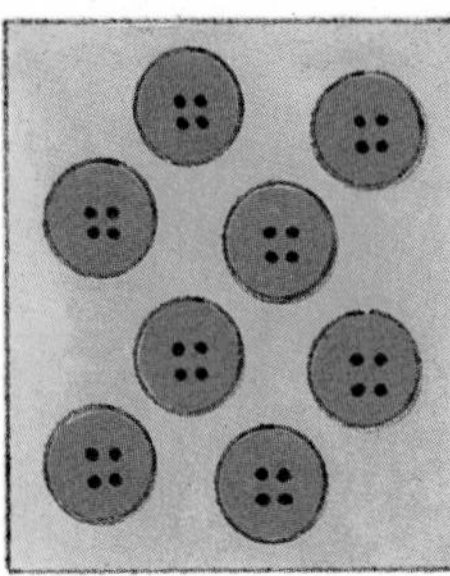 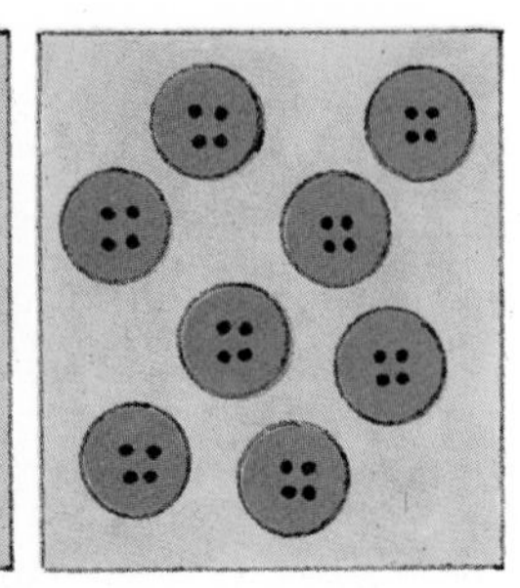

There are 3 groups of 8 in 24.
The students need 3 cards of buttons.

B. You can divide to find how many. Use what you know about multiplication to divide.

$24 \div 8 = $ 3 **Think:** 3 $\times 8 = 24.$
$3 \times 8 = 24$

So, $24 \div 8 = 3$.

Read this fact: Twenty-four divided by eight equals three.

Complete.

1. 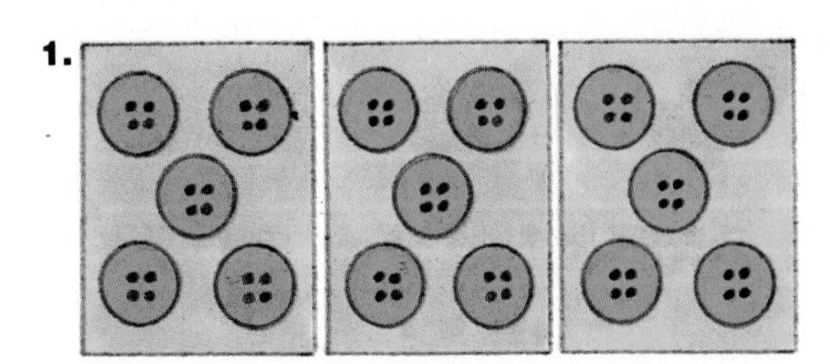

How many groups of 5?
_3_ groups of 5
$15 \div 5 = $ _3_

2. 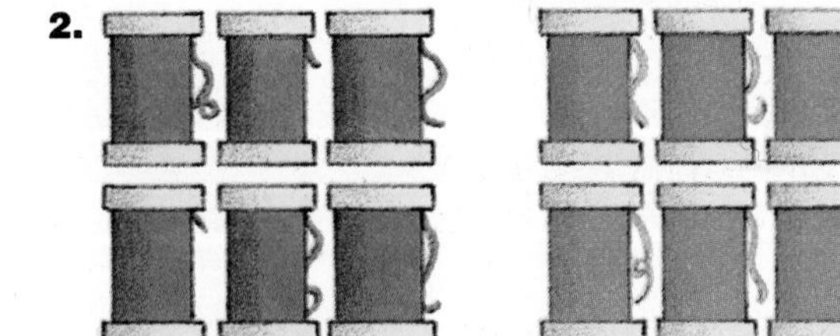

How many groups of 6?
_2_ groups of 6
$12 \div 6 = $ _2_

Complete.

3. __3__ groups of 6

$18 \div 6 =$ __3__

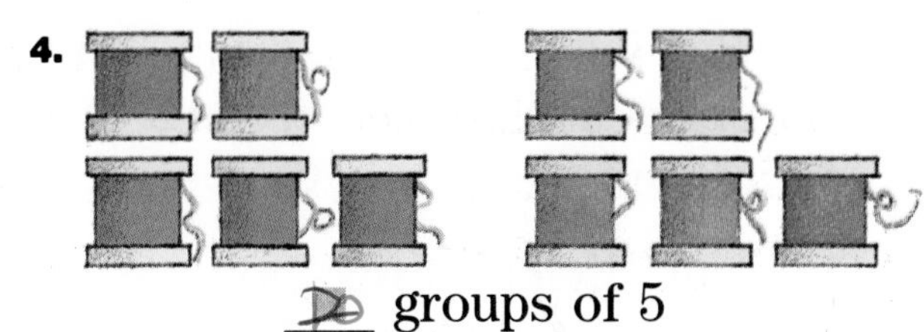

4. __2__ groups of 5

$10 \div 5 =$ __2__

Write a number sentence for each.

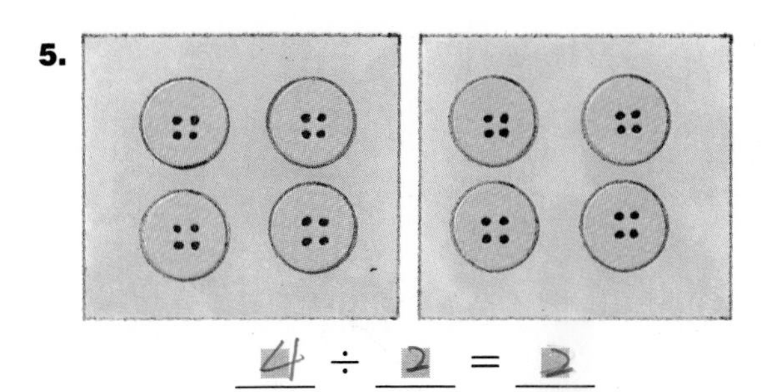

5. __4__ ÷ __2__ = __2__

2 groups of Nine

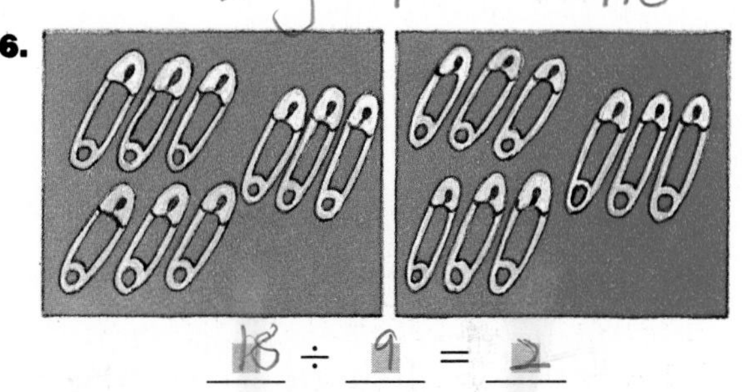

6. __18__ ÷ __9__ = __2__

7. __20__ ÷ __4__ = __5__

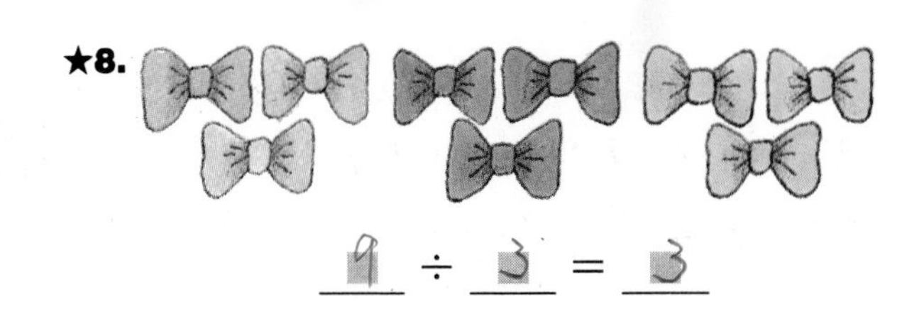

★8. __9__ ÷ __3__ = __3__

Divide.

9. __7__ $\times 2 = 14$

$14 \div 2 =$ __7__

10. __4__ $\times 6 = 24$

$24 \div 6 =$ __4__

11. __2__ $\times 8 = 16$

$16 \div 8 =$ __2__

12. __5__ $\times 2 = 10$

$10 \div 2 =$ __5__

13. __4__ $\times 3 = 12$

$12 \div 3 =$ __4__

14. __5__ $\times 5 = 20$

$20 \div 5 =$ __5__

CALCULATOR

You can use your calculator to solve.

1. Start with 18. Subtract 9 until you have 0. How many times did you subtract 9?

2 times

2. Start with 18. Subtract 6 until you have 0. How many times did you subtract 6?

3 times

Dividing by 2

Mr. Berry gives out 10 drumsticks to his music class. Each drummer needs 2 drumsticks. How many students can have drumsticks?

You can divide to find how many students can have drumsticks.

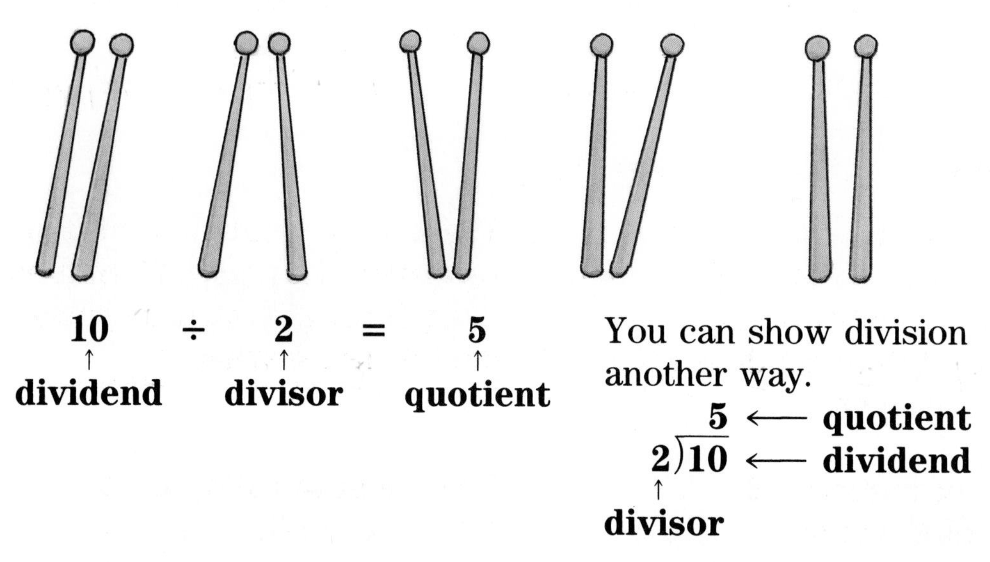

10	÷	2	=	5
↑		↑		↑
dividend		**divisor**		**quotient**

You can show division another way.

$$\begin{array}{r} 5 \\ 2\overline{)10} \end{array}$$

5 ⟵ **quotient**
2)10 ⟵ **dividend**
↑
divisor

5 students can have drumsticks.

Complete.

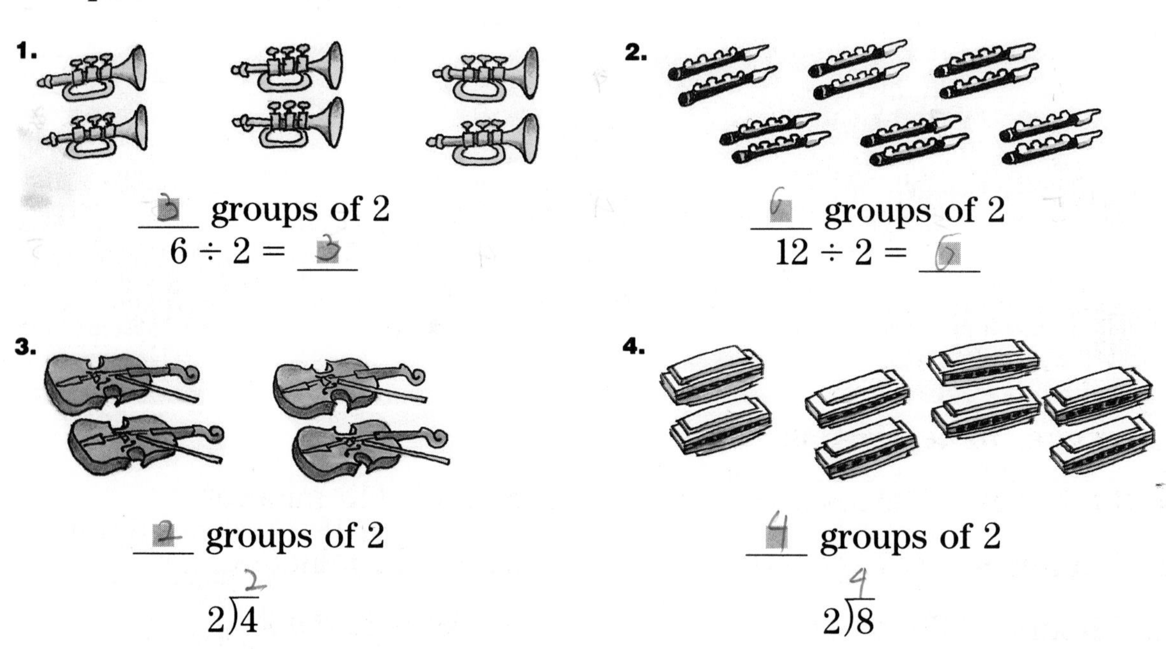

1. 3 groups of 2
   $6 \div 2 = 3$

2. 6 groups of 2
   $12 \div 2 = 6$

3. 2 groups of 2
   $2\overline{)4}$ 2

4. 4 groups of 2
   $2\overline{)8}$ 4

Divide.

5. $18 \div 2 =$ ____ 6. $14 \div 2 =$ ____ 7. $16 \div 2 =$ ____

8. $6 \div 2 =$ ____ 9. $12 \div 2 =$ ____ 10. $4 \div 2 =$ ____

11. $2\overline{)10}$ 12. $2\overline{)14}$ 13. $2\overline{)18}$ 14. $2\overline{)16}$ 15. $2\overline{)6}$

16. $2\overline{)16}$ 17. $2\overline{)12}$ 18. $2\overline{)4}$ 19. $2\overline{)10}$ 20. $2\overline{)8}$

21. $2\overline{)18}$ 22. $2\overline{)6}$ 23. $2\overline{)14}$ 24. $2\overline{)16}$ 25. $2\overline{)12}$

Solve. For Problem 29, use the Infobank.

26. There are 12 children playing violins, 2 children playing horns, and 1 child playing piano. How many children are playing instruments?

27. There are 4 children who play recorders. There are 2 children in each row. How many rows of children are there?

★28. At a school concert, 12 children play violins. Each pair of children shares a music stand. How many music stands do the children need?

★29. Use the information on page 402 to solve. Suppose each section of the orchestra is divided into 2 parts. How many players would there be in each part?

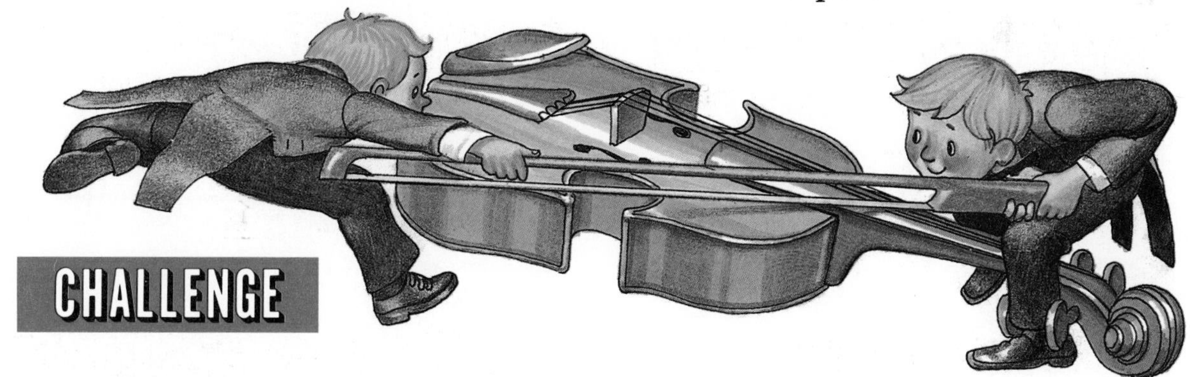

CHALLENGE

Compare. Write >, <, or = for ●.

1. 3 minutes ● 190 seconds
2. 2 hours ● 115 minutes
3. 2 minutes ● 105 seconds
4. 1 hour ● 60 minutes
5. 4 hours ● 230 minutes
6. 2 weeks ● 15 days

Dividing by 3

A group of students makes picture books for a crafts fair. They need 27 paintbrushes. Paintbrushes are sold in packs of 3. How many packs of paintbrushes do the students need?

You can divide to find how many packs are needed.

Think: 9 $\times 3 = 27$.

$9 \times 3 = 27$

So, $27 \div 3 = 9$, or $3\overline{)27}$ with 9 written above.

They need 9 packs of paintbrushes.

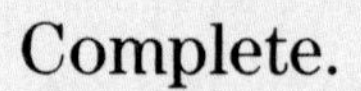

Complete.

1.

How many groups of 3?

$15 \div 3 =$ 5

2.

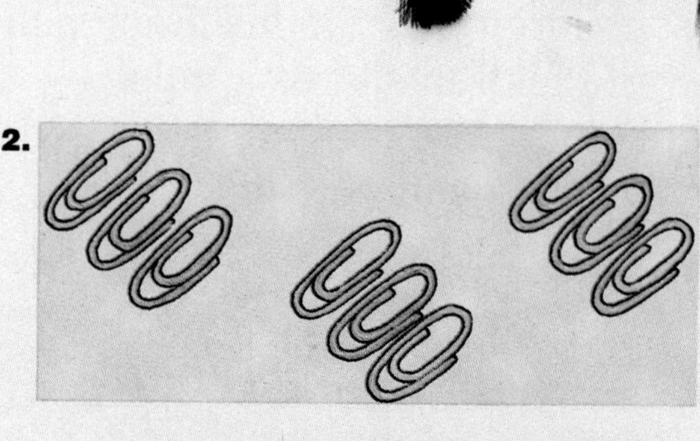

How many groups of 3?

$9 \div 3 =$ 3

3.

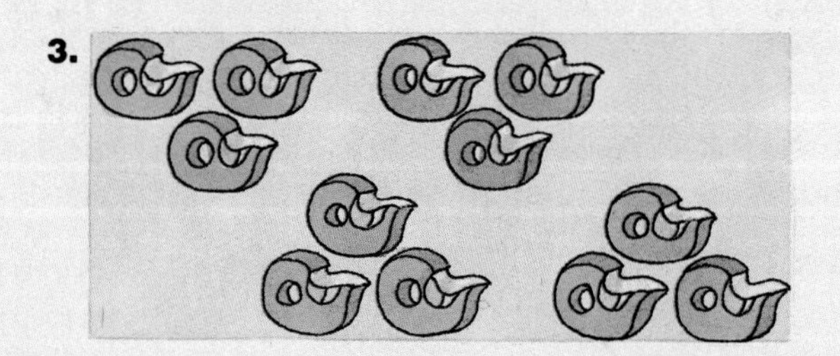

How many groups of 3?

$3\overline{)12}$ with 4 written above.

4.

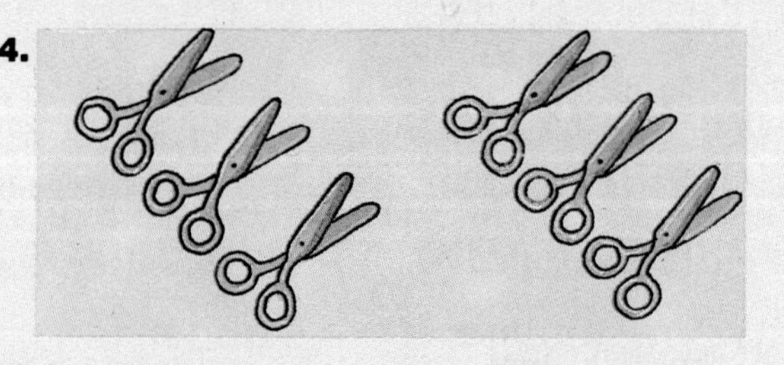

How many groups of 3?

$3\overline{)6}$ with 2 written above.

Find the quotient.

5. $21 \div 3 = \blacksquare$
6. $18 \div 3 = \blacksquare$
7. $24 \div 3 = \blacksquare$
8. $27 \div 3 = \blacksquare$
9. $15 \div 3 = \blacksquare$
10. $12 \div 3 = \blacksquare$
11. $6 \div 3 = \blacksquare$
12. $9 \div 3 = \blacksquare$
13. $12 \div 2 = \blacksquare$
14. $21 \div 3 = \blacksquare$
15. $18 \div 2 = \blacksquare$
16. $4 \div 2 = \blacksquare$

17. $3\overline{)6}$
18. $3\overline{)18}$
19. $3\overline{)12}$
20. $3\overline{)27}$
21. $3\overline{)15}$
22. $3\overline{)9}$
23. $3\overline{)24}$
24. $3\overline{)21}$
25. $3\overline{)6}$
26. $3\overline{)12}$
27. $3\overline{)18}$
28. $2\overline{)8}$
29. $3\overline{)27}$
30. $2\overline{)14}$
31. $2\overline{)16}$
32. $2\overline{)6}$
33. $3\overline{)15}$
34. $3\overline{)21}$
35. $3\overline{)24}$
36. $2\overline{)10}$

Solve.

37. The students need 12 tubes of paint. One box holds 3 tubes of paint. How many boxes do the children need?

38. Jenny's book is 27 pages long. Each story in it is 3 pages long. How many stories are there in her book?

39. Some students make book covers. They start with 18 sheets of heavy paper. They use 3 sheets of paper. How many sheets are left?

★40. To write a book, 21 students form groups. Each group has 1 artist and 2 writers. Write a division sentence to show how many groups there are.

CHALLENGE

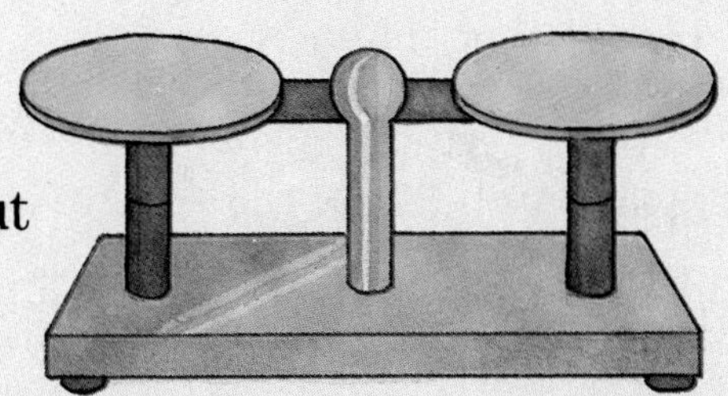

1. You put three 2-pound weights on the left pan. How many 3-pound weights do you put on the right pan to balance the scales?

2. You put six 3-pound weights on the right pan. How many 2-pound weights do you put on the left pan to balance the scales?

PROBLEM SOLVING

Choosing the Operation

Here are some hints to help you decide if you can multiply or divide to solve a problem.

A. For a book, an artist drew 12 princesses in 2 equal groups. How many princesses did the artist draw in each group?

12	÷	2	=	6
in all		groups		in each group

The artist drew 6 princesses in each group.

You know	how many in all. how many groups. that each group has the same number.
You want to find	how many in each group.
You can	DIVIDE.

B. The 12 princesses row across a lake. If the artist drew 3 princesses in each boat, how many boats would he draw?

12	÷	3	=	4
in all		in each group		groups

He would draw 4 boats.

You know	how many in all. how many in each group. that each group has the same number.
You want to find	how many groups.
You can	DIVIDE.

C. The artist drew 4 boats. Each boat held 3 princesses. How many princesses were there in the picture?

3	×	4	=	12
in each group		groups		princesses

There were 12 princesses in the picture.

You know	how many groups. how many in each group. that each group has the same number.
You want to find	how many in all.
You can	MULTIPLY.

Decide whether you would add, subtract, multiply, or divide to solve each problem. Write the letter of the correct answer.

1. Marian drew 18 pictures of machines for a book. She drew 3 pictures for each chapter. How many chapters did the book have?

   a. add b. subtract
   c. multiply d. divide

2. Lou has to draw 3 pictures for each chapter of a book. The book will have 6 chapters. How many pictures will he need to draw?

   a. add b. subtract
   c. multiply d. divide

Solve.

3. Rachel is doing the art for 3 stories in a book. She has 24 days to finish the art. If she spends the same amount of time on each story, how many days will it take to do the art for each?

4. Trina S. Hyman drew pictures for the book *Jane, Wishing.* She drew 12 pictures in black and white and 12 in color. How many pictures did the book have?

5. Between 1970 and 1980, 10 people won the Caldecott Medal for their illustrations in children's books. Another 31 people won Caldecott honors. How many people received Caldecott awards altogether?

6. In the book *The Nightingale,* 16 pages were used for Nancy Burkert's color pictures. Each picture covered 2 full pages. How many color pictures were there in *The Nightingale?*

★7. Ernesto has worked on 37 books. Of his books, 18 have black-and-white drawings. Another 5 books have photos. The rest have drawings in color. How many books have drawings in color?

★8. Nina drew 2 color pictures and 3 black-and-white pictures for each chapter of a book. She drew 27 black-and-white pictures. How many chapters were there? How many color pictures did she draw for the book?

Dividing by 4

For a children's show, Sandy has a bird act. She needs to carry 32 birds to the television station. Sandy wants to put 4 birds into each cage. How many cages does Sandy need?

You can divide to find how many cages she needs.

 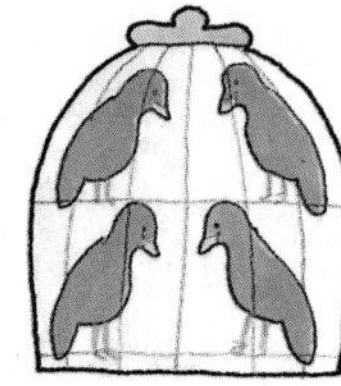  

Think: ■ $\times$ 4 = 32.
$8 \times 4 = 32$

So, $32 \div 4 = 8$, or $4\overline{)32}$ = 8.

Sandy needs 8 cages.

Divide.

1.     

$12 \div 4 =$ ■

2.     

$20 \div 4 =$ ■

3.  

$4\overline{)8}$

4.       

$4\overline{)24}$

Find the quotient.

5. $16 \div 4 = $ ___
6. $28 \div 4 = $ ___
7. $36 \div 4 = $ ___
8. $20 \div 4 = $ ___
9. $32 \div 4 = $ ___
10. $8 \div 4 = $ ___
11. $24 \div 4 = $ ___
12. $20 \div 4 = $ ___
13. $18 \div 3 = $ ___
14. $21 \div 3 = $ ___
15. $12 \div 4 = $ ___
16. $14 \div 2 = $ ___

17. $4\overline{)16}$
18. $4\overline{)8}$
19. $4\overline{)12}$
20. $4\overline{)24}$
21. $4\overline{)28}$
22. $4\overline{)36}$
23. $4\overline{)32}$
24. $4\overline{)20}$
25. $4\overline{)8}$
26. $4\overline{)16}$
27. $4\overline{)12}$
28. $3\overline{)24}$
29. $4\overline{)24}$
30. $2\overline{)6}$
31. $3\overline{)15}$
32. $4\overline{)28}$
33. $2\overline{)10}$
34. $3\overline{)12}$
35. $4\overline{)12}$
36. $2\overline{)4}$

Solve.

37. The third-grade classes sing 24 songs on a show. Each class sings 4 songs. How many third-grade classes sing songs on the show?

★38. A group of 18 children go to a show. Only 2 live close enough to walk. The others go by car. If 4 children ride in each car, how many cars are needed?

FOCUS: MENTAL MATH

You can skip-count backward to help you find division facts. Count the skips to find the quotient.

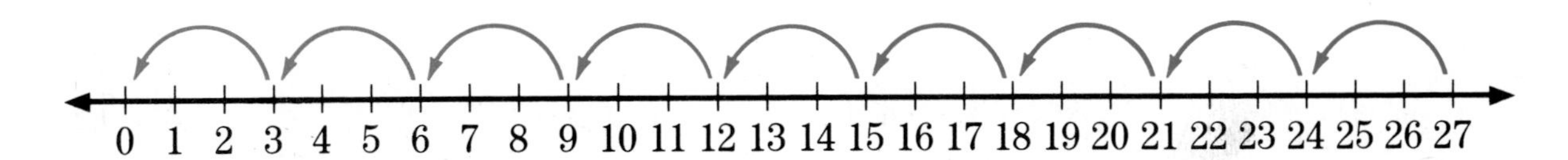

1. $27 \div 3 = $ ___
2. $12 \div 3 = $ ___
3. $15 \div 3 = $ ___
4. $9 \div 3 = $ ___
5. $21 \div 3 = $ ___
6. $6 \div 3 = $ ___
7. $18 \div 3 = $ ___
8. $24 \div 3 = $ ___

Dividing by 5

The children in Mr. Berg's sculpture class built frames. They needed 5 pieces of wire for each frame. There were 40 pieces of wire. How many frames were the children able to build?

You can divide to find out how many frames.

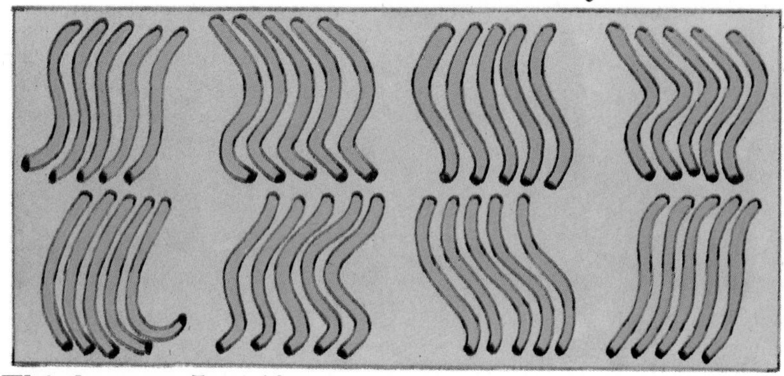

Think:  $\times 5 = 40$.

$8 \times 5 = 40$

So, $40 \div 5 = 8$, or $5\overline{)40}$ with quotient 8.

The children were able to build 8 frames.

Divide.

1.

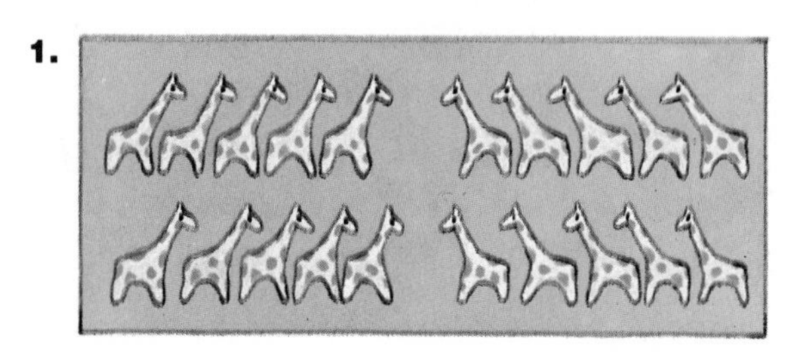

$20 \div 5 =$ 4

2.

$15 \div 5 =$ 3

3.

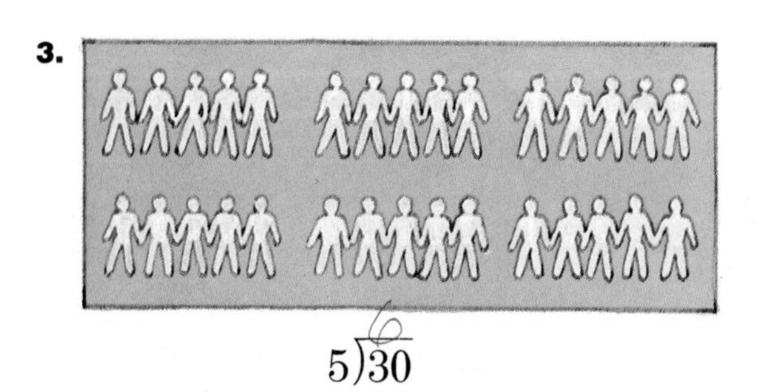

$5\overline{)30}$ 6

4.

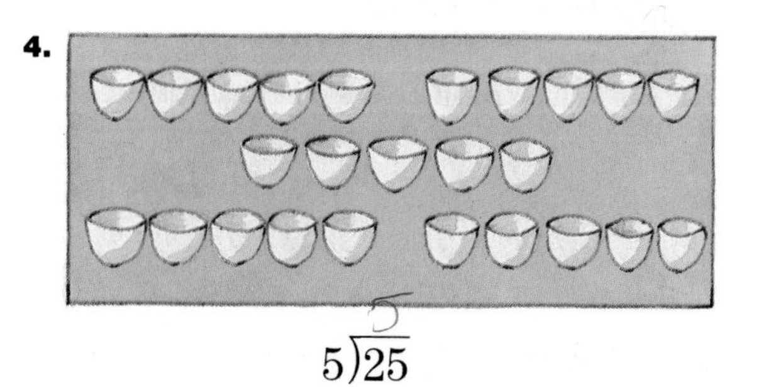

$5\overline{)25}$ 5

Find the quotient.

5. $5\overline{)45}$ **6.** $5\overline{)35}$ **7.** $5\overline{)10}$ **8.** $5\overline{)40}$ **9.** $5\overline{)15}$

10. $3\overline{)12}$ **11.** $4\overline{)16}$ **12.** $5\overline{)45}$ **13.** $5\overline{)20}$ **14.** $3\overline{)27}$

15. $5\overline{)30}$ **16.** $4\overline{)28}$ **17.** $3\overline{)15}$ **18.** $2\overline{)18}$ **19.** $4\overline{)36}$

20. $25 \div 5 =$ ____ **21.** $35 \div 5 =$ ____ **22.** $30 \div 5 =$ ____

23. $45 \div 5 =$ ____ **24.** $24 \div 3 =$ ____ **25.** $15 \div 5 =$ ____

26. $20 \div 4 =$ ____ **27.** $21 \div 3 =$ ____ **28.** $20 \div 5 =$ ____

29. $24 \div 4 =$ ____ **30.** $40 \div 5 =$ ____ **31.** $10 \div 5 =$ ____

Copy the chart. Divide to complete.

32.

Number	35	45	30	15	40	10	25	20
Divided by 5 =	**7**							

Solve.

33. One of Mr. Berg's classes made clay animals. He divided 35 pounds of clay into 5-pound chunks. Write a number sentence to show how many chunks he had.

★**34.** The children had 23 sculptures. There were 8 on one table. The rest were on 3 shelves. The same number of sculptures were on each shelf. How many sculptures were there on each shelf?

CHALLENGE

Complete.

1. $4 \times 4 \rightarrow$ ____ $\div 4 \rightarrow$ ____ $- 4 \rightarrow$ ____

2. $3 + 3 \rightarrow$ ____ $- 3 \rightarrow$ ____

3. $2 - 2 \rightarrow$ ____ $+ 2 \rightarrow$ ____ $\div 2 \rightarrow$ ____

4. $4 \times 9 \rightarrow$ ____ $- 12 \rightarrow$ ____

5. $3 \times 4 \rightarrow$ ____ $\div 2 \rightarrow$ ____ $\times 1 \rightarrow$ ____

6. $50 - 10 \rightarrow$ ____ $\div 5 \rightarrow$ ____

Dividing by 6

Roberta and Doreen went to an art show. They looked at 36 paintings. They saw 6 paintings in each room. How many rooms did Roberta and Doreen visit?

You can divide to find the number of rooms.

$36 \div 6 = 6$ $\quad$ $6\overline{)36}$ with quotient 6

Roberta and Doreen visited 6 rooms.

Divide.

1. $24 \div 6 =$ 4
2. $12 \div 6 =$ 2
3. $30 \div 6 =$ 5
4. $18 \div 6 =$ 3
5. $48 \div 6 =$ 8
6. $36 \div 6 =$ 6
7. $42 \div 6 =$ 7
8. $54 \div 6 =$ 9
9. $6\overline{)48}$ 8
10. $6\overline{)12}$ 2
11. $6\overline{)30}$ 5
12. $6\overline{)54}$ 7
13. $6\overline{)24}$ 4
14. $6\overline{)36}$ 6
15. $6\overline{)54}$ 9
16. $6\overline{)12}$ 2
17. $6\overline{)18}$ 3
18. $6\overline{)42}$ 7

Find the quotient.

19. 5)30 **20.** 3)15 **21.** 5)20 **22.** 6)24 **23.** 4)16

24. 2)18 **25.** 6)42 **26.** 4)8 **27.** 2)4 **28.** 6)30

29. 5)10 **30.** 3)12 **31.** 5)40 **32.** 5)45 **33.** 3)18

34. 12 ÷ 6 = ____ **35.** 36 ÷ 6 = ____ **36.** 42 ÷ 6 = ____

37. 24 ÷ 6 = ____ **38.** 30 ÷ 6 = ____ **39.** 8 ÷ 2 = ____

40. 54 ÷ 6 = ____ **41.** 32 ÷ 4 = ____ **42.** 24 ÷ 3 = ____

43. 18 ÷ 6 = ____ **44.** 48 ÷ 6 = ____ **45.** 12 ÷ 2 = ____

46. 9 ÷ 3 = ____ **47.** 36 ÷ 4 = ____ **48.** 14 ÷ 2 = ____

Solve.

49. A class makes shell pictures for the show. They use 54 shells. They use 6 shells for each picture. How many pictures do they make?

50. Another class makes 36 paper-bag masks for the show. They pack 6 masks in each box. How many boxes do they need?

51. Kevin and John make posters for the art show. They each make 6 posters. How many posters do they make?

★52. Postcards at the art gallery cost 6¢ each. Nat has 45¢. What is the greatest number of postcards he can buy?

MIDCHAPTER REVIEW

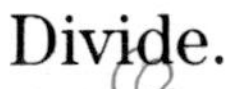

Divide.

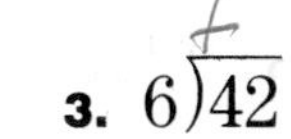

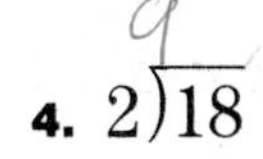

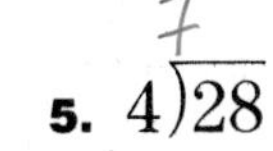

1. 3)18 **2.** 5)10 **3.** 6)42 **4.** 2)18 **5.** 4)28

6. 25 ÷ 5 = ____ **7.** 24 ÷ 3 = ____ **8.** 36 ÷ 6 = ____ **9.** 12 ÷ 4 = ____

More Practice, page 411

PROBLEM SOLVING

Practice

Write the letter of the number sentence that you could use to solve the problem.

1. The Louvre Museum in Paris opened in 1546. Only kings, queens, lords, and ladies were welcome. The museum was opened to everyone 247 years later. What year was that?

   a. $1546 - 247 = \_\_\_$
   b. $1546 + 247 = \_\_\_$

2. Rembrandt painted 79 major pictures. Dutch museums have 18 of them. The rest are in museums in other countries. How many are there in museums in other countries?

   a. $79 - 18 = \_\_\_$
   b. $79 + 18 = \_\_\_$

What information do you need to solve the problem? Write the letter of the correct answer.

3. Leonardo da Vinci lived from 1452 to 1519. He painted the *Mona Lisa* in 3 years. How old was he when he finished?

   a. his age when he started the painting
   b. his age in 1519

4. Hokusai painted more than 100 pictures of Mount Fuji. Goro has seen 17 of them. How many has Goro not seen?

   a. the number of paintings of Mount Fuji
   b. the size of Mount Fuji

Make a plan. Then solve.

5. Museum tickets cost $3.00 for adults and $2.00 for children. A group that has 1 adult pays $11.00. How many children are there in the group?

6. A museum is closed on Mondays. It is open from 10:00 A.M. to 9:00 P.M. on Tuesdays. It is open from 10:00 A.M. to 5:00 P.M. five days a week. For how many hours per week is it open?

Choose the operation. Write the letter of the correct answer.

7. The art show has 9 rows of paintings. Each row has 8 paintings. How many paintings are there in all?

   a. addition
   b. subtraction
   c. multiplication
   d. division

8. Drake buys 45 tubes of paint packed in boxes. Each box holds 5 tubes. How many boxes does Drake buy?

   a. addition
   b. subtraction
   c. multiplication
   d. division

Solve.

9. Hale uses 4 tubes of white paint in one month. How many tubes of white paint will he use in 8 months?

10. A group of 6 artists orders supplies. Each person pays $8.00. The bill amounts to $43.98. How much money is left?

11. An airbrush costs $13.50. How much change will Dan receive if he pays with a $20 bill?

12. Nan orders a sketch pad for $8.00, pencils for $6.95, and paints for $7.49. How much money does Nan need?

13. One drawing pad costs $4.00. How many drawing pads can Theodore buy for $24.00?

14. Drawing pens are packed 12 to a box. Joe has 3 pens left in one box and one full box. How many pens does Joe have?

15. One art store's catalog has 232 pages. Another store's catalog has 321 pages. How many more kinds of supplies does the second store offer in its catalog?

16. A company makes marking pens in 203 colors. Jan has the 22 red colors and the 36 blue colors. How many other pens must she buy to have all the colors?

Related Facts

In the morning, 12 students formed 4 groups for a singing contest. How many students were there in each group?

$$12 \div 4 = 3$$

These two number sentences are called **related facts.**

In the afternoon, 12 students formed groups for a dance contest. There were 3 students in each group. How many groups were there?

$$12 \div 3 = 4$$

Other examples:

$10 \div 5 = 2$ $\quad$ $18 \div 3 = 6$

$10 \div 2 = 5$ $\quad$ $18 \div 6 = 3$

Divide.

1. $20 \div 5 =$ ___
   $20 \div 4 =$ ___
2. $8 \div 4 =$ ___
   $8 \div 2 =$ ___
3. $6 \div 2 =$ ___
   $6 \div 3 =$ ___
4. $12 \div 6 =$ ___
   $12 \div 2 =$ ___
5. $15 \div 3 =$ ___
   $15 \div 5 =$ ___
6. $24 \div 4 =$ ___
   $24 \div 6 =$ ___
7. $30 \div 6 =$ ___
   $30 \div 5 =$ ___
8. $10 \div 2 =$ ___
   $10 \div 5 =$ ___

Find the quotient.

9. $4 \div 2 =$ ___
10. $18 \div 2 =$ ___
11. $40 \div 5 =$ ___
12. $36 \div 6 =$ ___
13. $35 \div 5 =$ ___
14. $21 \div 3 =$ ___
15. $16 \div 4 =$ ___
16. $48 \div 6 =$ ___
17. $28 \div 4 =$ ___
18. $9 \div 3 =$ ___
19. $45 \div 5 =$ ___
20. $32 \div 4 =$ ___

Dividing with 0 and 1

A. Dan, Sue, Nora, and Jerry went to find a box of drawing paper. The box was empty. How many sheets of paper did each person have if each person took the same number of sheets?

$0 \div 4 = 0$
0 divided by any number, except 0, is 0.

B. There are 4 sheets of paper. One student uses them. How many sheets does the student use?

$4 \div 1 = 4$
A number divided by 1 equals itself.

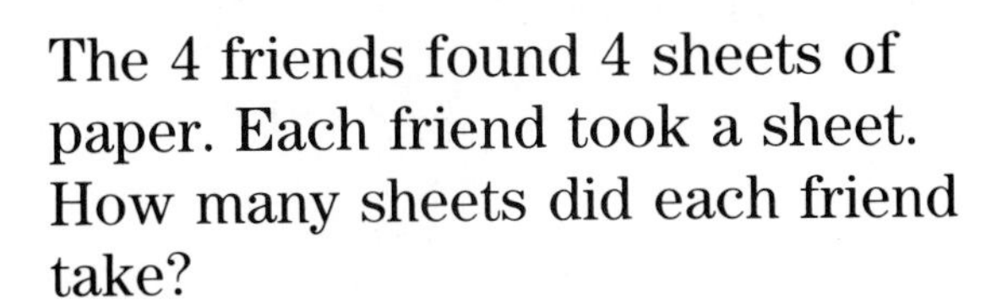

The 4 friends found 4 sheets of paper. Each friend took a sheet. How many sheets did each friend take?

$4 \div 4 = 1$

A number divided by itself, except 0, equals 1.

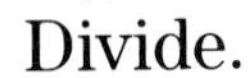

Divide.

1. $6 \div 1 =$ ____ **2.** $0 \div 1 =$ ____ **3.** $3 \div 3 =$ ____ **4.** $4 \div 1 =$ ____

5. $0 \div 5 =$ ____ **6.** $2 \div 2 =$ ____ **7.** $1 \div 1 =$ ____ **8.** $0 \div 4 =$ ____

9. $5 \div 5 =$ ____ **10.** $3 \div 1 =$ ____ **11.** $0 \div 2 =$ ____ **12.** $6 \div 6 =$ ____

13. $4\overline{)4}$ **14.** $3\overline{)0}$ **15.** $1\overline{)5}$ **16.** $6\overline{)0}$ **17.** $1\overline{)2}$

18. $1\overline{)6}$ **19.** $2\overline{)2}$ **20.** $4\overline{)0}$ **21.** $5\overline{)5}$ **22.** $3\overline{)3}$

23. $6\overline{)6}$ **24.** $1\overline{)4}$ **25.** $1\overline{)0}$ **26.** $2\overline{)0}$ **27.** $1\overline{)1}$

Dividing by 7

The students in third grade put on a play. They built 21 cardboard trees for the forest scene. The trees were arranged on the stage in 7 equal groups. How many trees were there in each group?

You can divide to find out how many there were in each group.

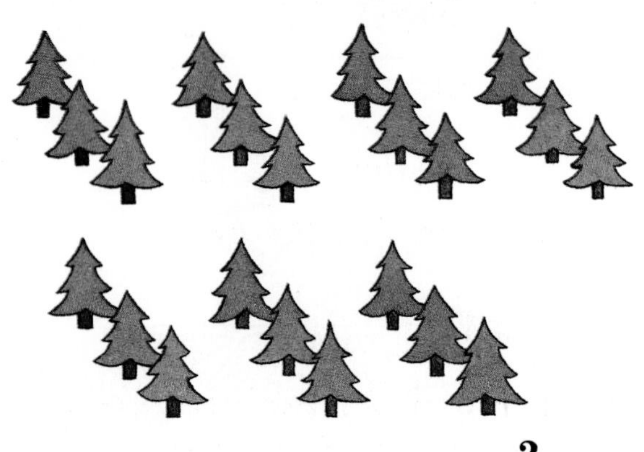

$21 \div 7 = 3$ $\qquad$ $7\overline{)21}$ = 3

There were 3 trees in each group.

Checkpoint Write the letter of the correct answer.

Divide.

1. $49 \div 7 =$ ____
a. 7
b. 17
c. 42
d. 56

2. $63 \div 7 =$ ____
a. 3
b. 8
c. 9
d. 70

3. $7\overline{)14}$
a. 2
b. 3
c. 20
d. 21

4. $7\overline{)35}$
a. 5
b. 6
c. 35
d. 50

Divide.

1. $14 \div 7 =$ ____ **2.** $35 \div 7 =$ ____ **3.** $28 \div 7 =$ ____ **4.** $49 \div 7 =$ ____

5. $63 \div 7 =$ ____ **6.** $21 \div 7 =$ ____ **7.** $42 \div 7 =$ ____ **8.** $56 \div 7 =$ ____

9. $7\overline{)49}$ **10.** $7\overline{)35}$ **11.** $7\overline{)21}$ **12.** $7\overline{)56}$ **13.** $7\overline{)14}$

Find the quotient.

14. $7\overline{)28}$ **15.** $7\overline{)42}$ **16.** $7\overline{)63}$ **17.** $7\overline{)35}$ **18.** $7\overline{)56}$

19. $1\overline{)5}$ **20.** $7\overline{)14}$ **21.** $2\overline{)10}$ **22.** $3\overline{)9}$ **23.** $5\overline{)15}$

24. $6\overline{)12}$ **25.** $2\overline{)12}$ **26.** $3\overline{)3}$ **27.** $6\overline{)0}$ **28.** $4\overline{)8}$

29. $4\overline{)24}$ **30.** $7\overline{)49}$ **31.** $4\overline{)32}$ **32.** $1\overline{)1}$ **33.** $5\overline{)35}$

34. $7 \div 7 =$ ____ **35.** $35 \div 7 =$ ____ **36.** $21 \div 7 =$ ____

37. $42 \div 7 =$ ____ **38.** $28 \div 7 =$ ____ **39.** $30 \div 5 =$ ____

40. $24 \div 6 =$ ____ **41.** $0 \div 7 =$ ____ **42.** $63 \div 7 =$ ____

43. $8 \div 2 =$ ____ **44.** $36 \div 4 =$ ____ **45.** $8 \div 1 =$ ____

Solve.

46. A play was written by 28 children. They worked in groups of 7. Each group wrote one act. How many acts did the play have?

47. Each of 7 children read part of a 42-line poem. Each child had the same number of lines. How many lines did each child read?

48. Jo bought 7 oranges as a treat for the actors. She cut each orange into 4 pieces. How many pieces were there?

★49. The class printed 60 programs. They posted 4 and stacked the rest in 7 equal piles. How many programs were in each pile?

ANOTHER LOOK

Write the time.

1. 

2. 

3. 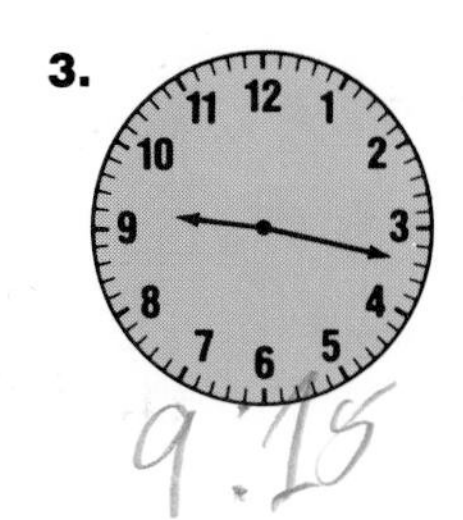

4. 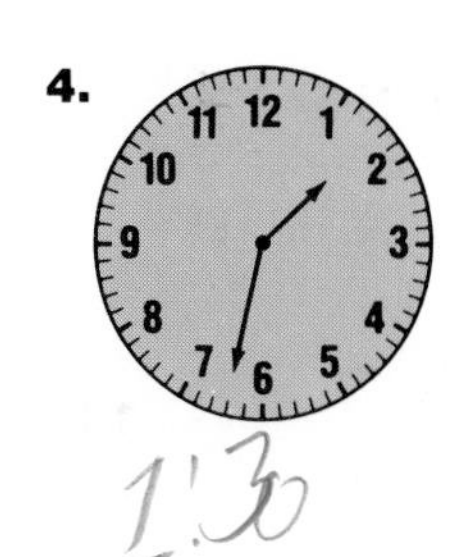

More Practice, page 411

Dividing by 8 and 9

A. There are 24 third graders who learn to square dance. There are 8 dancers in each square. How many squares are there?

You can divide to find how many squares.

$$24 \div 8 = 3, \text{ or } 8\overset{3}{\overline{)24}}.$$

There are 3 squares.

B. Each dancer has the same partner for 9 minutes, then they change partners. If they dance for 36 minutes, how many partners will each dancer have?

You can divide to find how many partners.

$$36 \div 9 = 4, \text{ or } 9\overset{4}{\overline{)36}}.$$

Each dancer will have 4 partners.

Checkpoint Write the letter of the correct answer.

Divide.

1. $81 \div 9 =$ ____
   - a. 1
   - b. 8
   - c. 9
   - d. 90
2. $48 \div 8 =$ ____
   - a. 5
   - b. 6
   - c. 8
   - d. 56
3. $9\overline{)54}$
   - a. 6
   - b. 7
   - c. 45
   - d. 60
4. $8\overline{)64}$
   - a. 6
   - b. 8
   - c. 18
   - d. 72

Divide.

1. $56 \div 8 =$ ____
2. $72 \div 9 =$ ____
3. $18 \div 9 =$ ____
4. $40 \div 8 =$ ____
5. $24 \div 8 =$ ____
6. $16 \div 8 =$ ____
7. $45 \div 9 =$ ____
8. $63 \div 9 =$ ____
9. $9\overline{)81}$
10. $8\overline{)32}$
11. $9\overline{)54}$
12. $8\overline{)48}$
13. $9\overline{)36}$

Find the quotient.

14. 8)64 **15.** 8)16 **16.** 8)24 **17.** 8)72 **18.** 8)56

19. 9)72 **20.** 9)45 **21.** 9)81 **22.** 9)27 **23.** 9)18

24. 8)32 **25.** 7)28 **26.** 4)12 **27.** 9)54 **28.** 5)5

29. $18 \div 9 =$ ___ **30.** $45 \div 9 =$ ___ **31.** $36 \div 9 =$ ___

32. $40 \div 8 =$ ___ **33.** $32 \div 8 =$ ___ **34.** $48 \div 8 =$ ___

35. $72 \div 8 =$ ___ **36.** $27 \div 9 =$ ___ **37.** $0 \div 7 =$ ___

38. $54 \div 9 =$ ___ **39.** $9 \div 1 =$ ___ **40.** $56 \div 8 =$ ___

Copy the chart. Divide to complete.

41.

Number	40	32	64	16	56	24	72	48
Divided by 8 =	___	___	___	___	___	___	___	___

42.

Number	54	63	27	81	45	18	72	36
Divided by 9 =	___	___	___	___	___	___	___	___

Solve.

43. One person tells the dancers which steps to do. He works for 45 minutes. After every 9 minutes, he calls a different dance. How many dances does he call?

44. There were 32 fiddlers in a music contest. They played in groups of 8. Write a division sentence to show how many fiddlers played in each group.

45. Don brought 9 records to the dance. Each record had 2 songs on it. How many songs were there on Don's records?

★46. There were 72 children square dancing. There were 8 children in each square. After the first dance, 8 more dancers joined in. How many squares were there in all?

More Practice, page 411

Fact Families

Division and multiplication are related. You can use the numbers 4, 3, and 12 to write four true sentences.

4 kinds
3 of each kind
12 in all

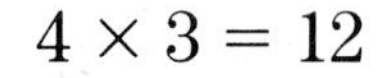

$4 \times 3 = 12$

12 in all
3 of each kind
4 kinds

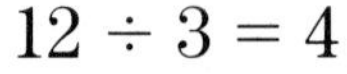

$12 \div 3 = 4$

3 kinds
4 of each kind
12 in all

$3 \times 4 = 12$

12 in all
4 of each kind
3 kinds

$12 \div 4 = 3$

These four number sentences are called a **family of facts.**

Complete each number sentence.

1. $7 \times 4 = 28$
$4 \times 7 = 28$
$28 \div 4 = 7$
$28 \div 7 =$ 4

2. $15 \div 3 = 5$
$15 \div 5 = 3$
$3 \times 5 =$ 15
$5 \times 3 =$ 15

3. $12 \div 6 = 2$
$12 \div 2 =$ 6
$2 \times 6 =$ 12
$6 \times 2 =$ 12

4. $3 \times 9 = 27$
$9 \times 3 =$ 27
$27 \div 9 =$ 3
$27 \div 3 =$ 9

5. $36 \div 9 =$ 4
$36 \div 4 =$ 9
$9 \times 4 =$ 36
$4 \times 9 =$ 36

6. $35 \div 7 =$ 5
$35 \div 5 =$ 7
$7 \times 5 =$ 35
$5 \times 7 =$ 35

7. $4 \times 8 =$ 32
$8 \times 4 =$ 32
$32 \div 8 =$ 4
$32 \div 4 =$ 8

8. $6 \times 3 =$ 18
$3 \times 6 =$ 18
$18 \div 3 =$ 6
$18 \div 6 =$ 3

Complete each number sentence.

9. $3 \times 8 = \_\_\_$
$8 \times 3 = \_\_\_$
$24 \div 3 = \_\_\_$
$24 \div 8 = \_\_\_$

10. $6 \times 5 = \_\_\_$
$5 \times 6 = \_\_\_$
$30 \div 6 = \_\_\_$
$30 \div 5 = \_\_\_$

11. $7 \times 3 = \_\_\_$
$3 \times 7 = \_\_\_$
$21 \div 7 = \_\_\_$
$21 \div 3 = \_\_\_$

12. $16 \div \_\_\_ = 2$
$\_\_\_ \div 2 = 8$
$8 \times \_\_\_ = 16$
$2 \times \_\_\_ = 16$

13. $\_\_\_ \div 9 = 8$
$72 \div \_\_\_ = 9$
$\_\_\_ \times 8 = 72$
$8 \times 9 = \_\_\_$

14. $\_\_\_ \times 6 = 48$
$6 \times 8 = \_\_\_$
$48 \div 6 = \_\_\_$
$\_\_\_ \div 8 = 6$

Complete each family of facts.

15. $3 \times 2 = \_\_\_$
$2 \times 3 = \_\_\_$
$6 \div 3 = \_\_\_$

16. $6 \times 9 = \_\_\_$
$9 \times 6 = \_\_\_$
$54 \div 9 = \_\_\_$

17. $42 \div 7 = \_\_\_$
$42 \div 6 = \_\_\_$
$7 \times 6 = \_\_\_$

Write four number sentences for each set of numbers.

18. 5, 8, 40
19. 45, 9, 5
20. 14, 2, 7
21. 4, 24, 6
22. 7, 8, 56
23. 2, 18, 9
24. 4, 5, 20
25. 7, 63, 9

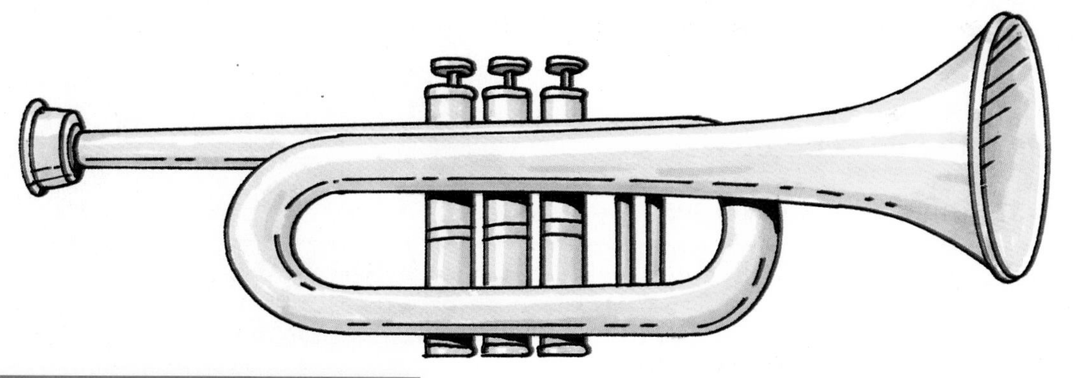

FOCUS: REASONING

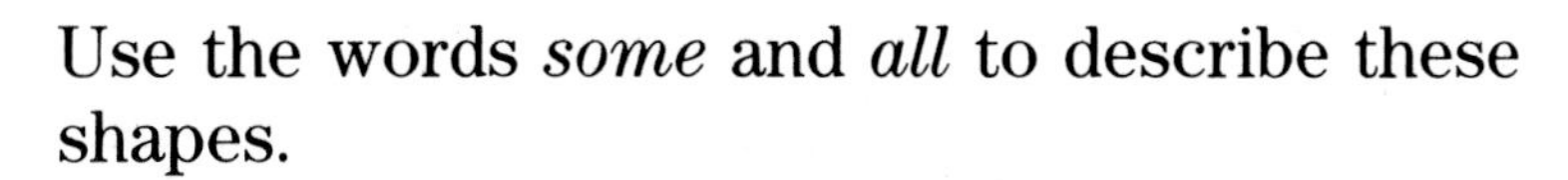

Some means at least one but not all.
All means every single one.

Use the words *some* and *all* to describe these shapes.

Examples: Some of these shapes are blue.
All the squares are red.

PROBLEM SOLVING

Checking That the Solution Answers the Question

Some word problems ask for answers that are not just numbers with labels. Pay special attention to the question that is asked. Be sure you answer it.

A ballad is a folk song that tells a story. One famous ballad has 28 verses. There are 7 singers who will sing the ballad on a radio show. They will take turns singing the verses. Each singer is to sing the same number of verses. Each singer will sing alone. Will any singer sing more than 6 verses?

Which is the correct answer.

a. Each singer sings 4 verses.
b. No, no singer will sing more than 6 verses.
c. Those singers must get very tired.

The correct answer is *b.* If you are not sure why, carefully reread the question that was asked in the problem.

Which sentence answers the question? Write the letter of the correct answer.

1. The 6 members of a folk group sing 3 songs each in a show. Are there more than 15 songs in the show?

   a. There are 18 songs.
   b. Yes, there are.
   c. No, there are not.

2. Ralph bought 25 strings for 5 new banjos that his group just bought. How many strings does a banjo have?

   a. A banjo has 5 strings.
   b. Yes, they were all used.
   c. No strings were left.

Solve.

3. Singer John Jacob Niles had 110 old folk songs printed in a book. Ralph Vaughan Williams's book had 59 old folk songs. Which man's book had more songs?

4. The *Grand Ole Opry* is a radio show of country music. From 1925 to 1974, it was broadcast from a building in Nashville, Tennessee. How many years is that?

5. Guitars are often used to play folk and country music. One kind of guitar has 6 strings. If Jess buys 30 strings, will he have enough for 4 of these guitars?

6. One country-music record featured 14 men and 16 women singers. Another 28 people played musical instruments. Were there more singers or more instrument players?

7. Sara's new record has 8 songs on side A and 12 songs on side B. Which side has more songs on it?

8. Each of the 4 singers in a folk group sings 3 songs on their new record. How many songs are on the record?

9. One record company has 98 folk singers who record for it. Of this number, 53 belong to groups. The others sing alone. How many sing by themselves?

★10. Martha takes folk-singing lessons. She buys sheet music for $12.50 and a record for $7.99. Will a $10 bill and two $5 bills pay for both?

READING MATH

Read this sentence.

 **careful when** U X **the** 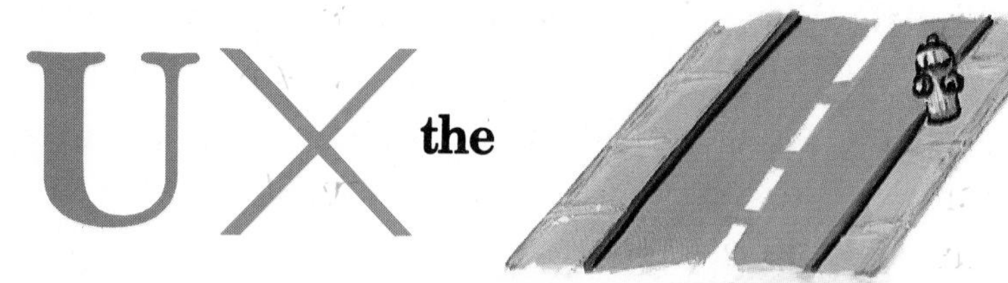

The symbols in this sentence stand for words. When we use words for the symbols the sentence can be read this way: "Be careful when you cross the street."

We use symbols in math every day. Look at the symbols below. Write the meaning of each symbol.

1. $+$ **2.** $-$

3. $<$ **4.** $\div$

5. $=$ **6.** $\times$

Use the symbols to write each sentence.

7. Six plus two equals eight.

8. Nineteen is less than twenty-four.

9. Six multiplied by nine is fifty-four.

10. Twenty divided by two equals ten.

11. Eighteen minus three is fifteen.

Look at the problems below. What is missing in each?

12. $6 \blacksquare 6 = 36$ **13.** $400 \blacksquare 100 = 300$

14. $36 \blacksquare 71 = 107$ **15.** $81 \blacksquare 9 = 9$

GROUP PROJECT

Talent Show

The problem: Your school is going to put on a talent show. The different acts will include singing, dancing, playing music, acting, telling jokes, and gymnastics. You have to make up the schedule.

Key Questions

- Should each student, or each class, have a set time?
- Should every student be in the show?
- How many classes are there in the school?
- How many students are there in each class?
- How long should the whole show last?

CHAPTER TEST

Find the quotient. (pages 192, 194, 198, 200, 202, 208, and 210)

1. $24 \div 3 =$ ____ **2.** $27 \div 3 =$ ____ **3.** $24 \div 4 =$ ____

4. $32 \div 4 =$ ____ **5.** $16 \div 4 =$ ____ **6.** $30 \div 5 =$ ____

7. $21 \div 3 =$ ____ **8.** $24 \div 6 =$ ____ **9.** $20 \div 4 =$ ____

10. $40 \div 5 =$ ____ **11.** $6 \div 6 =$ ____ **12.** $21 \div 7 =$ ____

13. $35 \div 7 =$ ____ **14.** $81 \div 9 =$ ____ **15.** $54 \div 6 =$ ____

16. $7\overline{)28}$ **17.** $9\overline{)72}$ **18.** $8\overline{)72}$ **19.** $8\overline{)64}$

20. $7\overline{)49}$ **21.** $4\overline{)24}$ **22.** $7\overline{)42}$ **23.** $6\overline{)48}$

Complete each family of facts. (page 212)

24. ____ $\times 4 = 20$
$4 \times 5 =$ ____
$20 \div$ ____ $= 5$
____ $\div 5 = 4$

25. $4 \times 8 =$ ____
$8 \times$ ____ $= 32$
$32 \div 8 =$ ____
$32 \div 4 =$ ____

26. $9 \times$ ____ $= 54$
$6 \times$ ____ $= 54$
$54 \div$ ____ $= 9$
____ $\div 9 = 6$

Write four number sentences for each set of numbers. (page 212)

27. 36, 9, 4 **28.** 56, 8, 7 **29.** 63, 7, 9

Write the letter of the word that tells how you would solve each problem. (page 196)

30. There are 6 baseball cards in a pack. Doris buys 4 packs. How many cards does she buy in all?

a. add **b.** subtract
c. multiply **d.** divide

31. Klee has 27 postcards. He pastes 3 cards on each page of his album. How many pages does he fill?

a. add **b.** subtract
c. multiply **d.** divide

Which answers the question? Write *a*, *b*, or *c*. (page 214)

32. Delores bought 2 tickets for the acrobat show at $7.00 each. She had three $5.00 bills. Does she receive any change?

a. $1.00
b. No, she receives no change.
c. Yes, she receives change.

33. Humbert tried 14 stunts. He did 6 cartwheels and 6 headstands. He fell on his other tries. How many stunts did he complete?

a. He fell twice.
b. He completed 12 stunts.
c. He tried 14 stunts.

BONUS

Compare. Write $>$, $<$, or $=$ for ●.

1. $63 \div 7$ ● $45 \div 5$ **2.** $64 \div 8$ ● $56 + 5$ **3.** 7×6 ● $99 - 25$

4. $18 + 9$ ● $68 - 44$ **5.** 4×6 ● 7×2 **6.** $81 \div 9$ ● $67 - 56$

7. $53 + 37$ ● 9×9 **8.** 8×8 ● $134 - 59$ **9.** $82 - 47$ ● $163 - 124$

RETEACHING

Knowing one multiplication fact can help you remember three other facts.

3 rows
5 children
in each row
15 in all
$3 \times 5 = 15$

15 in all
3 rows
5 children
in each row
$15 \div 3 = 5$

5 rows
3 children
in each row
15 in all
$5 \times 3 = 15$

15 in all
5 rows
3 children
in each row
$15 \div 5 = 3$

These four number sentences are called a **family of facts.**

Complete each number sentence.

1. $6 \times 4 =$ ___	**2.** $5 \times 2 =$ ___	**3.** $8 \times 3 =$ ___	**4.** $7 \times 4 =$ ___
$4 \times 6 =$ ___	$2 \times 5 =$ ___	$3 \times 8 =$ ___	$4 \times 7 =$ ___
$24 \div 4 =$ ___	$10 \div 2 =$ ___	$24 \div 3 =$ ___	$28 \div 4 =$ ___
$24 \div 6 =$ ___	$10 \div 5 =$ ___	$24 \div 8 =$ ___	$28 \div 7 =$ ___

Write four number sentences for each set of numbers.

5. 2, 3, 6 **6.** 7, 3, 21 **7.** 3, 4, 12 **8.** 9, 5, 45

ENRICHMENT

Order of Operations

Sometimes, a problem includes more than one operation. To solve this kind of problem, follow these rules.

1. **Do the operation in parentheses first.**
2. **Multiply or divide in order from left to right.**
3. **Add or subtract in order from left to right.**

Solve: $3 \times (8 - 2) + 3 =$ Do the operation in parentheses.
$3 \times 6 + 3 =$ Multiply.
$18 + 3 =$
$18 + 3 = 21$ Add.

Other examples:

$31 - (6 \times 4) + 19 =$
$31 - 24 + 19 =$
$7 + 19 =$
$7 + 19 = 26$

$(36 \div 9) + 2 \times 4 =$
$4 + 2 \times 4 =$
$4 + 8 =$
$4 + 8 = 12$

Solve.

1. $(6 + 3) \times 2 =$ ____
2. $18 - (3 \times 3) =$ ____
3. $8 \div (2 \times 2) =$ ____
4. $(5 \times 3) - 9 =$ ____
5. $(7 \times 3) + 6 =$ ____
6. $5 + (35 \div 5) =$ ____
7. $49 - (2 \times 7) + 2 =$ ____
8. $(6 \times 6) - 6 - 5 =$ ____
9. $(6 \times 7) - 8 \times 3 =$ ____
10. $81 - 64 + (9 \div 3) =$ ____
11. $57 - (7 \times 8) + 24 =$ ____
12. $(5 \times 5) + 10 - 5 =$ ____

★13. $4 \times 4 - 6 + (72 \div 9) \times 2 =$ ____

★14. $2 \times (30 \div 6) + 48 - 4 \times 2 =$ ____

TECHNOLOGY

The LOGO turtle can move in any direction and draw any shape. In order for the turtle to draw the shape you want, you must write the commands in the proper order.

The FD command will make the turtle move forward the number of steps shown.

The BK command will make the turtle move in the opposite direction the number of steps shown.

REMEMBER:

FD is short for FORWARD. RT is short for RIGHT.

BK is short for BACK. LT is short for LEFT.

1. This procedure should draw three sides of a square. The list of commands is written in the wrong order. Rewrite the list of commands in the correct order.

TO THREESIDES

LT 90
FD 30
FD 30
LT 90
FD 30
END

2. These commands should draw the shape shown. Some commands are in the wrong order. Write the commands in the correct order.

RT 90
FD 40
LT 72
LT 72
FD 40
LT 72
FD 40
LT 72
FD 40
FD 40

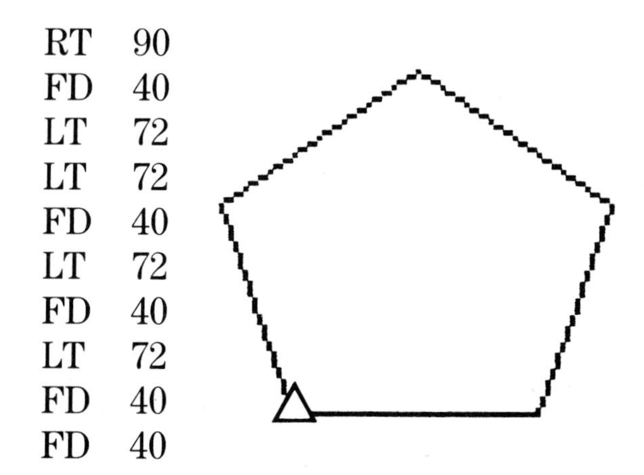

★3. Read the list of commands. Draw what the commands tell you to draw. When the command is RT 45, turn the turtle this far.

FD 40
RT 45
FD 40
RT 45
FD 40
RT 45
FD 40
RT 45
FD 40
RT 45
FD 40
RT 45
FD 40
RT 45
FD 40

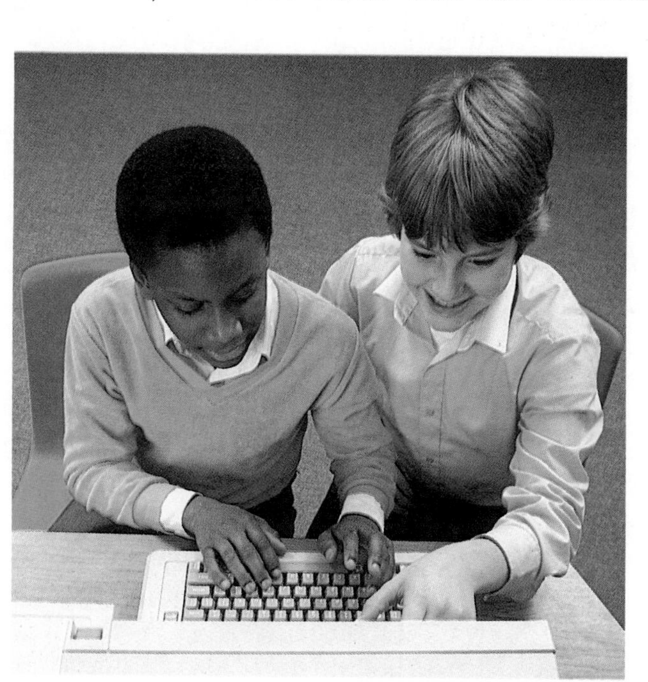

CUMULATIVE REVIEW

Write the letter of the correct answer.

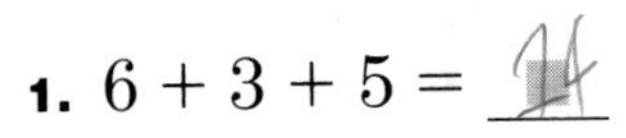

1. $6 + 3 + 5 = \_\_\_$

   a. 8 b. 9 c. 14 d. not given

2. Compare. 427 ● 319

   a. $<$ b. $>$ c. $=$ d. not given

3. How much money?

   a. 41¢ b. 51¢ c. 56¢ d. not given

4. $\begin{array}{r} 6{,}478 \\ +\,3{,}568 \\ \hline \end{array}$

   a. 9,046 b. 10,046 c. 11,046 d. not given

5. $\begin{array}{r} \$20.75 \\ -\;\;9.80 \\ \hline \end{array}$

   a. \$19.95 b. \$29.15 c. \$30.55 d. not given

6. $8 \times 0 = \_\_\_$

   a. 0 b. 1 c. 8 d. not given

7. $\begin{array}{r} 5 \\ \times 8 \\ \hline \end{array}$

   a. 13 b. 40 c. 58 d. 85

8. $\begin{array}{r} 7 \\ \times 9 \\ \hline \end{array}$

   a. 53 b. 54 c. 69 d. not given

9. $7 \times 8 = \_\_\_$

   a. 56 b. 65 c. 72 d. not given

10. $3 \times 1 \times 4 = \_\_\_$

    a. 8 b. 12 c. 341 d. not given

11. Nat had 24 worms. He sold 22 worms for 10¢. How many worms does he have now?

    a. 2 worms b. 24 worms c. 220¢ d. not given

12. Anne has 3 bags of marbles. There are 9 marbles in each bag. How many marbles does she have in all?

    a. 12 marbles b. 27 marbles c. 93 marbles d. not given

8 FRACTIONS AND DECIMALS

How do you spend the first two hours after school each day? Think of the best way to keep track of how you spend your time. Then keep a record of your time for five days. Decide if you would like to change your schedule in some way.

Fractional Parts of a Whole

Frank washes the window of his booth at the school fair. The window has 4 windowpanes. Frank washes 3 of the panes. What part of the window does he wash?

You can use **fractions** to name parts of a whole.

3 ⟶ **numerator** ⟶ **panes washed**
4 ⟶ **denominator** ⟶ **panes in all**

$\frac{3}{4}$ means 3 of 4 equal parts.

Frank washes $\frac{3}{4}$ of the window.

Other examples:

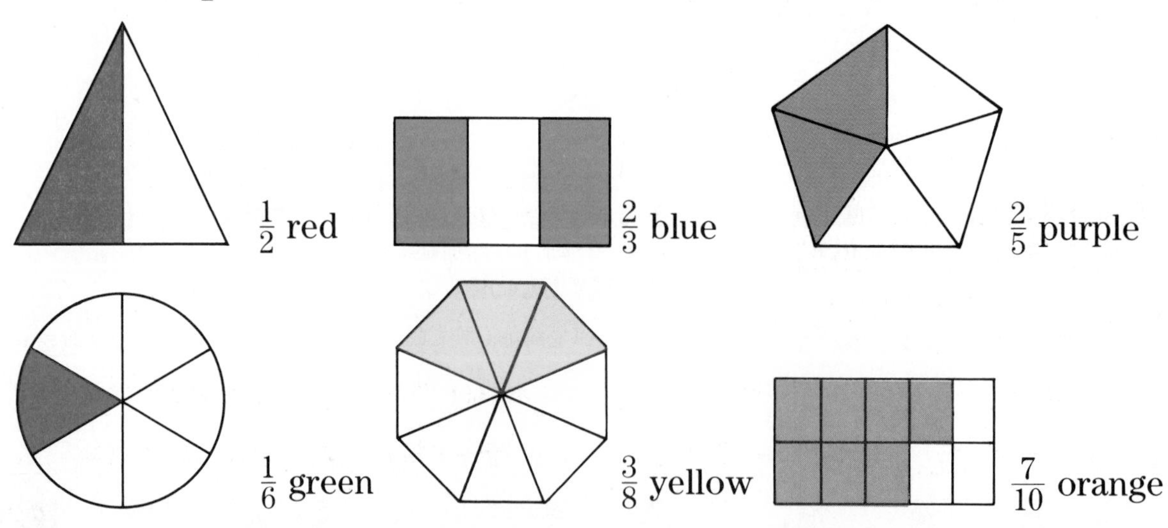

Write the fraction that names the shaded part.

1. 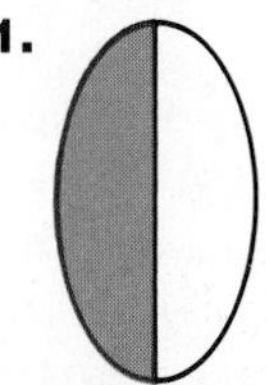

2. 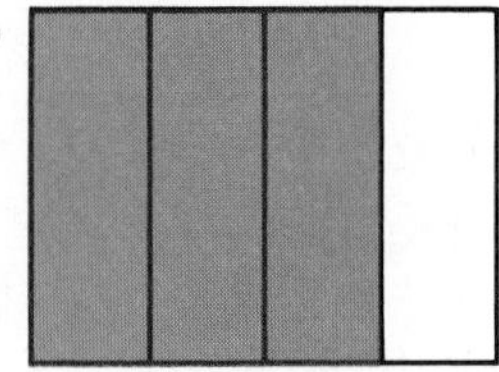

3. 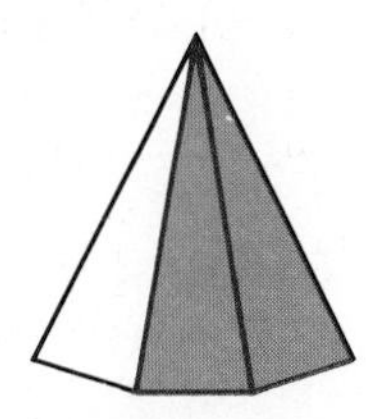

4. 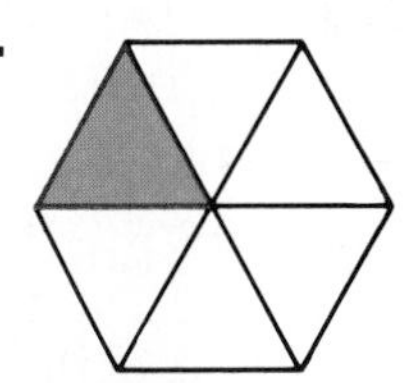

5. 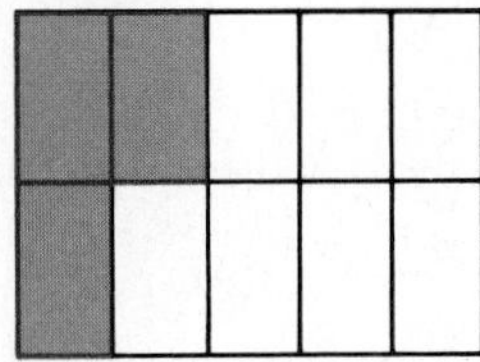

6. 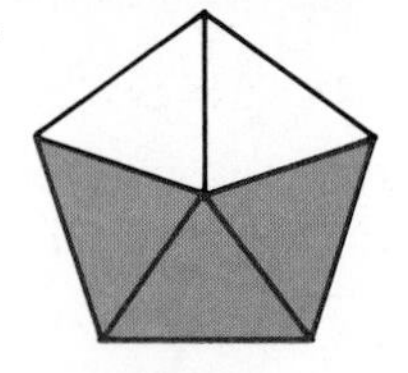

7. 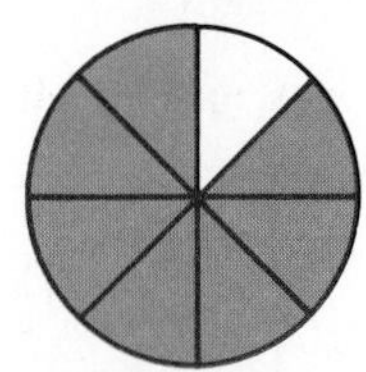

8. 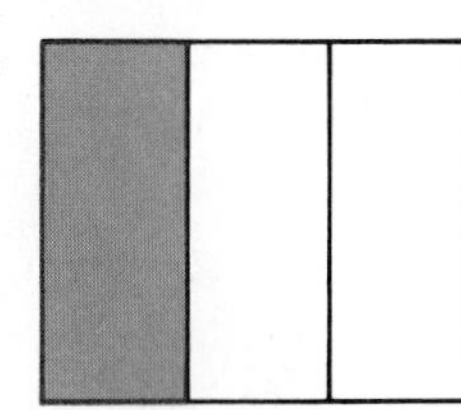

9. 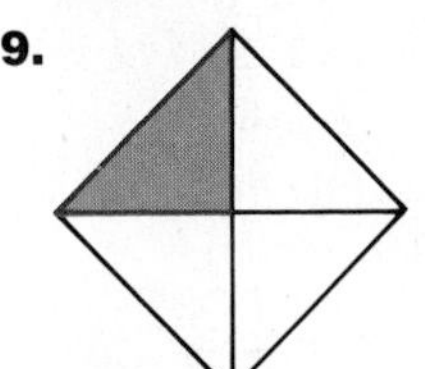

10. 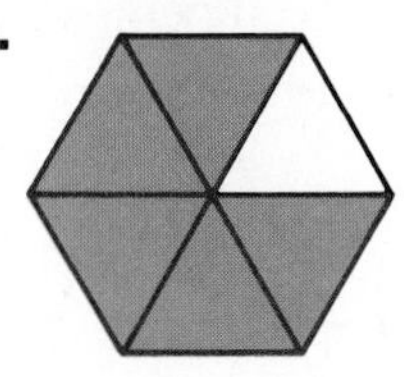

11. 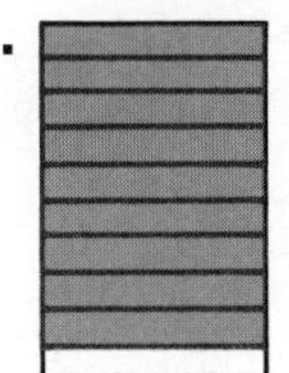

12. 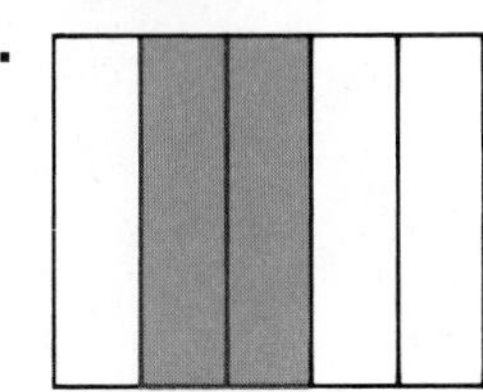

Solve.

13. Hans builds a table for a booth. He cuts 1 board into 3 equal parts. Write the fraction for one part of the board.

14. Shelves for a booth were made from 1 board that was cut into 10 equal pieces. Write the fraction for 2 shelves.

15. Sue paints a spinner for a school-fair game. The spinner has 6 equal parts. She paints 3 parts blue. Write the fraction that names the 3 blue parts.

★16. A spinner for the fair has 6 equal parts. It has 2 green parts, 2 red parts, and the rest are white. Write the fraction for each color.

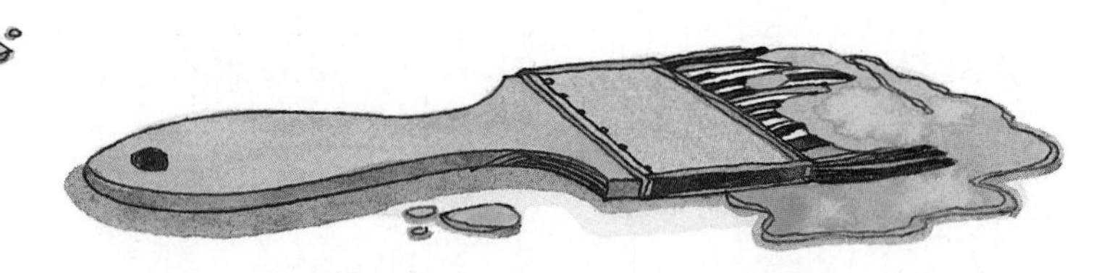

ANOTHER LOOK

Divide.

1. $8 \div 2 =$ ___ **2.** $12 \div 3 =$ ___ **3.** $24 \div 6 =$ ___ **4.** $45 \div 5 =$ ___

5. $56 \div 8 =$ ___ **6.** $63 \div 7 =$ ___ **7.** $48 \div 6 =$ ___ **8.** $72 \div 8 =$ ___

Fractional Parts of a Set

Many students brought their sticker collections to the school fair. Cloe had a set of heart stickers. What fraction of the set was purple?

You can use fractions to name parts of a set.

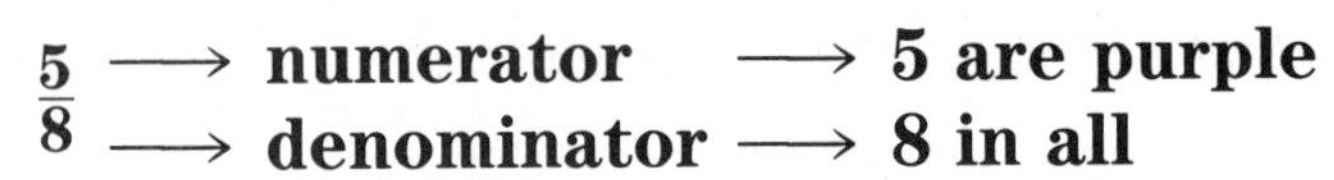

$\frac{5}{8}$ → **numerator** → **5 are purple**
→ **denominator** → **8 in all**

$\frac{5}{8}$ means 5 of 8 equal parts.

$\frac{5}{8}$ of the set was purple.

Other examples:

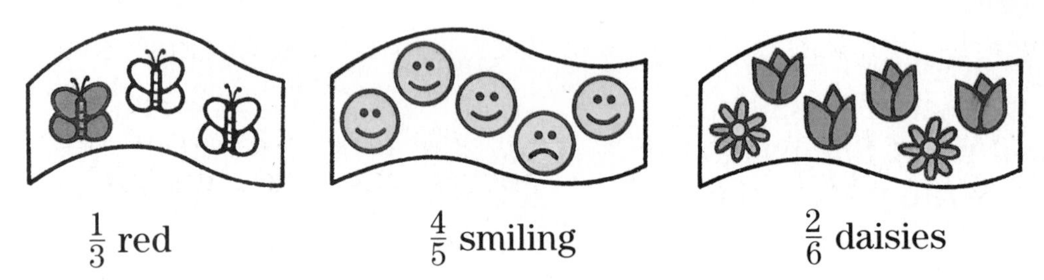

$\frac{1}{3}$ red $\quad$ $\frac{4}{5}$ smiling $\quad$ $\frac{2}{6}$ daisies

Checkpoint Write the letter of the correct answer.

Choose the fraction that names the shaded part of the set.

1.

a. $\frac{1}{5}$ **b.** $\frac{1}{4}$ **c.** 1 **d.** $\frac{5}{1}$

2.

a. $\frac{1}{3}$ **b.** $\frac{2}{3}$ **c.** 2 **d.** 3

3.

a. $\frac{2}{6}$ **b.** $\frac{4}{6}$ **c.** $\frac{4}{2}$ **d.** $\frac{6}{2}$

Write the fraction that names the shaded part of the set.

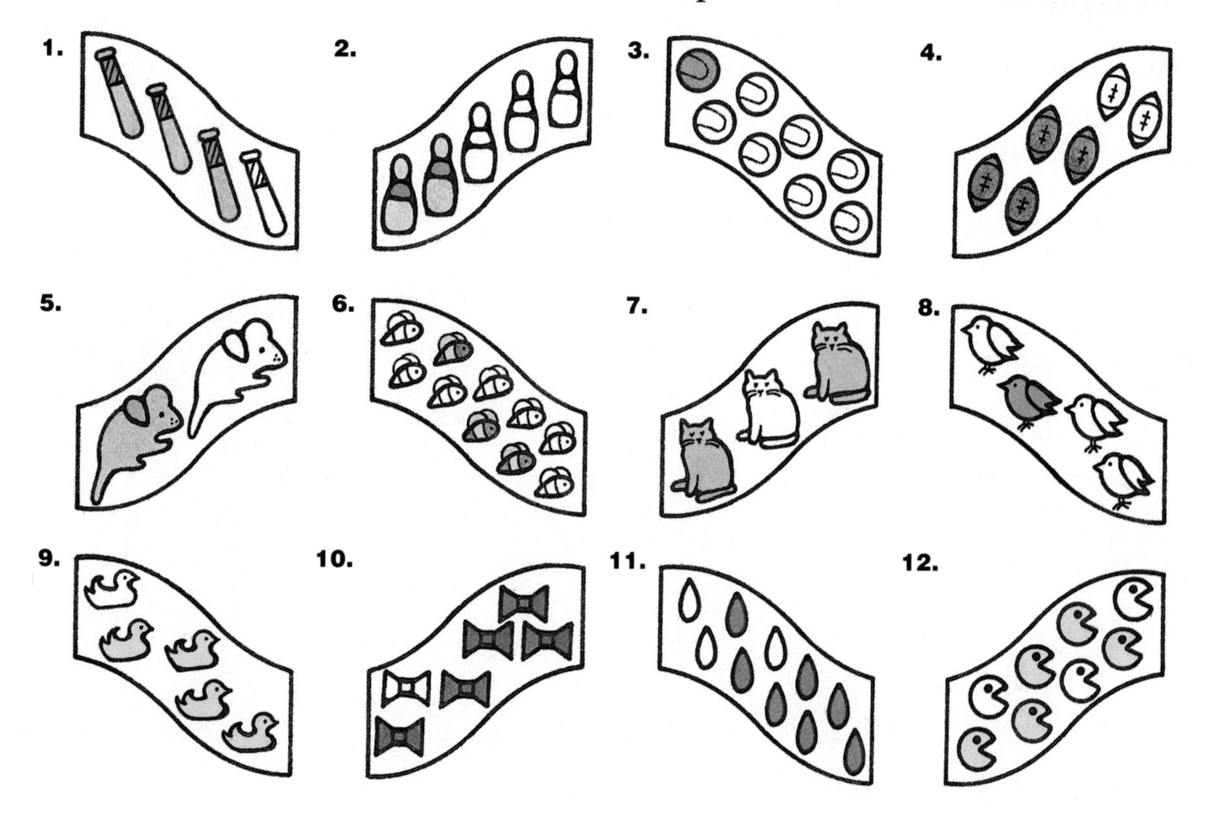

Solve.

13. Willie has 4 lamb stickers. He trades 3 of them for a whale sticker. What fraction of the stickers does he trade?

★14. Elsie has 10 star stickers that glow in the dark. She trades 3 of them. What fraction of the stickers does she have left?

FOCUS: ESTIMATION

Is more or less than $\frac{1}{2}$ of the figure shaded? Write *yes* or *no*.

1. 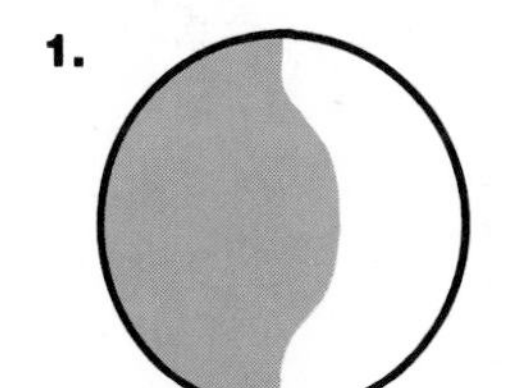

2. 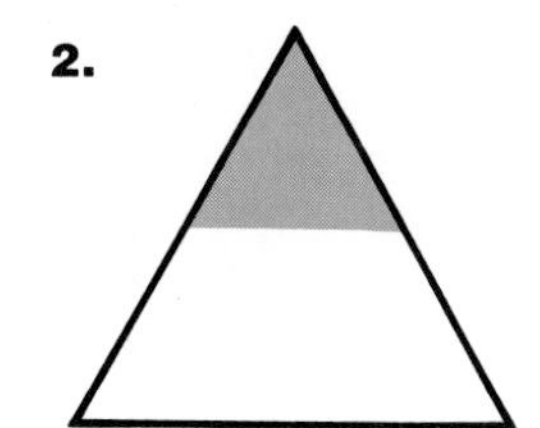

3. 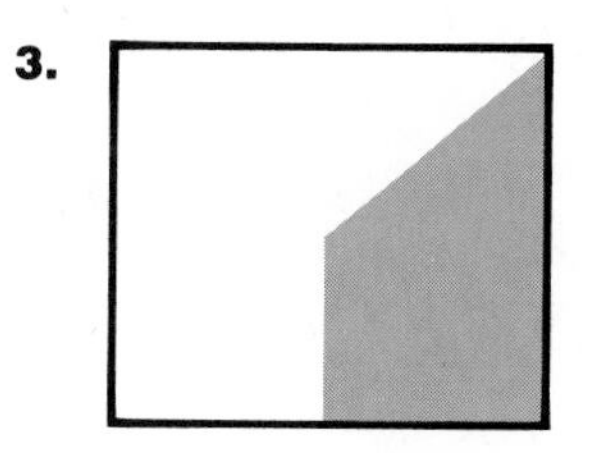

4. 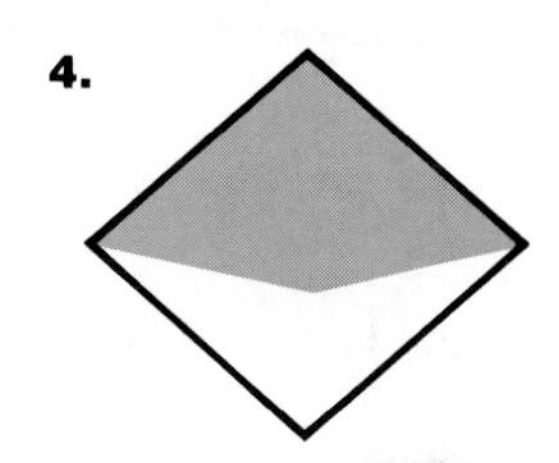

Finding Fractions of a Set

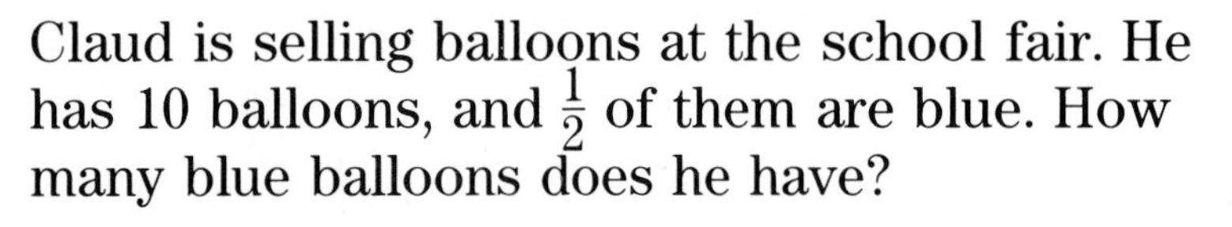

Claud is selling balloons at the school fair. He has 10 balloons, and $\frac{1}{2}$ of them are blue. How many blue balloons does he have?

You can separate the balloons into 2 equal sets to find how many blue balloons Claud has.

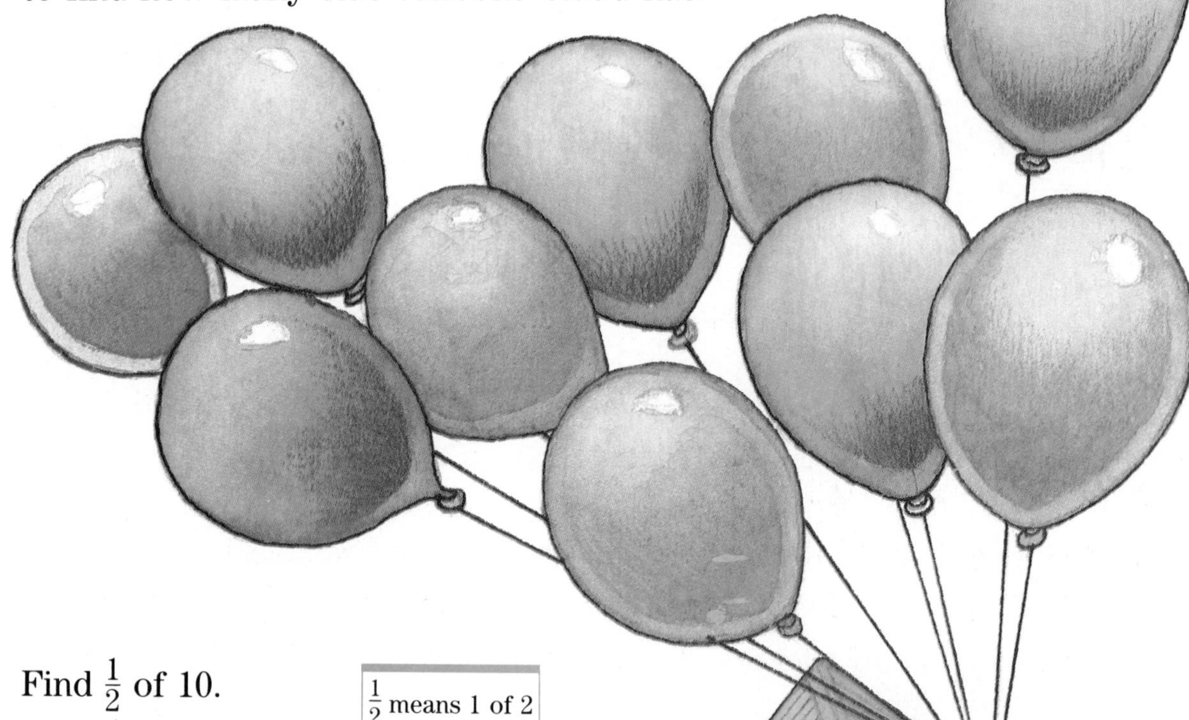

Find $\frac{1}{2}$ of 10.

Think: $10 \div 2 = 5$.

$\frac{1}{2}$ of 10 is 5.

Claud has 5 blue balloons.

> $\frac{1}{2}$ means 1 of 2 equal sets.

Other examples:

Find $\frac{1}{3}$ of 9.
$9 \div 3 = 3$
$\frac{1}{3}$ of 9 is 3.

> $\frac{1}{3}$ means 1 of 3 equal sets.

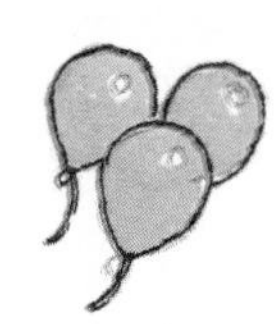
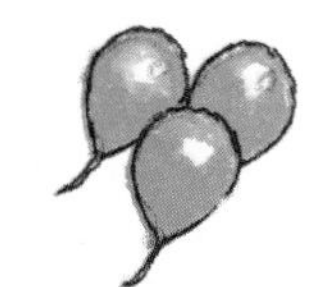
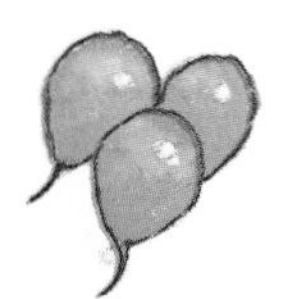

Find $\frac{1}{5}$ of 10.
$10 \div 5 = 2$
$\frac{1}{5}$ of 10 is 2.

> $\frac{1}{5}$ means 1 of 5 equal sets.

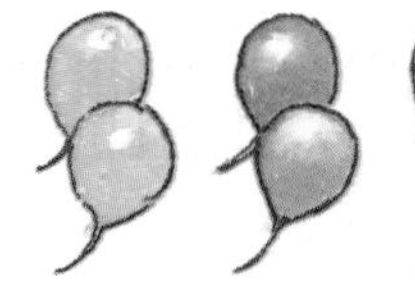
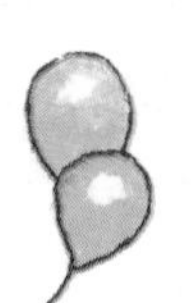
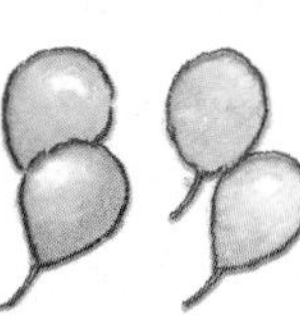

Solve.

1. $\frac{1}{3}$ of 6 = ___

2. $\frac{1}{6}$ of 6 = ___

3. $\frac{1}{5}$ of 5 = ___

4. $\frac{1}{4}$ of 12 = ___

5. $\frac{1}{4}$ of 8 = ___

6. $\frac{1}{8}$ of 16 = ___

7. $\frac{1}{2}$ of 6 = ___

8. $\frac{1}{8}$ of 8 = ___

9. $\frac{1}{6}$ of 18 = ___

10. $\frac{1}{2}$ of 16 = ___

11. $\frac{1}{5}$ of 20 = ___

12. $\frac{1}{5}$ of 25 = ___

13. $\frac{1}{3}$ of 12 = ___

14. $\frac{1}{3}$ of 27 = ___

15. $\frac{1}{4}$ of 24 = ___

16. $\frac{1}{2}$ of 14 = ___

17. $\frac{1}{6}$ of 42 = ___

18. $\frac{1}{5}$ of 40 = ___

19. $\frac{1}{6}$ of 24 = ___

20. $\frac{1}{4}$ of 32 = ___

21. $\frac{1}{6}$ of 12 = ___

22. $\frac{1}{8}$ of 40 = ___

23. $\frac{1}{4}$ of 20 = ___

24. $\frac{1}{3}$ of 24 = ___

25. $\frac{1}{5}$ of 35 = ___

26. $\frac{1}{8}$ of 56 = ___

Solve. Use the Infobank on page 402.

27. Marlene rides on $\frac{1}{5}$ of the rides. How many rides does she ride?

28. Whistles are sold in $\frac{1}{4}$ of the souvenir booths. How many souvenir booths sell whistles?

29. Food from other countries is sold in $\frac{1}{2}$ of the food booths. How many food booths sell food from other countries?

★30. George wins prizes from $\frac{1}{3}$ of the games and from $\frac{1}{3}$ of the raffle drawings. How many prizes does he win altogether?

CHALLENGE

Davy spent $6.00 at the school fair. The $6.00 Davy spent was $\frac{1}{3}$ of the money he had. How much money did Davy start with? How much money does he have left?

PROBLEM SOLVING
Writing a Number Sentence

In most problems, you are given certain facts. You use those facts to answer the question in the problem. Writing a number sentence can help you find the answer.

> The World Day Fair has booths, shows, games, and much more. All 5 grades in Murray School take part in the fair. There are 3 classes in each grade. How many classes are there at the fair?

1. List what you know and what you need to find.

know There are 5 grades. There are 3 classes in each grade.

find How many classes are there at the fair?

2. Think about what your list tells you about whether to +, −, ×, or ÷. Write a number sentence about this problem. Use ■ to stand for the number you need to find.

5	×	3	=	■
grades		classes in each grade		total number of classes

3. Solve. Write the answer.

$5 \times 3 = ■$
$5 \times 3 = 15$

There are 15 classes at the fair.

Write the letter of the correct number sentence.

1. One class sets up a Japanese puppet theater. Each giant puppet needs 3 people to work it. If 6 puppets are onstage at once, how many people are needed to work them?

   a. $6 + 3 = \blacksquare$
   b. $6 \times 3 = \blacksquare$

2. The class that chose Peru set up a marketplace. In the marketplace, 12 children worked in the booths. Each booth was run by 2 children. How many booths were there?

   a. $12 \times 2 = \blacksquare$
   b. $12 \div 2 = \blacksquare$

Write a number sentence, and solve.

3. In one booth, people can play the African game Ayo. On the board, there are 2 rows of 6 cups each. How many cups are there?

4. Jay needs to put 4 seeds or pebbles into each cup on the Ayo board. How many seeds will he need to fill 6 cups in 1 row?

5. There are 25 students in Mr. Yo's class. If 5 students are needed to help in each booth, in how many booths can Mr. Yo's students help?

6. Barney buys gifts in 3 booths. He buys 4 gifts in the first booth, 5 gifts in the second booth, and 3 gifts in the third booth. How many gifts did Barney buy?

Use the price list for Questions 7–9.

Price List	
Brazilian Bean Soup	$0.85
French Bread & Cheese	$0.65
Chinese Beef & Beans	$1.25
Nigerian Peanut Stew	$1.10
Indonesian Rice	$0.95
Orange Juice	$0.35
Cherry Juice	$0.45
Pineapple Punch	$0.40

7. Kim buys soup, rice, and punch. How much does he pay?

8. Fran has $2.75. If she buys stew and cherry juice, how much will she have left?

★9. Ed buys stew. Ann buys cherry juice and 1 other dish. Together, her 2 items cost what Ed's stew costs. What other dish did Ann buy?

Equivalent Fractions

Children made pictures to display at the crafts fair. Joy and Bill drew boxes and painted them to show a fraction. Joy's box showed $\frac{1}{2}$. Bill's box showed $\frac{4}{8}$. Both fractions looked the same. Are they the same amount?

You need to find out if $\frac{1}{2}$ is the same as $\frac{4}{8}$.

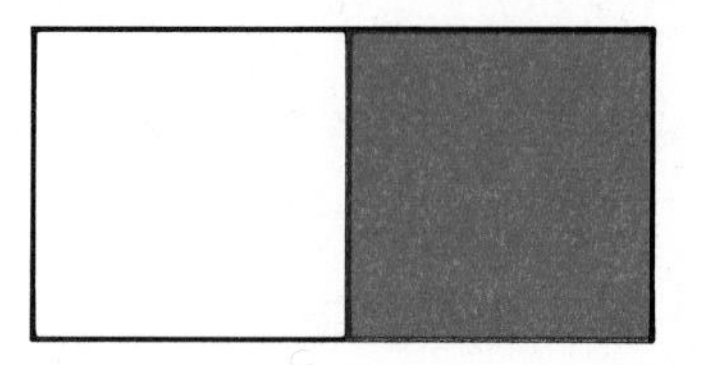

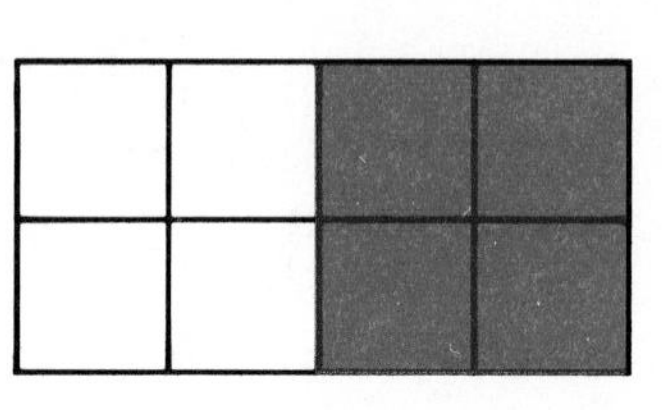

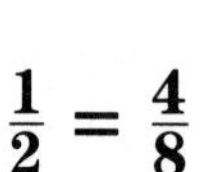

$\frac{1}{2} = \frac{4}{8}$

$\frac{1}{2}$ names the same amount as $\frac{4}{8}$.

$\frac{1}{2}$ and $\frac{4}{8}$ are **equivalent fractions.**

Other examples:

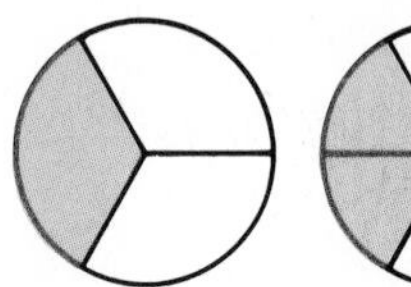

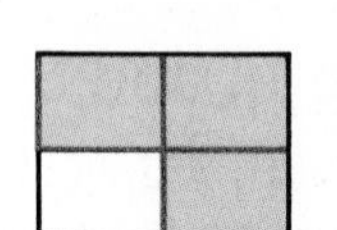

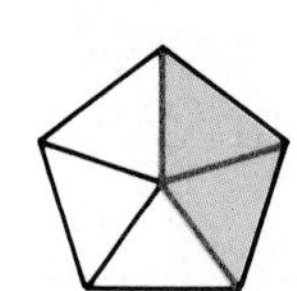

$\frac{1}{3} = \frac{2}{6}$ $\frac{3}{4} = \frac{6}{8}$

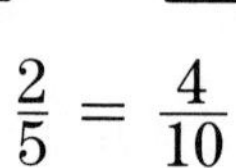

$\frac{2}{5} = \frac{4}{10}$

Checkpoint Write the letter of the correct answer.

Name the equivalent fractions.

1.

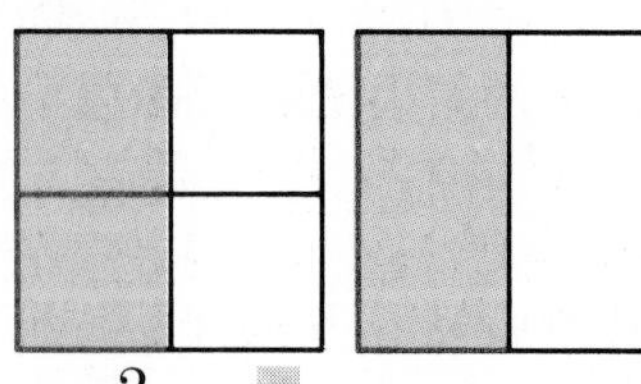

$\frac{2}{4} = \frac{\blacksquare}{\blacksquare}$

a. $\frac{1}{2}$ **b.** 1

c. $\frac{2}{1}$ **d.** 2

2.

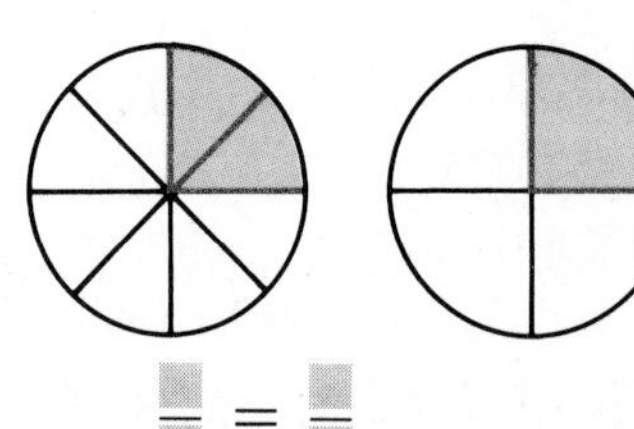

$\frac{\blacksquare}{\blacksquare} = \frac{\blacksquare}{\blacksquare}$

a. $\frac{1}{4} = \frac{2}{8}$ **b.** $\frac{2}{8} = \frac{1}{4}$

c. $\frac{2}{6} = \frac{1}{3}$ **d.** $\frac{6}{8} = \frac{2}{4}$

Name the equivalent fraction.

1. 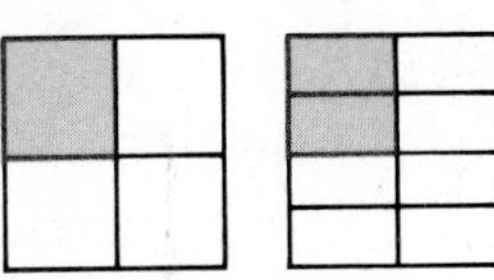

$\frac{1}{4} = \frac{\square}{8}$

2. 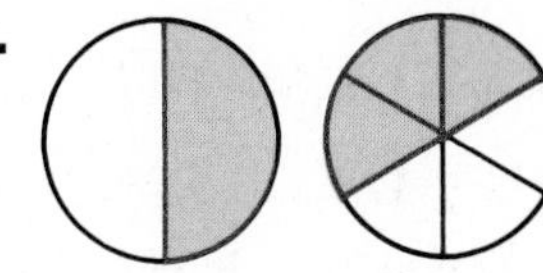

$\frac{1}{2} = \frac{3}{\square}$

3. 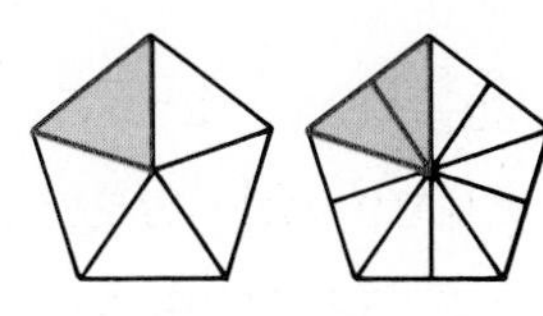

$\frac{1}{5} = \frac{\square}{10}$

4. 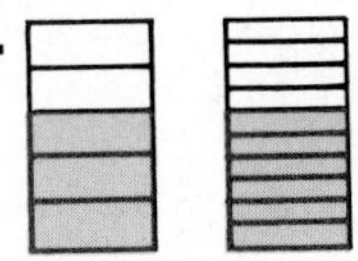

$\frac{3}{5} = \frac{\square}{\square}$

5. 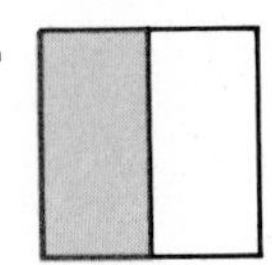

$\frac{1}{2} = \frac{\square}{\square}$

6. 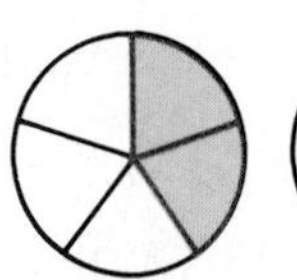

$\frac{2}{5} = \frac{\square}{\square}$

Write two equivalent fractions.

7. 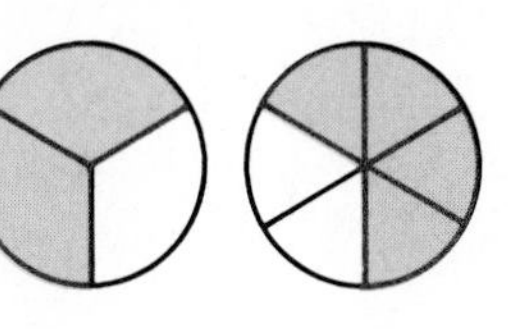

$\frac{\square}{\square} = \frac{\square}{\square}$

8. 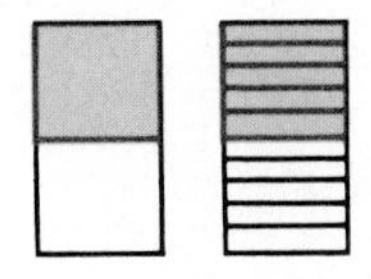

$\frac{\square}{\square} = \frac{\square}{\square}$

9. 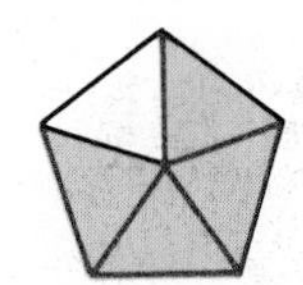

$\frac{\square}{\square} = \frac{\square}{\square}$

10. 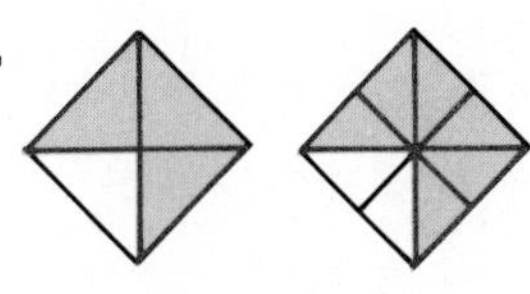

$\frac{\square}{\square} = \frac{\square}{\square}$

11. 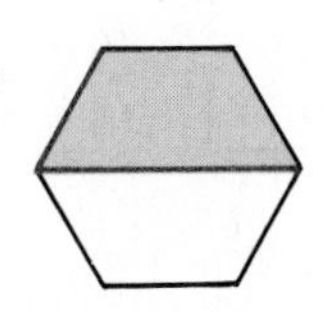

$\frac{\square}{\square} = \frac{\square}{\square}$

★12. 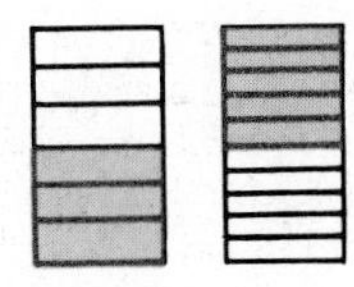

$\frac{\square}{\square} = \frac{\square}{\square}$

FOCUS: MENTAL MATH

$\frac{1}{4} \times \frac{4}{4} = \frac{4}{16}$

An easy way to find equivalent fractions is to multiply. Multiply the numerator and the denominator by the same number.

1. $\frac{1}{2} \times \frac{\square}{\square} = \frac{\square}{\square}$ **2.** $\frac{2}{3} \times \frac{\square}{\square} = \frac{\square}{\square}$ **3.** $\frac{3}{4} \times \frac{\square}{\square} = \frac{\square}{\square}$ **4.** $\frac{2}{5} \times \frac{\square}{\square} = \frac{\square}{\square}$

Comparing Fractions

A. Ray had $\frac{3}{4}$ jar of paint. Tina had $\frac{1}{4}$ jar of paint. Who had more paint?

You can compare fractions to find who had more paint.

$\frac{3}{4} > \frac{1}{4}$

The denominators are like.
Compare the numerators.
$3 > 1$
$\frac{3}{4}$ is greater than $\frac{1}{4}$.

Ray had more paint.

B. You can use a picture to help you compare fractions that have unlike denominators.

Which is less, $\frac{2}{3}$ or $\frac{5}{6}$?

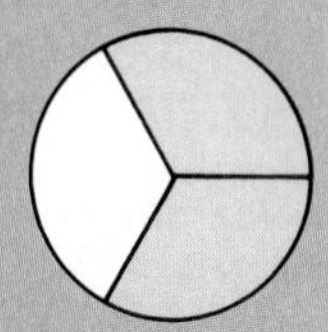

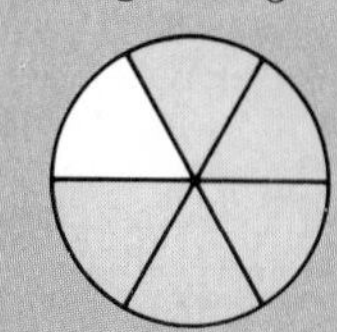

$\frac{2}{3}$ is less than $\frac{5}{6}$. $\frac{2}{3} < \frac{5}{6}$

Which is more, $\frac{4}{5}$ or $\frac{9}{10}$?

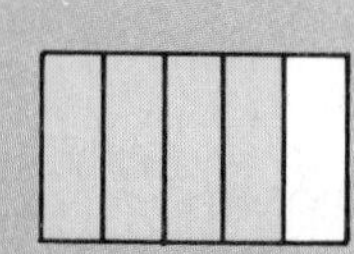

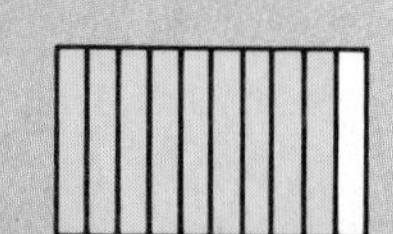

$\frac{9}{10}$ is more than $\frac{4}{5}$. $\frac{9}{10} > \frac{4}{5}$

Checkpoint Write the letter of the correct answer.

Compare.

1.

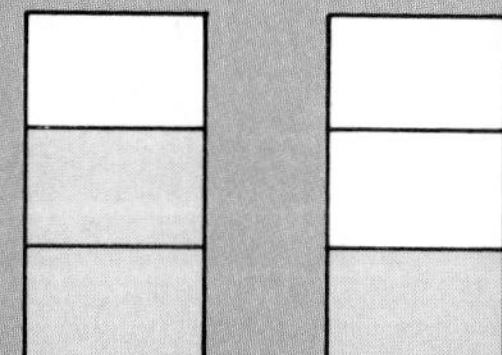

a. $\frac{2}{1} > \frac{1}{2}$ b. $\frac{2}{3} > \frac{1}{3}$
c. $\frac{2}{3} < \frac{1}{3}$ d. $\frac{1}{2} < \frac{2}{3}$

2.

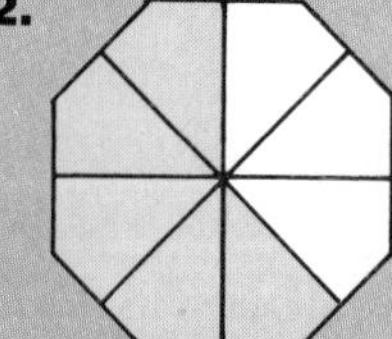

a. $\frac{3}{8} > \frac{1}{4}$ b. $\frac{5}{8} > \frac{3}{4}$
c. $\frac{3}{5} > \frac{1}{3}$ d. $\frac{5}{8} < \frac{3}{4}$

Write >, <, or = for ●.

1. 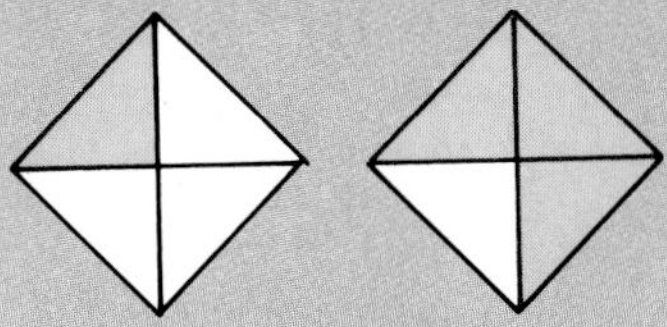

$\frac{1}{4}$ ● $\frac{3}{4}$

2. 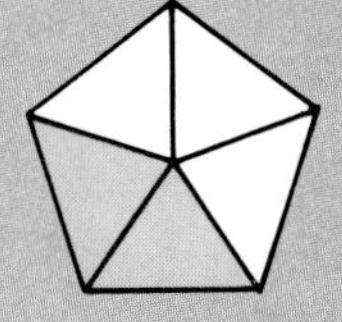 

$\frac{2}{5}$ ● $\frac{1}{5}$

3. 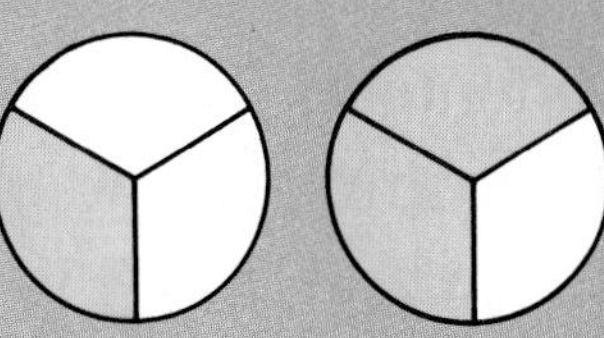

$\frac{1}{3}$ ● $\frac{2}{3}$

4. $\frac{5}{6}$ ● $\frac{2}{6}$ **5.** $\frac{2}{4}$ ● $\frac{3}{4}$ **6.** $\frac{1}{8}$ ● $\frac{3}{8}$ **7.** $\frac{3}{5}$ ● $\frac{2}{5}$ **8.** $\frac{6}{10}$ ● $\frac{9}{10}$

9. 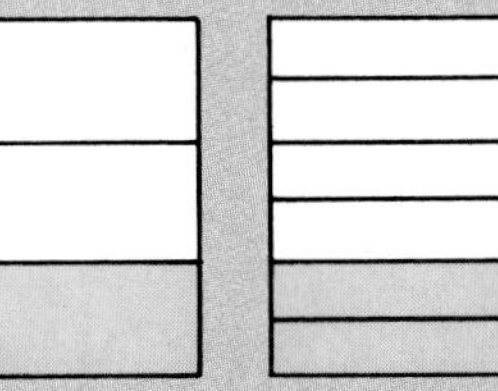 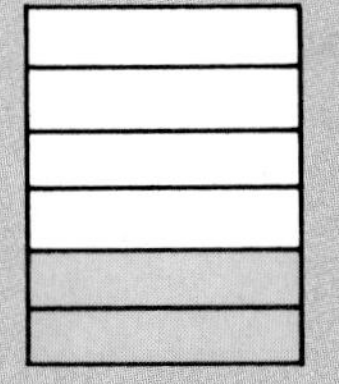

$\frac{1}{3}$ ● $\frac{2}{6}$

10. 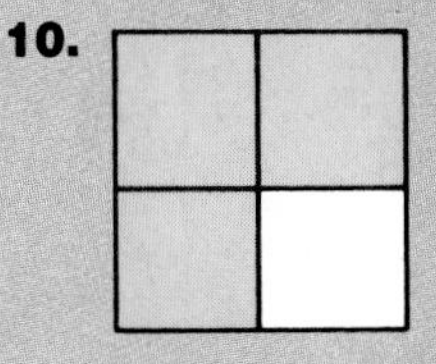

$\frac{3}{4}$ ● $\frac{5}{8}$

11.  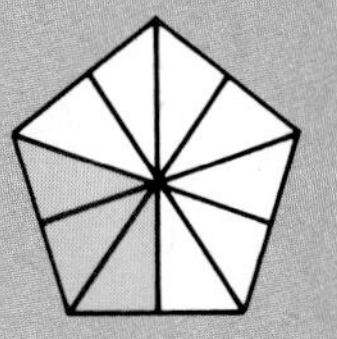

$\frac{3}{10}$ ● $\frac{2}{5}$

★**12.** $\frac{1}{8}$ ● $\frac{3}{4}$ ★**13.** $\frac{1}{2}$ ● $\frac{2}{4}$ ★**14.** $\frac{5}{6}$ ● $\frac{1}{3}$ ★**15.** $\frac{8}{10}$ ● $\frac{4}{5}$ ★**16.** $\frac{2}{3}$ ● $\frac{1}{6}$

Solve.

17. Merida painted a picture of a costume parade. She used $\frac{4}{5}$ jar of orange paint and $\frac{3}{5}$ jar of black paint. Which color did she use more of?

★**18.** Cheyenne painted a picture of a snowman. She used $\frac{4}{5}$ jar of white paint, $\frac{1}{5}$ jar of green, and $\frac{4}{10}$ jar of blue. Which color did she use the most? Which color did she use the least?

CHALLENGE

Complete the sequence.

1. $\frac{1}{2}, \frac{2}{4}, \frac{3}{6}, \frac{\blacksquare}{\blacksquare}, \frac{5}{10}, \frac{\blacksquare}{\blacksquare}, \frac{7}{14}, \frac{\blacksquare}{\blacksquare}, \frac{9}{18}, \frac{\blacksquare}{\blacksquare}$

2. $\frac{1}{3}, \frac{2}{6}, \frac{3}{9}, \frac{\blacksquare}{\blacksquare}, \frac{5}{15}, \frac{\blacksquare}{\blacksquare}, \frac{7}{21}, \frac{\blacksquare}{\blacksquare}, \frac{9}{27}, \frac{\blacksquare}{\blacksquare}$

More Practice, page 412

Whole Numbers and Mixed Numbers

Students in the school are making costumes for the costume parade. Sammy is going to be a space captain. He needs 1 yard of cloth to sew the pants and $\frac{3}{4}$ yard to sew the top. How many yards of cloth does he need?

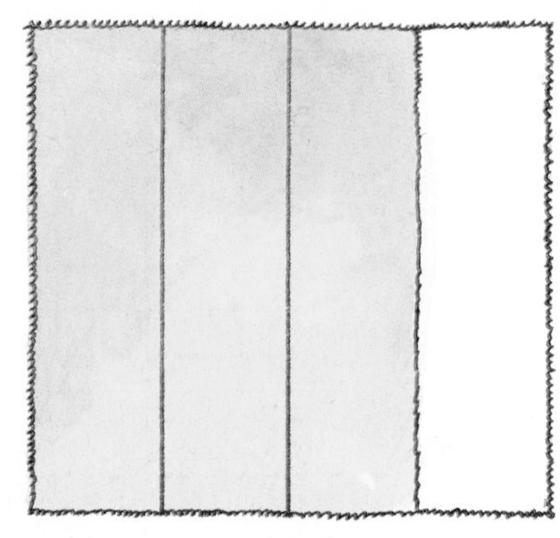

1 **and** $\mathbf{\frac{3}{4}} \longrightarrow \mathbf{1\frac{3}{4}}$

↑
mixed number

Sammy needs $1\frac{3}{4}$ yards of cloth.

Other examples:

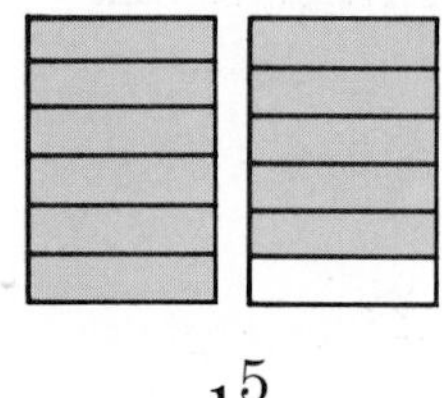

$1\frac{5}{6}$

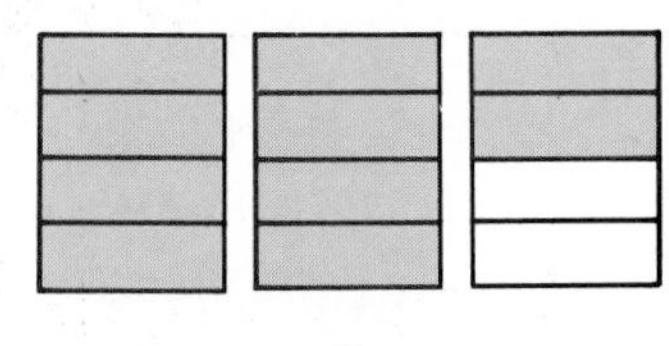

$2\frac{2}{4}$

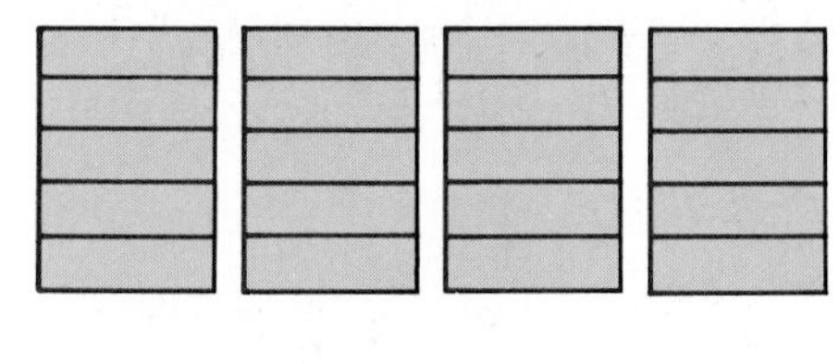

4

Write the whole number or the mixed number to show how much is shaded.

1. 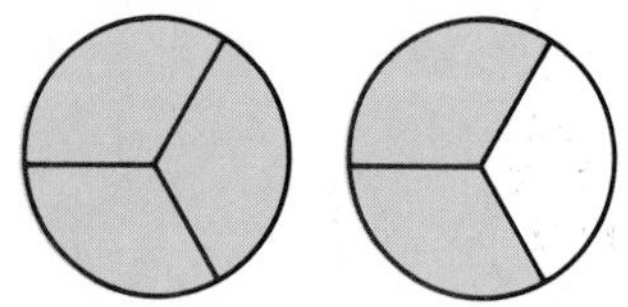

2. 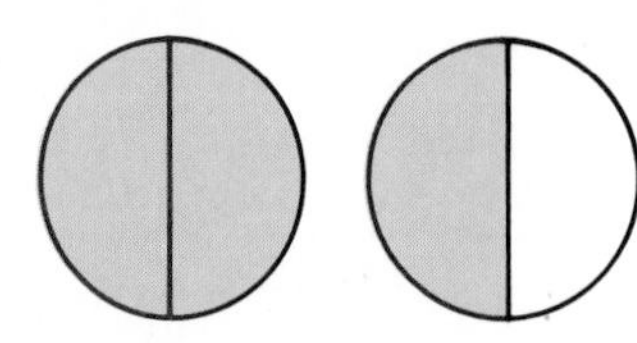

3. 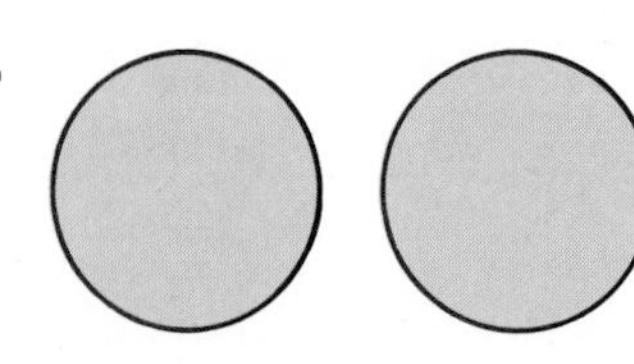

Write the whole number or the mixed number to show how much is shaded.

4. 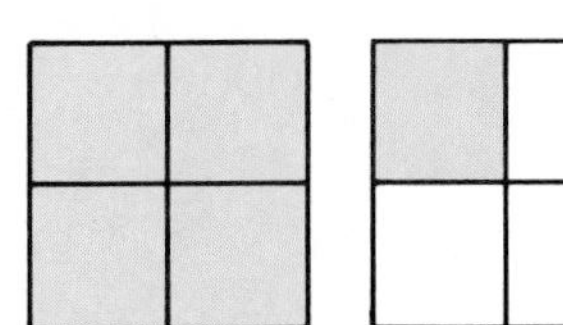

5. 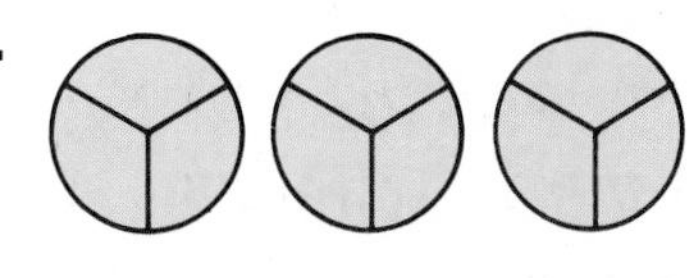

6. 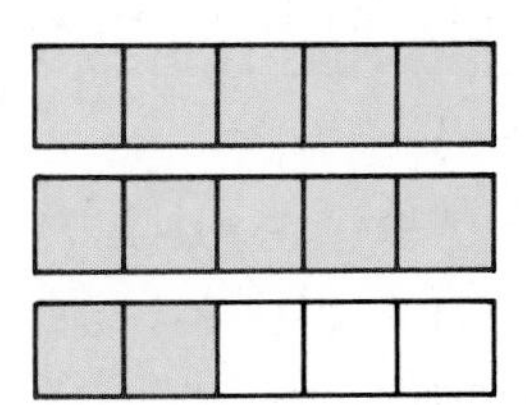

7. 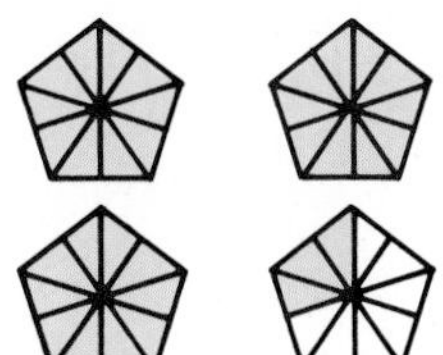

8. 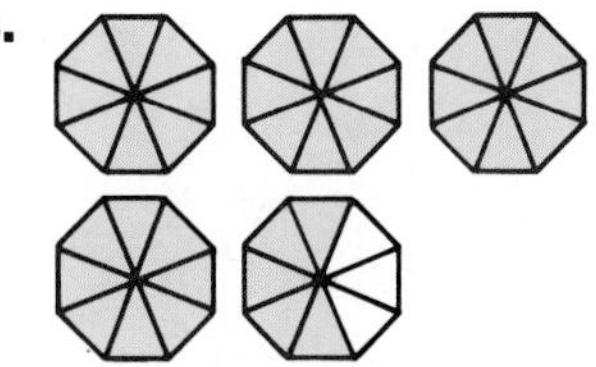

9. 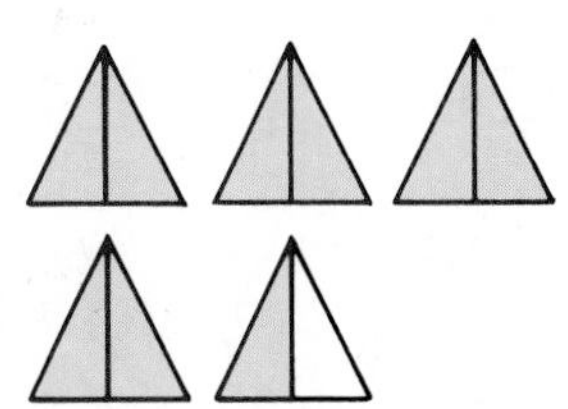

10. 

11. 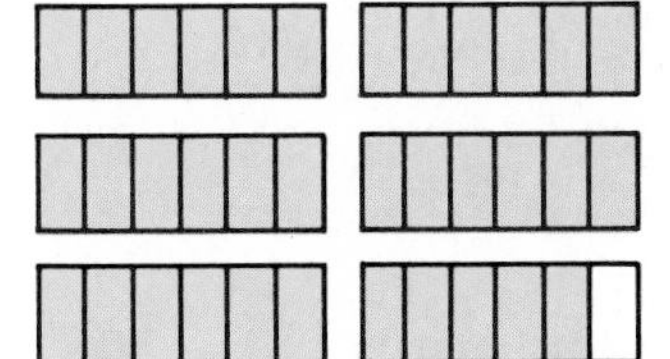

12. 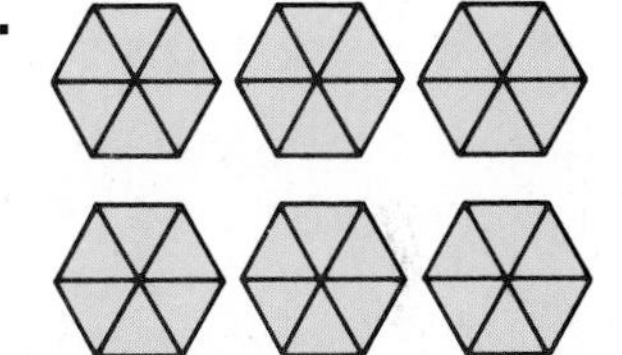

13. 

14. 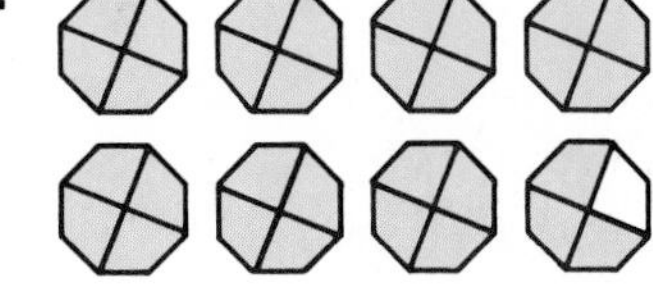

15. 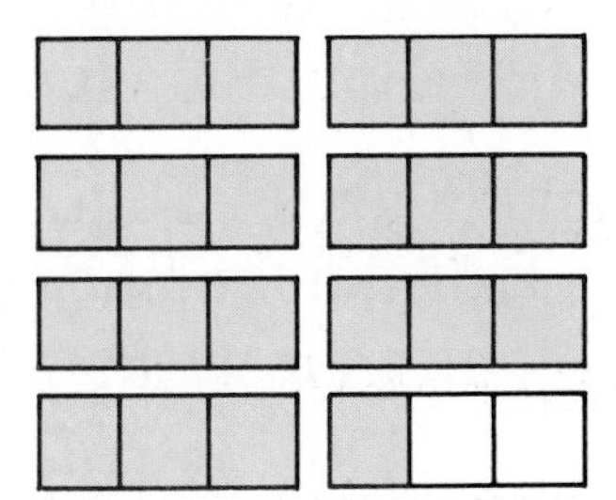

MIDCHAPTER REVIEW

Complete.

1. $\frac{1}{8}$ of 24 = $\square$ **2.** $\frac{1}{6}$ of 36 = $\square$ **3.** $\frac{1}{5}$ of 30 = $\square$ **4.** $\frac{1}{3}$ of 21 = $\square$

5. 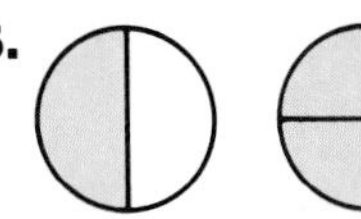

$\frac{1}{2} = \frac{4}{\square}$

6. 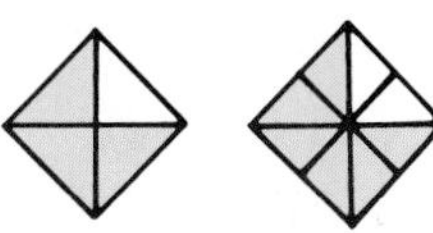

$\frac{3}{4} = \frac{\square}{8}$

7. 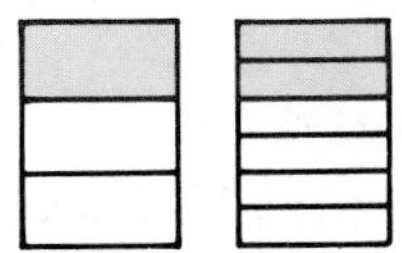

$\frac{1}{3} = \frac{\square}{\square}$

8. 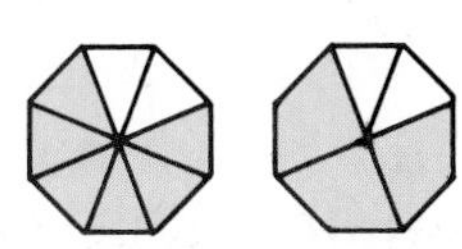

$\frac{\square}{\square} = \frac{\square}{\square}$

Write >, <, or = for ●.

9. $\frac{3}{4}$ ● $\frac{2}{4}$ **10.** $\frac{4}{10}$ ● $\frac{5}{10}$ **11.** $\frac{1}{2}$ ● $\frac{3}{6}$ **12.** $\frac{5}{8}$ ● $\frac{3}{4}$

PROBLEM SOLVING

Guessing and Checking

Solving some problems is like playing detective. Using the clues, you guess the answer. Then, you check to see if your guess is correct. If it isn't correct, try to make a better guess. Then check it.

> Benny, Paco, and Phil make 11 signs for the school fair. Each boy makes a different number of signs. Each boy makes fewer than 6 signs. How many signs does each boy make?

Think about what you know about the 3 numbers you must find. Think about the clues you have.

- Their sum is 11.
- Each number is less than 6.
- There are 3 numbers of signs.

Make a guess. You know that each of the 3 numbers is fewer than 6. Start your guess with 5.

$$\begin{array}{r} 11 \\ -\ 5 \\ \hline 6 \end{array}$$

When you have subtracted 5 from 11, you have 6 left. If 5 is one of the 3 numbers, the other 2 numbers must add up to 6. Which numbers add up to 6?

You know the numbers can't be 3 and 3. Each number must be different. The numbers can't be 5 and 1 either. You already have a 5. That leaves 4 and 2.

$3 + 3 = 6$
$5 + 1 = 6$
$4 + 2 = 6$

You have found the numbers that match the clues.

The boys make 5, 4, and 2 signs.

Stanley the Great has a booth at the fair. Help him guess the numbers.

1. Vera has opened a book. The sum of the 2 facing page numbers is 91. Neither number is greater than 50. Since the pages face each other, they are numbers that come one after the other in order. What are the 2 page numbers?

2. Ralph is thinking of 3 numbers that add up to 15. The numbers come one after another in order. Each number is less than 8. What are the numbers?

3. Pauline has 9 coins in her pocket. They add up to $0.51. What are they?

★4. Jo is thinking of 3 numbers. Their sum is 15. None of the numbers is greater than 10. The numbers are all different. If you multiply the smaller 2 numbers, you can add that product to the largest number and get 17. What are the numbers?

★5. Four children want Stanley to guess their ages without looking at them. Their ages are all even numbers. The sum of their ages is 34. None of them is older than 11. There are 2 possible answers. What are both sets of numbers?

Probability

A coin has 2 sides.

This side is **heads.** This side is **tails.**

When you toss a coin, it must come up heads or tails.

There is an equal chance of the coin coming up heads or tails.

The chance of heads coming up is 1 out of 2.
The chance of tails coming up is 1 out of 2.

1. Toss a coin. Does it come up heads or tails?

2. Copy the chart below. Then toss a coin 20 times. Mark the chart with a slash to show which side of the coin comes up.

	1	2	3	4	5	6	7	8	9	10	11	12	13	14	15	16	17	18	19	20
Heads																				
Tails																				

3. Toss a coin 50 times. Do you get heads about as often as you get tails?

4. Suppose you tossed a coin 100 times. About how many times do you think you would get heads?

★5. Suppose you toss a coin. The coin comes up heads. What is the chance it will come up heads on the next toss?

Chris has 2 pairs of socks in his drawer. One pair is brown and the other pair is blue. Without looking, Chris reaches into the drawer to pull out a pair of socks.

Answer each question.

1. How many colors are there?
2. What are the chances that Chris will pick the brown pair of socks?
3. What are the chances that he will pick the blue pair of socks?

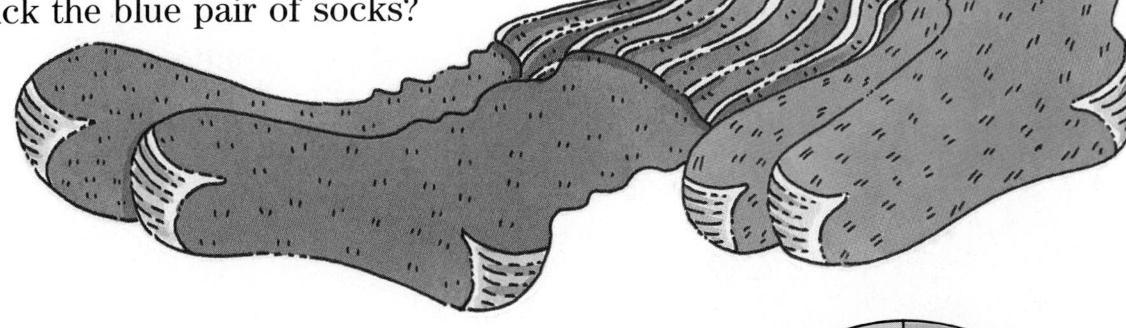

Peg's class forms 2 teams for the spelling bee. The class uses a spinner to choose the teams. Each student spins once.

Answer each question.

4. What are the chances that Peg will be on team A?
5. What are the chances that she will be on team B?

Copy the chart. Each student in the class can spin a spinner to choose which team he or she will be on. After each student spins, record the results.

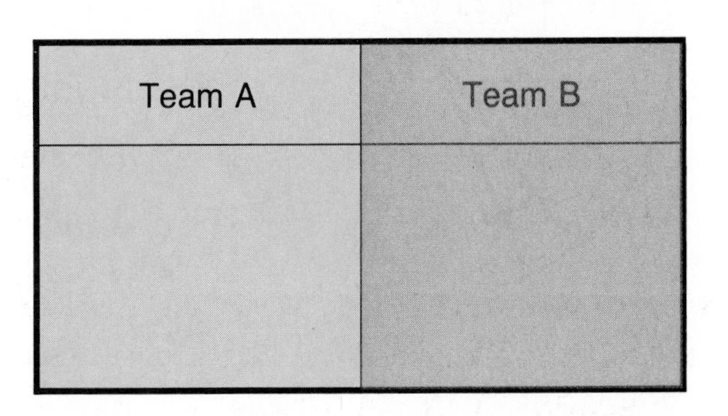

Team A	Team B

6. Count your marks. Is the number of students on team A about the same as the number on team B?

More Probability

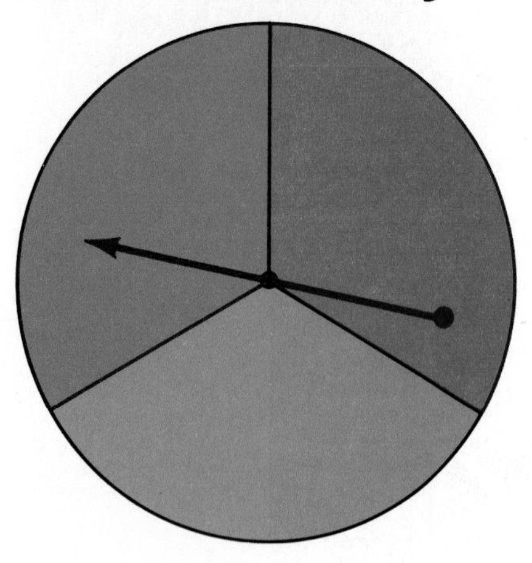

A. Suppose you make a spinner that has 3 equal parts. What is the chance that the arrow will point to red?

1 out of 3 parts is red. The chance of the arrow pointing to red is 1 out of 3. 1 out of 3 is the probability.

You can also show probability as a fraction.

$\frac{\mathbf{1}}{\mathbf{3}}$ ← number of red parts / ← total number of parts

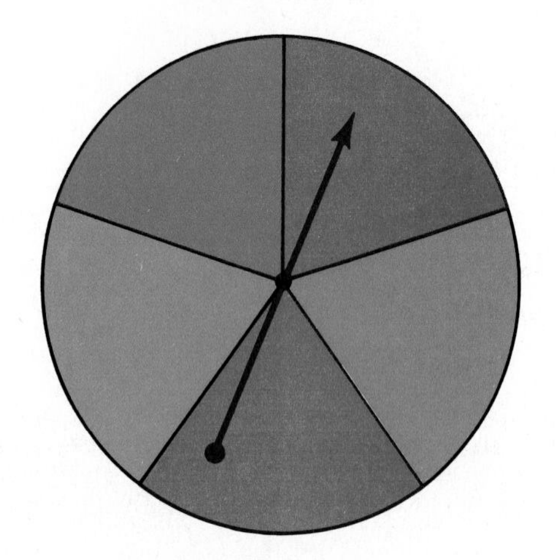

B. What is the chance that the arrow on this spinner will point to blue?

$\frac{\mathbf{2}}{\mathbf{5}}$ ← number of blue parts / ← total number of parts

2 out of 5 parts are blue. The probability of the arrow pointing to blue is $\frac{2}{5}$.

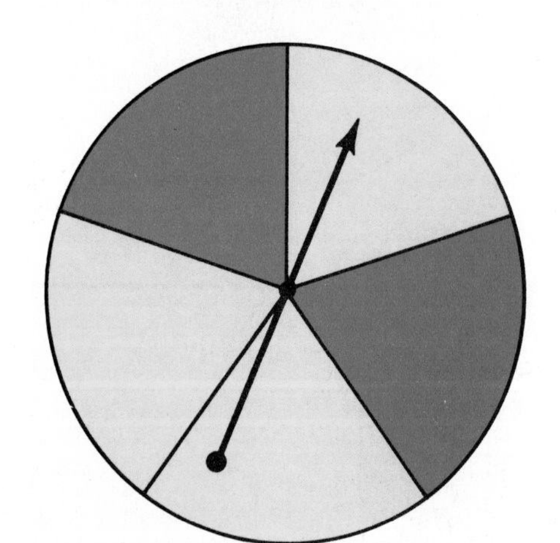

C. Which color is the arrow on this spinner most likely to point to?

To find out, compare the probabilities for each color.

3 out of 5 parts are yellow. The probability of the arrow pointing to yellow is $\frac{3}{5}$.

2 out of 5 parts are green. The probability of the arrow pointing to green is $\frac{2}{5}$.

$\frac{3}{5}$ is greater than $\frac{2}{5}$. So, the arrow is more likely to point to yellow.

What is the probability of spinning

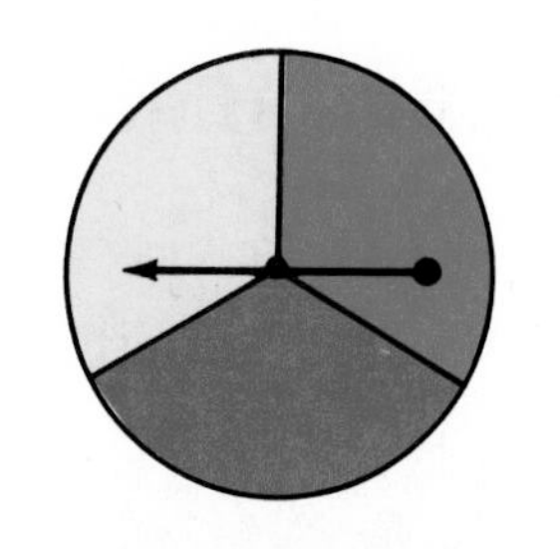

1. yellow?
2. blue?
3. green?
★4. yellow or blue?

What is the probability of picking

5. a red marble?
6. an orange marble?
7. a silver marble?

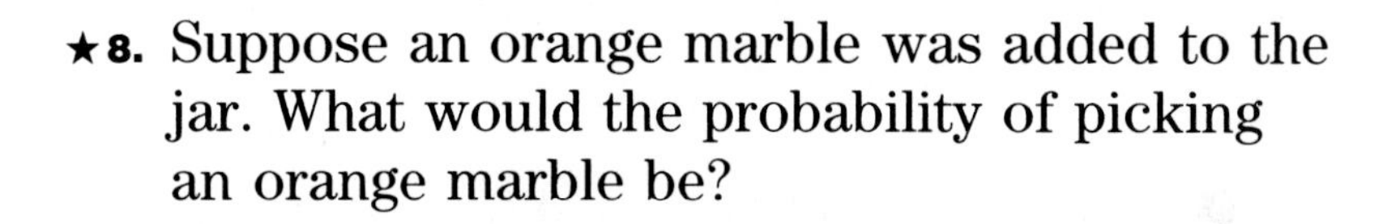

★8. Suppose an orange marble was added to the jar. What would the probability of picking an orange marble be?

Put 3 red crayons, 2 blue crayons, 1 white crayon, and 2 green crayons in a box.

9. Guess the color that you are most likely to pick.
10. Guess the color that you are least likely to pick.
11. List each color and write the probability of picking that color.

Shake the box of crayons. Then pick 1 crayon at a time without looking. Put each crayon back in the box and pick again. Do this 20 times. Keep track of the number of times that you pick each color.

12. Which color did you pick most often?
13. Which color did you pick least often?
14. Do your results agree with the probability for each color?

Decimals and Fractions

Judy made 10 bookmarks for the school crafts fair. She decorated 4 of them with faces. You can show 4 of the 10 as a fraction or a decimal.

$\frac{4}{10}$ ⟶ **0.4**

fraction **decimal**

You read both $\frac{4}{10}$ and 0.4 as **"four tenths."**

Another example:

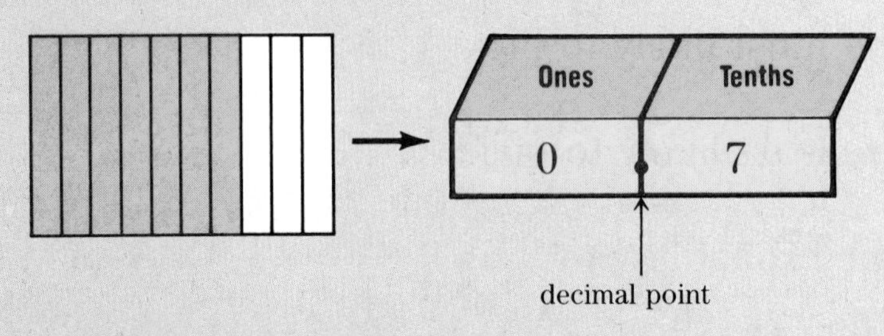

Read: seven tenths.
Write: 0.7.

Checkpoint Write the letter of the correct answer.

Choose the decimal.

1.

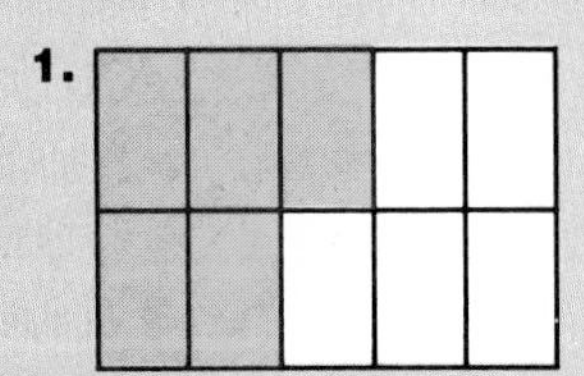

a. $\frac{5}{10}$

b. 0.5

c. 05

d. 5.0

2. one tenth

a. 0.01

b. $\frac{1}{10}$

c. 0.1

d. 110

3. $\frac{3}{10}$

a. 0.03

b. 0.3

c. 03

d. 3.1

Write the decimal.

1. 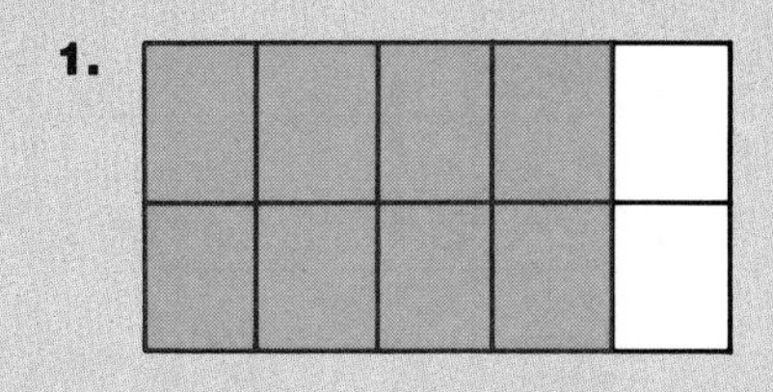

2. 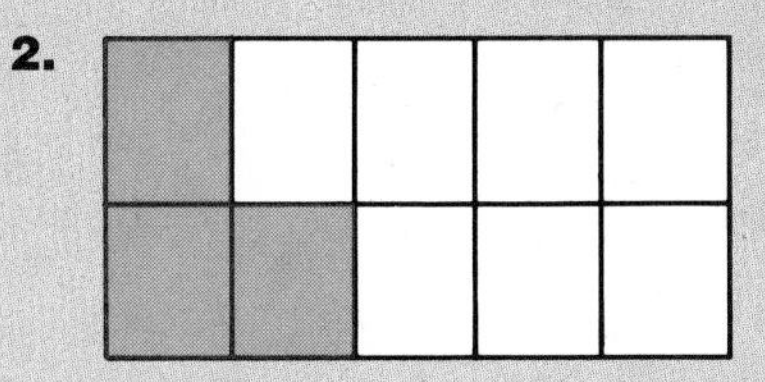

3. 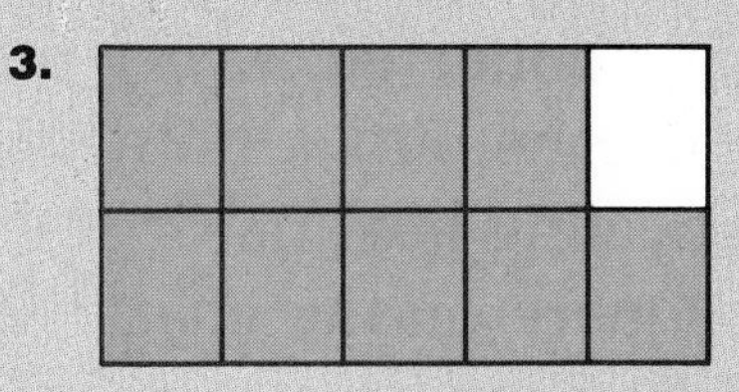

4.

Ones	Tenths
0 .	2

5.

Ones	Tenths
0 .	5

6. 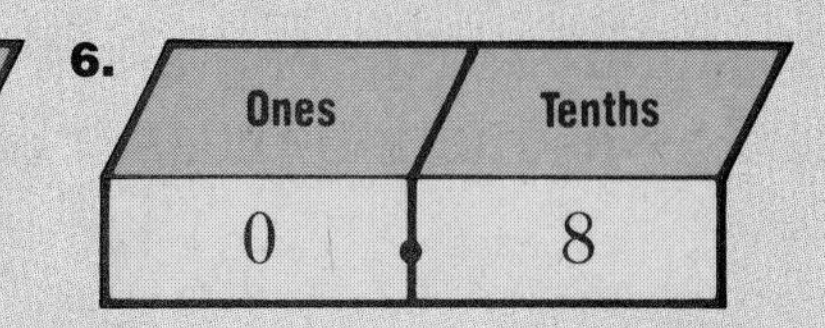

7. $\frac{3}{10}$

8. $\frac{2}{10}$

9. $\frac{6}{10}$

10. $\frac{9}{10}$

11. four tenths

12. five tenths

13. two tenths

14. one tenth

Write the word name for the decimal.

15. 0.8

16. 0.2

17. 0.6

18. 0.1

Use the picture to answer each question.
Write the answer as a decimal.

19. What part of the set of bookmarks is hands?

20. What part of the set of bookmarks is feet?

21. What part of the set of bookmarks is animals?

22. What part of the set of bookmarks is red?

23. What part of the set of bookmarks is blue?

24. What part of the set of bookmarks is green?

Tenths and Hundredths

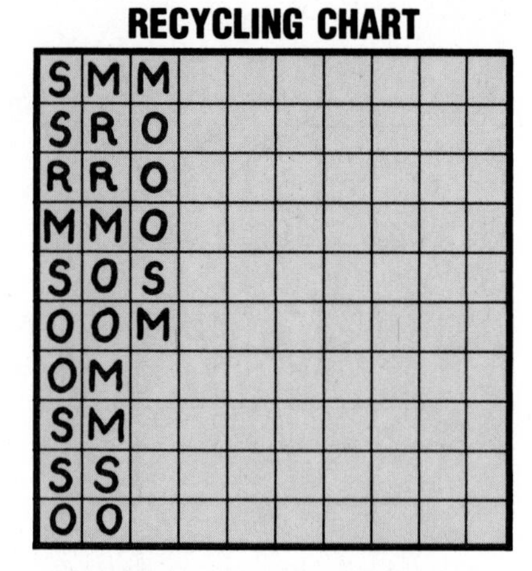

A. Sean's class draws a chart to show how many bottles they have collected. Every time someone puts a bottle into the recycling box, he or she marks a square on the chart. At the end of one week, Sean counts 26 squares. You can show 26 of 100 as a fraction or as a decimal.

$\frac{26}{100}$ ⟶ 0.26

fraction **decimal**

Ones	Tenths	Hundredths
0	. 2	6

↑ decimal point

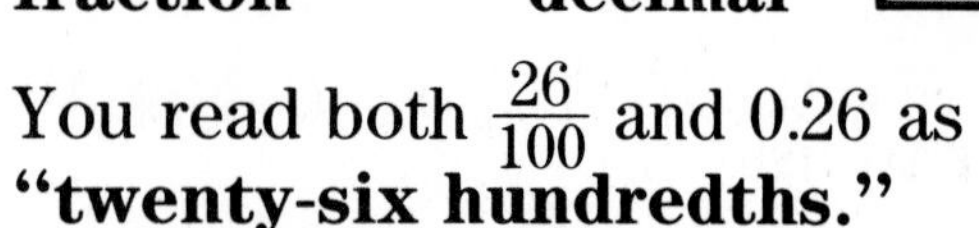

You read both $\frac{26}{100}$ and 0.26 as **"twenty-six hundredths."**

B. You can think of cents as hundredths of a dollar.

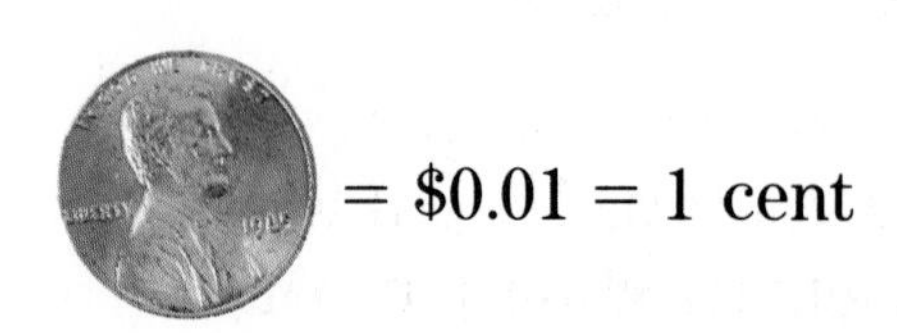

= \$0.01 = 1 cent

one hundredth of a dollar

= \$0.25 = 25 cents

twenty-five hundredths of a dollar

Checkpoint Write the letter of the correct answer.

Choose the decimal.

1. $\frac{73}{100}$
- **a.** 0.073
- **b.** 0.73
- **c.** 073
- **d.** 73.01

2. seventy-one hundredths
- **a.** $\frac{71}{100}$
- **b.** 0.17
- **c.** 0.71
- **d.** 71.01

3. 35 cents
- **a.** \$0.35
- **b.** 0.35
- **c.** $\frac{35}{100}$
- **d.** \$35.00

Write the decimal.

1. 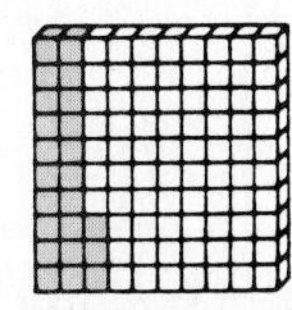

2. 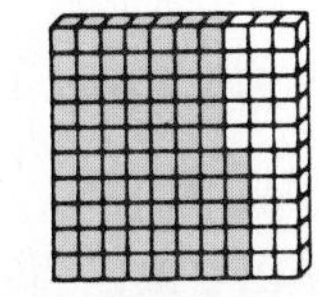

3. 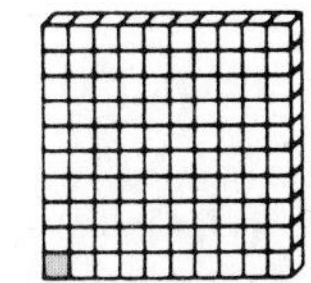

4. 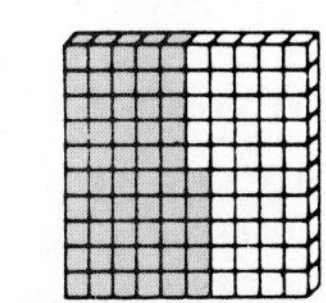

5.

Ones	Tenths	Hundredths
0 .	7	9

6.

Ones	Tenths	Hundredths
0 .	0	8

7. forty-eight hundredths

8. nine hundredths

9. nineteen hundredths

10. thirty-three hundredths

Write the word name for each decimal.

11. 0.91

12. 0.15

13. 0.73

14. 0.08

15. 0.42

Write each amount as a decimal. Remember to use the dollar sign and the cents point.

16. twenty-five cents

17. seventy-nine cents

18. three cents

Use the chart on the previous page to answer each question.

19. Each student marked the squares with an initial letter. What part of the chart did Sean fill? Write the answer as a decimal.

★20. Mona finds 6 bottles. Oscar finds 3 bottles. They add them to the chart. What part of the chart is filled in now? Write the answer as a decimal.

CALCULATOR

Use your calculator to write these fractions as decimals.

1. $\frac{1}{4} = 1 \div 4 =$ ____

2. $\frac{2}{4} = 2 \div 4 =$ ____

3. $\frac{3}{4} = 3 \div 4 =$ ____

4. Look at your answers to Examples 1–3. How would you describe the pattern?

Decimals

A. Emma's class makes a quilt. Each student makes one block. When they finish, there are enough blocks for 1 quilt and for $\frac{3}{10}$ of another. You can show $1\frac{3}{10}$ as a mixed number or as a decimal.

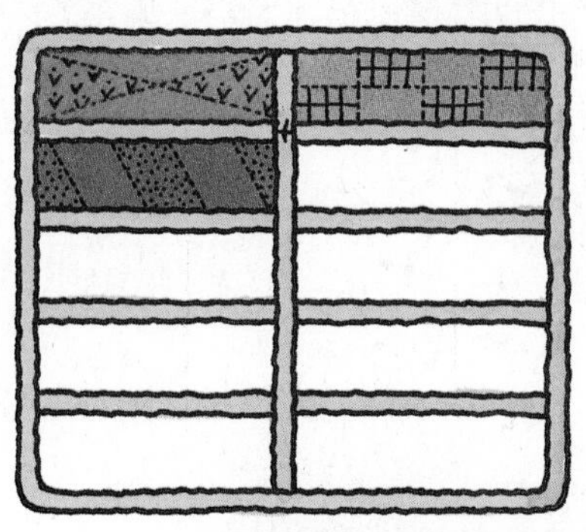

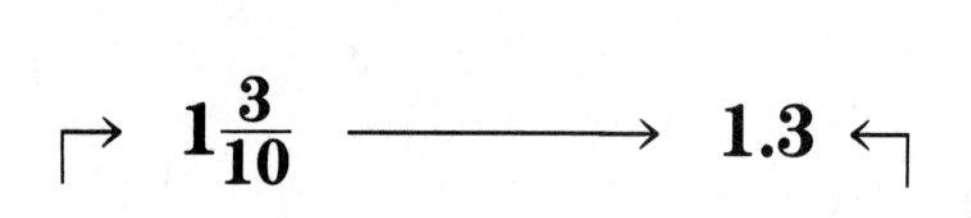

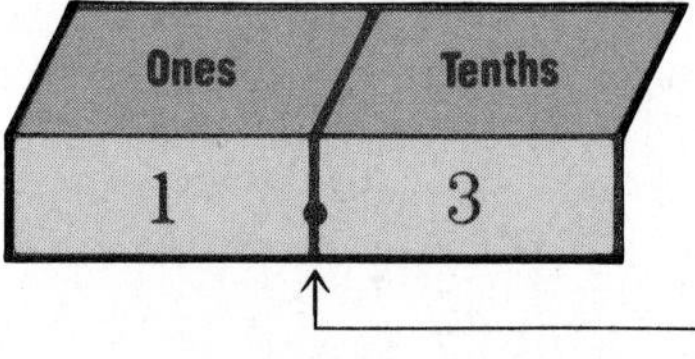

mixed number **decimal** **one and three tenths**

B. You can think of dollars and cents as decimals.

= $1.26

one and twenty-six hundredths

= $5.05

five and five hundredths

Checkpoint

Write the letter of the correct answer.

Choose the decimal.

1. $1\frac{1}{10}$

a. 0.11 **b.** 1.1
c. 11 **d.** 110

2. three and four hundredths

a. 3.04 **b.** $3\frac{4}{100}$
c. 3.40 **d.** 304

3. two dollars and twelve cents

a. $0.212 **b.** $2.12
c. $2.2 **d.** $212

Write the decimal.

1.

2. 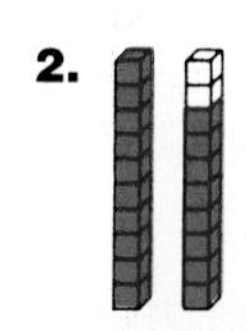

3.  

4. 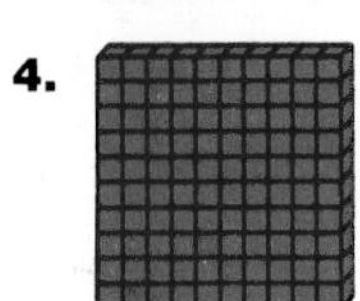  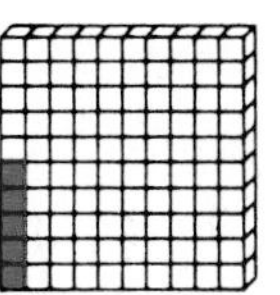

5. $5\frac{6}{10}$ 6. $9\frac{9}{10}$ 7. $2\frac{8}{10}$ 8. $4\frac{1}{10}$ 9. $6\frac{4}{10}$

10. $9\frac{93}{100}$ 11. $4\frac{20}{100}$ 12. $2\frac{6}{100}$ 13. $1\frac{1}{100}$ 14. $7\frac{15}{100}$

15. $7\frac{35}{100}$ 16. $8\frac{6}{10}$ 17. $3\frac{4}{100}$ 18. $6\frac{75}{100}$ 19. $5\frac{7}{10}$

20. one and two tenths

21. twelve and one hundredth

22. twenty-seven and five tenths

Write each amount as a decimal.

23. five dollars and forty-seven cents

24. seven dollars and eight cents

25. twenty-one dollars and seventeen cents

CHALLENGE

Write the missing numbers in each pattern.

1. $1\frac{1}{2}$, 2, $2\frac{1}{2}$, 3, ____, 4, $4\frac{1}{2}$, ____

2. \$6.25, \$____, \$7.25, \$7.75, \$8.25, \$____

3. ____, $\frac{1}{6}$, $\frac{1}{12}$, ____, $\frac{1}{48}$, $\frac{1}{96}$

4. 3.7, 4.0, 4.3, ____, 4.9, ____, 5.5

PROBLEM SOLVING
Checking for a Reasonable Answer

When you do a problem, think about your answer. Is it reasonable? If it isn't, you will need to correct your answer.

Farrell School is having a winter fair. In the parade, the students will carry large flags. It takes 3 students to carry each flag. If there are 6 flags, how many students will be needed?

5 students

18 students

60 students

How can you tell that Wendy's answer is too small?
How can you tell that George's answer is too big?

Without finding the exact answer, write the letter of the most reasonable answer.

1. This year, 2 schools hold a winter fair together. There are 249 children in Farrell School and 312 children in Reed School. How many children are there in both schools?
   a. 63 children
   b. 110 children
   c. 561 children

2. The snow-sculpture contest has 49 sculptures. Of them, 12 were chosen to be in the final round of judging. How many sculptures did not make the final round?
   a. 37 sculptures
   b. 61 sculptures
   c. 169 sculptures

Without finding the exact answer, write the letter of the most reasonable answer.

3. Each ice-skating race can have only 5 skaters. If 20 people want to race, how many races must there be?

   a. 4 races
   b. 15 races
   c. 25 races

4. The snow-castle contest has 27 people in it. They are on 3 teams of equal size. How many are there on each team?

   a. 3 people
   b. 9 people
   c. 30 people

5. One Sno Delight costs $0.25. You pay with $1.00. How much change do you receive?

   a. $0.75
   b. $7.50
   c. $75

6. How many paper snowflakes can 8 children make if each child makes 6?

   a. 2 flakes
   b. 48 flakes
   c. 60 flakes

7. The Food Committee spends $67.98 on food. They earn $84.55 by selling this food. How much more do they earn than they spend?

   a. $16.57
   b. $52.59
   c. $152.53

8. The teachers make 627 links for a paper chain. The parents make 451 links. The students make 809 links. How many links does the chain have?

   a. 97 links
   b. 1,887 links
   c. 8,087 links

★9. It costs $0.10 per person to take the sleigh ride. If 4 people go on the ride twice, how much does it cost them altogether?

   a. $0.08
   b. $0.40
   c. $0.80

★10. One newspaper will print a 210-line article about the fair. Jo wrote 84 lines about the races. Bob wrote 47 lines about the contests. How many lines are left?

   a. 37 lines
   b. 79 lines
   c. 131 lines

Comparing and Ordering Decimals

A. Reiko and Fritz build ant farms for the science fair. Reiko fills 0.9 of her jar with sand. Fritz fills 0.6 of his jar with sand. Which ant farm has more sand?

REMEMBER: The number line shows numbers in order from the least to the greatest.

You can use a number line to help you compare decimals.

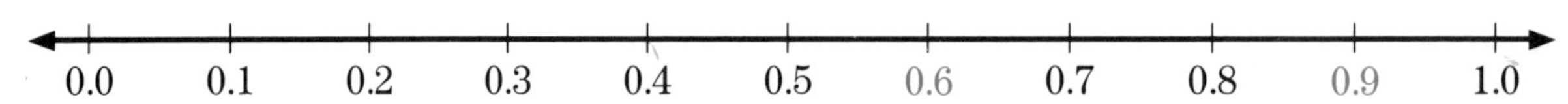

Since 0.9 is farther away from zero than 0.6, 0.9 is greater than 0.6.

$0.9 > 0.6$

Reiko's ant farm has more sand.

B. Compare 1.4 and 1.7.

Line up the decimal points.

1.4
1.7

So, $1.4 < 1.7$.

Begin by comparing digits at the left.

1.4
1.7

1 = 1

Continue comparing digits.

1.4
1.7

4 < 7

C. You can write numbers in order from the least to the greatest: 2.6, 2, 2.3.

Line up the decimal points.

2.6
2.0
2.3

Write zeros if you need them.
2 = 2.0

Compare to find the least number.

2.6 > 2.0
2.0 < 2.3

Find the next least number.

2.3 < 2.6

So, 2.0 is the least number.

The order from the least to the greatest is 2, 2.3, 2.6.
The order from the greatest to the least is 2.6, 2.3, 2.

Use the number line to answer each problem.
Write >, <, or = for ●.

1.0 1.1 1.2 1.3 1.4 1.5 1.6 1.7 1.8 1.9 2.0

1. 1.2 ● 2.0 **2.** 1.4 ● 1.1 **3.** 2.0 ● 1.8 **4.** 1.5 ● 1.6

Write >, <, or = to complete each problem.

5. 0.9 ● 0.1 **6.** 0.3 ● 0.3 **7.** 0.8 ● 0.7 **8.** 0.1 ● 0.2

9. 3.1 ● 4.9 **10.** 4.0 ● 5.9 **11.** 15 ● 15.0 **12.** 14.3 ● 12.7

13. 24.0 ● 30.8 **14.** 21.5 ● 21.9 **15.** 36.8 ● 36.80 **16.** 56.2 ● 59.5

Write the missing decimal.

17. **18.**

Write the decimals in order from the least to the greatest.

19. 1.0, 1.3, 1.1 **20.** 0.7, 0.5, 0.6 **21.** 1.9, 1.7, 1.8

Write the decimals in order from the greatest to the least.

22. 1.3, 2.1, 0.8 **23.** 6.7, 5.8, 6.4 **24.** 3.2, 4.3, 2.4

Use the chart to answer each question.

25. For the science fair, Pablo made an instrument that measures the speed of the wind. On which day was the wind the fastest?

26. Write the wind speeds in order from the least to the greatest.

SPEED OF THE WIND (miles per hour)	
Monday	18.1
Tuesday	12.7
Wednesday	8.9

Adding Decimals

A. It was physical-fitness week at school. Jason's class walked 1.2 kilometers on Monday and 1.5 kilometers on Thursday. How many kilometers did the class walk that week?

Add 1.2 + 1.5.

You can add decimals to find how many kilometers they walked. Add decimals the same way you add whole numbers.

Line up the decimal points.	Add the tenths.	Add the ones. Write the decimal point.
$\begin{array}{r} 1.2 \\ +1.5 \\ \hline \end{array}$	$\begin{array}{r} 1.2 \\ +1.5 \\ \hline 7 \end{array}$	$\begin{array}{r} 1.2 \\ +1.5 \\ \hline 2.7 \end{array}$

The class walked 2.7 kilometers that week.

B. You can regroup decimals the same way you regroup whole numbers.

Add 4.67 + 2.65.

Line up the decimal points. Add the hundredths. Regroup.	Add the tenths. Regroup.	Add the ones. Write the decimal point.
$\begin{array}{r} ^{1} \\ 4.67 \\ +2.65 \\ \hline 2 \end{array}$	$\begin{array}{r} ^{1\ 1} \\ 4.67 \\ +2.65 \\ \hline 32 \end{array}$	$\begin{array}{r} ^{1\ 1} \\ 4.67 \\ +2.65 \\ \hline 7.32 \end{array}$

Other examples:

3.13 + 2.48 = 5.61

$\begin{array}{r} ^{1\ 1} \\ 0.75 \\ +0.88 \\ \hline 1.63 \end{array}$

$\begin{array}{r} ^{1} \\ \$4.17 \\ +\ 1.69 \\ \hline \$5.86 \end{array}$

\$0.19 + \$0.45 = \$0.64

Add.

1. 0.8 + 0.1	**2.** 0.7 + 0.2	**3.** 0.34 + 0.24	**4.** \$0.23 + 0.62	**5.** \$0.04 + 0.53
6. 6.5 + 2.6	**7.** \$3.82 + 1.56	**8.** 6.2 + 5.3	**9.** \$5.89 + 4.01	**10.** 4.7 + 4.6
11. \$1.66 + 1.76	**12.** 4.5 + 8.7	**13.** \$5.31 + 8.94	**14.** 3.98 + 8.23	**15.** \$1.87 + 3.45
16. \$0.48 + 0.70	**17.** 9.6 + 2.7	**18.** 0.6 + 0.5	**19.** 4.1 + 3.8	**20.** 0.57 + 0.44

21. 0.2 + 0.6 = ___ **22.** \$0.73 + \$0.04 = ___ **23.** \$0.15 + \$0.63 = ___

24. \$4.54 + \$3.91 = ___ **25.** 6.4 + 2.8 = ___ **26.** 8.7 + 5.1 = ___

27. 8.3 + 6.9 = ___ **28.** \$7.50 + \$4.50 = ___ **29.** 7.9 + 5.7 = ___

30. \$0.98 + \$0.72 = ___ **31.** 0.7 + 0.4 = ___ **32.** \$8.40 + \$1.49 = ___

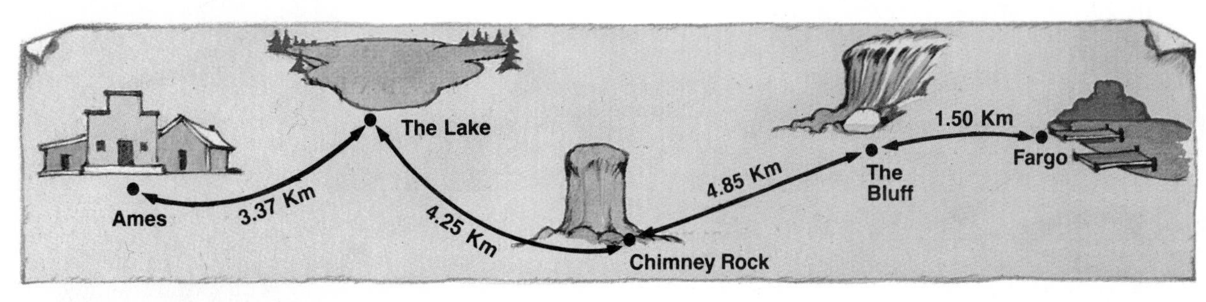

Use the map to answer each question.

33. Mr. Miller's class takes a hiking trip. They begin at Ames. When they get to Chimney Rock, how far have they hiked?

34. It takes the class two days to hike from Chimney Rock to Fargo. How many kilometers do they hike?

35. Which is closer to Chimney Rock, the Lake or the Bluff?

★36. Steve walks from the Lake to Fargo. How many kilometers does he walk?

More Practice, page 412

Subtracting Decimals

A. Jason jumped 1.2 meters on Monday of Sports Week. He jumped 1.5 meters on Friday. How much farther did he jump on Friday?

Subtract 1.5 − 1.2.

You can subtract decimals the same way you subtract whole numbers.

Line up the decimal points.

$$\begin{array}{r} 1.5 \\ -1.2 \\ \hline \end{array}$$

Subtract the tenths.

$$\begin{array}{r} 1.5 \\ -1.2 \\ \hline 3 \end{array}$$

Subtract the ones. Write the decimal point.

$$\begin{array}{r} 1.5 \\ -1.2 \\ \hline 0.3 \end{array}$$

Jason jumped 0.3 meters farther on Friday.

B. You can regroup decimals the same way you regroup whole numbers.

Subtract 7.48 − 4.92.

Line up the decimal points. Subtract the hundredths.

$$\begin{array}{r} 7.48 \\ -4.92 \\ \hline 6 \end{array}$$

Regroup. Subtract the tenths.

$$\begin{array}{r} \scriptstyle 6\ \ 14\ \ \ \\ \not{7}.\not{4}8 \\ -4.92 \\ \hline 56 \end{array}$$

Subtract the ones. Write the decimal point.

$$\begin{array}{r} \scriptstyle 6\ \ 14\ \ \ \\ \not{7}.\not{4}8 \\ -4.92 \\ \hline 2.56 \end{array}$$

Other examples:

$$\begin{array}{r} \scriptstyle 14\ \ \ \\ \scriptstyle 5\ \ \not{4}\ 15 \\ \$\not{6}.\not{5}\not{5} \\ -\ \ 1.98 \\ \hline \$4.57 \end{array}$$

1.36 − 0.53 = 0.83

$$\begin{array}{r} \scriptstyle 3\ \ 10\ \ \\ \$\not{4}.\not{0}0 \\ -\ \ 2.70 \\ \hline \$1.30 \end{array}$$

Subtract.

1. $0.7 - 0.5$
2. $0.8 - 0.4$
3. $0.29 - 0.17$
4. $\$0.84 - 0.24$
5. $0.46 - 0.31$
6. $7.65 - 3.47$
7. $\$6.49 - 1.87$
8. $5.2 - 2.9$
9. $\$8.33 - 7.90$
10. $4.5 - 2.8$
11. $8.83 - 6.96$
12. $3.45 - 2.56$
13. $9.23 - 3.64$
14. $7.67 - 5.88$
15. $\$5.41 - 4.92$
16. $5.48 - 2.62$
17. $4.6 - 3.9$
18. $\$0.23 - 0.14$
19. $7.34 - 2.86$
20. $8.8 - 6.7$
21. $\$0.89 - \$0.20 =$ ____
22. $0.5 - 0.3 =$ ____
23. $0.76 - 0.70 =$ ____
24. $8.49 - 3.55 =$ ____
25. $6.7 - 3.9 =$ ____
26. $\$9.35 - \$1.43 =$ ____
27. $\$5.27 - \$1.39 =$ ____
28. $\$4.31 - \$2.84 =$ ____
29. $7.36 - 6.57 =$ ____
30. $0.36 - 0.28 =$ ____
31. $5.63 - 5.43 =$ ____
32. $6.10 - 3.72 =$ ____

Use the chart to answer each question.

33. For fitness week, Jim's class has a rope-climbing contest. Willie climbs 8.5 meters. How much higher does he climb than Sean?

34. How much higher does Cheyenne climb than Lily?

35. What is the difference between Wendy's and Willie's climbs?

★36. Who wins the contest?

ROPE-CLIMBING CONTEST

Willie 8.5 meters
Sean 7.8 meters
Wendy 6.4 meters
Cheyenne 8.5 meters
Lily 8.1 meters
Jim 6.8 meters

The red dots mark how high each person climbs.

More Practice, page 412

CALCULATOR

The display on your calculator will not show fractions. But you can write a fraction by writing it as a division.

Write a decimal for $\frac{1}{10}$

$\frac{1}{10}$ is one tenth. $\frac{1}{10}$ also means $1 \div 10$.

Press: [1] [÷] [1] [0] [=]

The display should show 0.1. This is a decimal.
$\frac{1}{10} = 0.1$

Copy and complete the table.

	Fraction	Calculator Keys	Display
Example	$\frac{1}{20}$	[1] [÷] [2] [0] [=]	0.05
1.	$\frac{1}{5}$	[ ] [ ] [ ] [ ]	■
2.	$\frac{1}{8}$	[ ] [ ] [ ] [ ]	■
3.	$\frac{5}{20}$	[ ] [ ] [ ] [ ] [ ]	■

Your calculator can help you decide whether two fractions are equivalent. If two fractions can be written as the same decimal, they are equivalent.

Copy and complete the table.

	Fraction	Calculator Keys	Display	Equivalent?
4.	$\frac{3}{6}$ $\frac{1}{2}$	[ ] [ ] [ ] [ ] [ ] [ ] [ ] [ ]	■ ■	■
5.	$\frac{6}{10}$ $\frac{8}{20}$	[ ] [ ] [ ] [ ] [ ] [ ] [ ] [ ] [ ] [ ]	■ ■	■

GROUP PROJECT

Library Hours

The problem: Many people in your town enjoy using the library. How can library hours be arranged so that the greatest number of people can use the library?

Key Facts

- The town has only one librarian. The town cannot pay more than one librarian.
- The librarian can work only 40 hours a week.
- School ends at 3:00 on weekdays.
- Retired people and people who have very young children like to use the library during the day.
- Many people work and go to the library in the evening.
- Some people are busy during the week and use the library on the weekend.

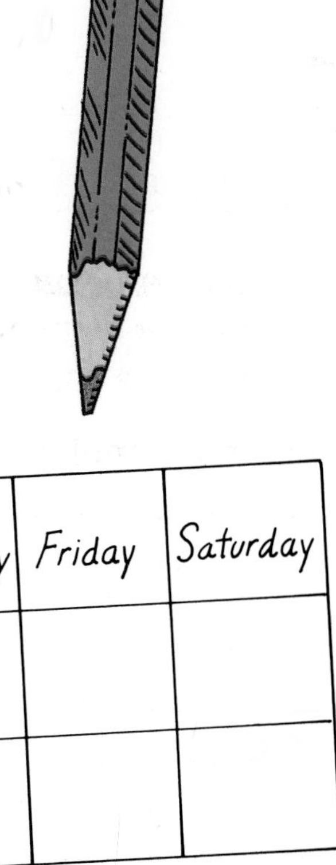

Making a Schedule

How can you solve the problem? How can you arrange the librarian's hours to give everyone a chance to use the library? Copy the chart. Make a schedule.

	Sunday	Monday	Tuesday	Wednesday	Thursday	Friday	Saturday
From							
To							

CHAPTER TEST

Write the fraction that names the shaded part. (pages 226 and 228)

1. 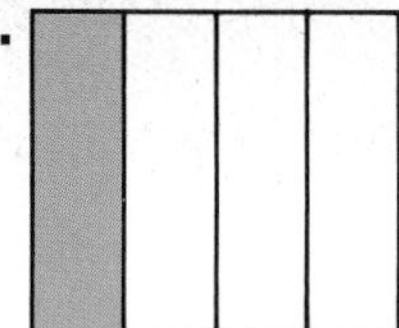

2. 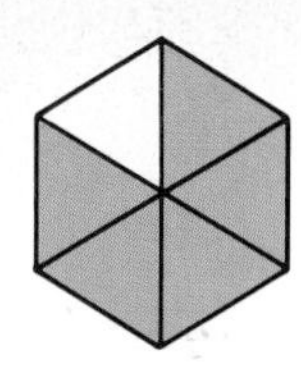

3. 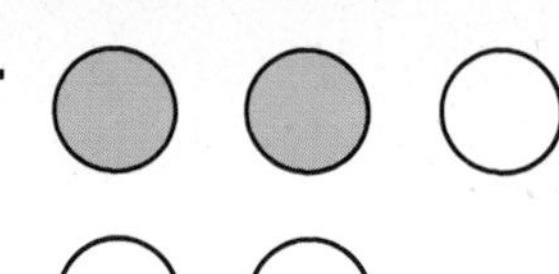

4. 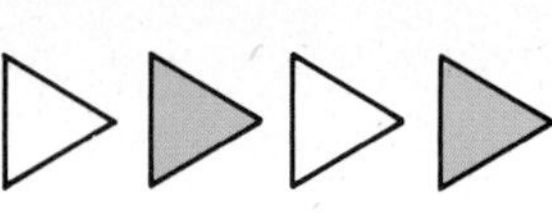
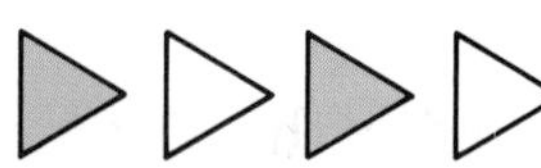

Solve. (page 230)

5. $\frac{1}{2}$ of 6 = ___ **6.** $\frac{1}{2}$ of 8 = ___ **7.** $\frac{1}{3}$ of 6 = ___ **8.** $\frac{1}{3}$ of 9 = ___

Write the equivalent fractions. (page 234)

9. 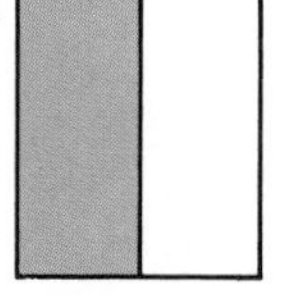

$\frac{\square}{\square} = \frac{\square}{\square}$

10. 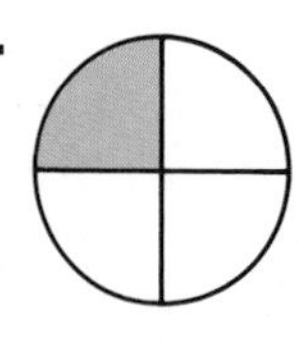

$\frac{\square}{\square} = \frac{\square}{\square}$

11. 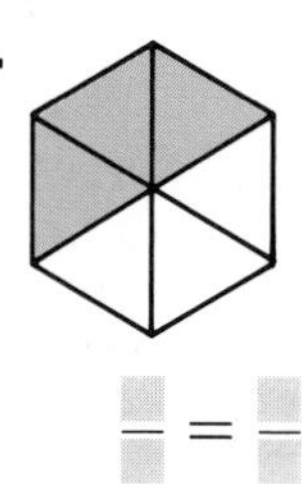 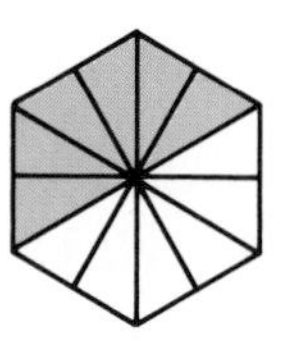

$\frac{\square}{\square} = \frac{\square}{\square}$

Write >, <, or = for ●. (page 236)

12. $\frac{1}{3}$ ● $\frac{2}{3}$ **13.** $\frac{5}{8}$ ● $\frac{3}{8}$ **14.** $\frac{8}{10}$ ● $\frac{7}{10}$

15. 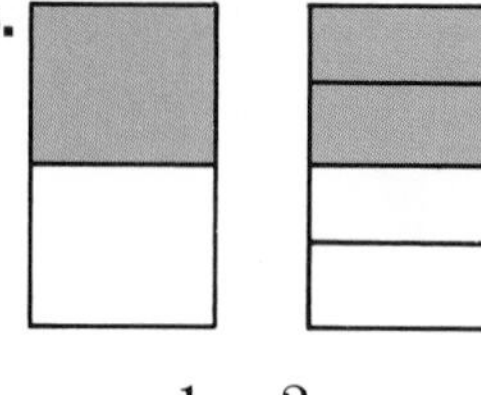

$\frac{1}{2}$ ● $\frac{3}{4}$

16. 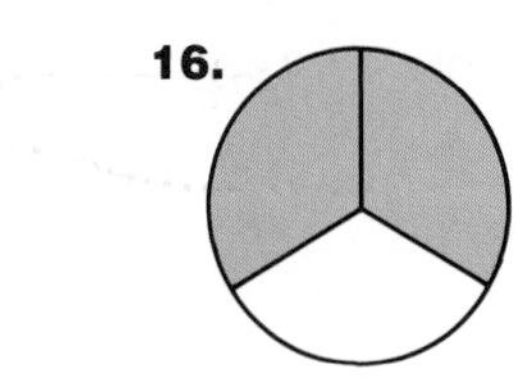 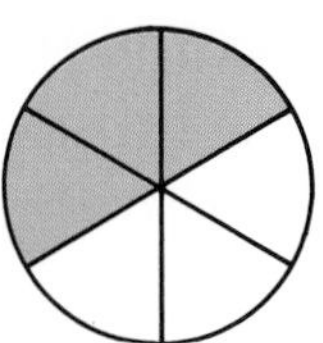

$\frac{2}{3}$ ● $\frac{3}{6}$

17. 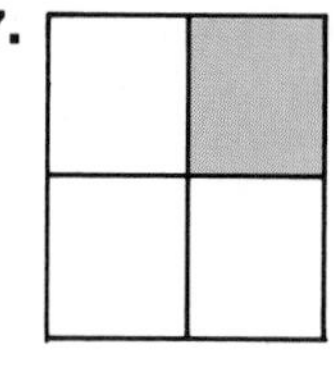 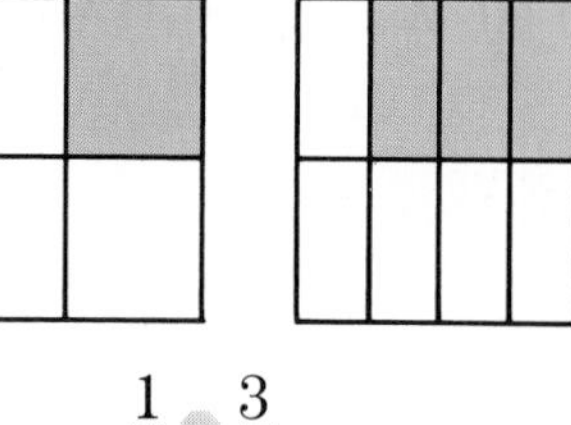

$\frac{1}{4}$ ● $\frac{3}{8}$

Write the whole or the mixed number to show how much is shaded. (page 238)

18.  

19. 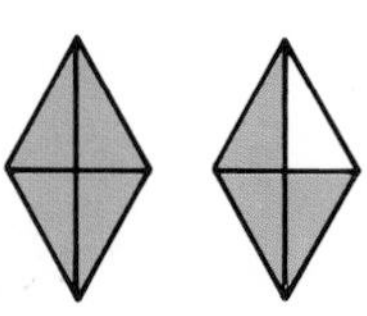

20. 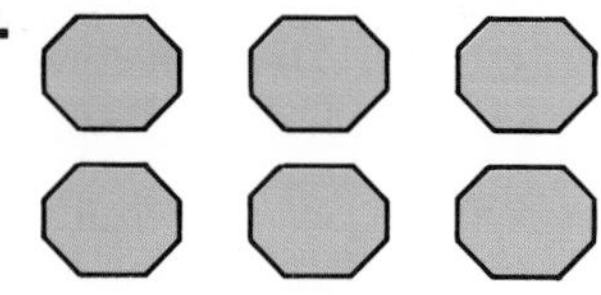

Write the decimal for each. (page 246)

21. three tenths

22. fifty-four hundredths

23. one and seven tenths

24. eleven dollars and sixteen cents

Order the decimals from the least to the greatest. (page 254)

25. 0.2, 0.7, 0.6

26. 7.4, 8.4, 7.7

27. 5.6, 6.5, 5.7

Add or subtract. (pages 256 and 258)

28. $\begin{array}{r} 6.8 \\ +3.2 \\ \hline \end{array}$

29. $\begin{array}{r} 0.6 \\ -0.4 \\ \hline \end{array}$

30. $\begin{array}{r} \$2.75 \\ +\ 3.79 \\ \hline \end{array}$

31. $\begin{array}{r} 0.57 \\ +0.97 \\ \hline \end{array}$

Write a number sentence and solve. (page 232)

32. Ty buys five packs of juice. Each pack has 6 cans. How many cans does Ty buy?

Write the letter of the most reasonable answer. (page 252)

33. Danny is 14 years old. His grandfather is 70 years old. His grandfather is about

a. twice as old as Danny.
b. half as old as Danny.
c. 5 times Danny's age.

BONUS

Suppose you have 12 eggs. You can place from 1 to 12 of them in the carton at a time. Think of all the fractions that can describe how much of the carton is full. There are 27 possible answers.

Example: There are 2 eggs in the carton. The carton is $\frac{2}{12}$ full. The carton is $\frac{1}{6}$ full.

RETEACHING

A. You can add and regroup decimals the same way you add and regroup whole numbers.

Add 4.88 + 5.73.

Line up the decimal points. Add the hundredths. Regroup.	Add the tenths. Regroup.	Add the ones. Regroup. Write the decimal point in the answer.
$\begin{array}{r} ^{1} \\ 4.88 \\ +5.73 \\ \hline 1 \end{array}$	$\begin{array}{r} ^{1\ 1} \\ 4.88 \\ +5.73 \\ \hline 61 \end{array}$	$\begin{array}{r} ^{1\ 1} \\ 4.88 \\ +5.73 \\ \hline 10.61 \end{array}$

B. You can subtract decimals the same way you subtract whole numbers.

Subtract 7.31 − 4.53.

Line up the decimal points. Regroup. Subtract the hundredths.	Regroup. Subtract the tenths.	Subtract the ones. Write the decimal point in the answer.
$\begin{array}{r} ^{2\ 11} \\ 7.\not{3}\not{1} \\ -4.53 \\ \hline 8 \end{array}$	$\begin{array}{r} ^{12} \\ ^{6\ \not{2}\ 11} \\ \not{7}.\not{3}\not{1} \\ -4.53 \\ \hline 78 \end{array}$	$\begin{array}{r} ^{12} \\ ^{6\ \not{2}\ 11} \\ \not{7}.\not{3}\not{1} \\ -4.53 \\ \hline 2.78 \end{array}$

Complete.

1. $\begin{array}{r} 3.17 \\ +1.99 \\ \hline \end{array}$ **2.** $\begin{array}{r} 6.54 \\ +2.47 \\ \hline \end{array}$ **3.** $\begin{array}{r} 4.04 \\ +0.98 \\ \hline \end{array}$ **4.** $\begin{array}{r} 3.21 \\ +6.09 \\ \hline \end{array}$ **5.** $\begin{array}{r} \$2.14 \\ +\ 2.99 \\ \hline \end{array}$

6. $\begin{array}{r} 4.96 \\ -1.01 \\ \hline \end{array}$ **7.** $\begin{array}{r} \$1.95 \\ -\ 1.07 \\ \hline \end{array}$ **8.** $\begin{array}{r} 8.09 \\ -3.92 \\ \hline \end{array}$ **9.** $\begin{array}{r} 4.40 \\ -3.99 \\ \hline \end{array}$ **10.** $\begin{array}{r} \$5.07 \\ -\ 2.68 \\ \hline \end{array}$

11. \$6.26 + \$2.29 = ▇ **12.** 5.06 + 3.48 = ▇ **13.** 3.35 − 1.73 = ▇

14. 0.32 + 0.83 = ▇ **15.** \$5.92 + \$1.89 = ▇ **16.** \$9.02 − \$7.35 = ▇

ENRICHMENT

Adding and Subtracting Fractions

To add or subtract fractions that have like denominators, add or subtract the numerators. Then write the sum as a fraction. The denominator remains the same.

A. Add $\frac{2}{5} + \frac{1}{5}$.

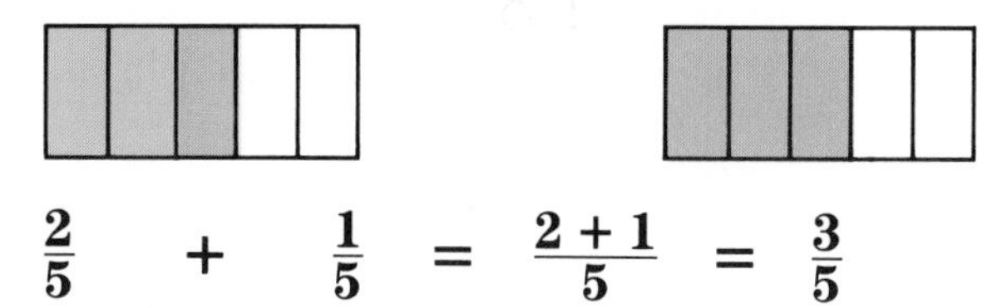

$$\frac{2}{5} + \frac{1}{5} = \frac{2+1}{5} = \frac{3}{5}$$

B. Subtract $\frac{3}{5} - \frac{1}{5}$.

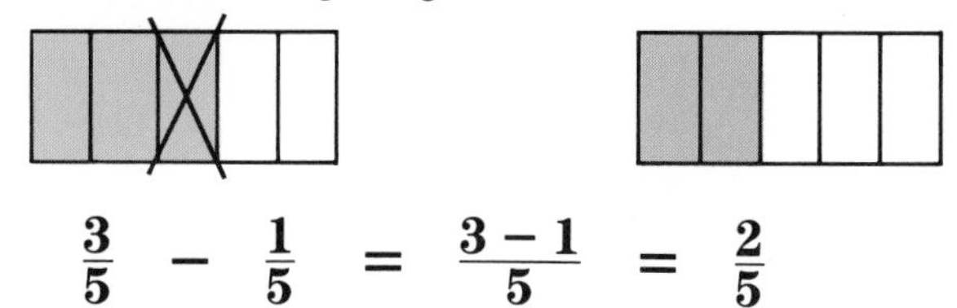

$$\frac{3}{5} - \frac{1}{5} = \frac{3-1}{5} = \frac{2}{5}$$

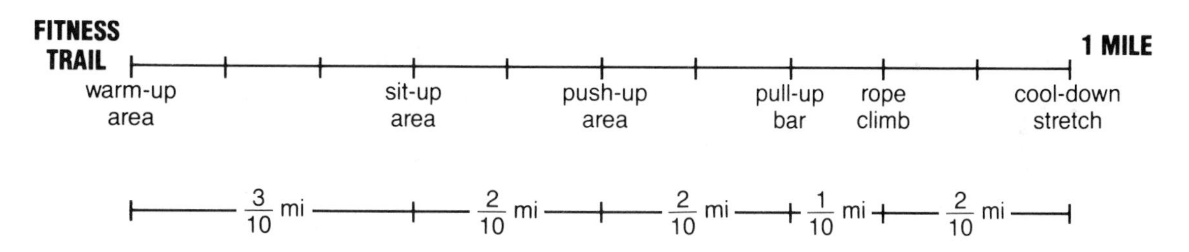

Use the diagram to solve.

1. Bob jogs from the push-up area to the pull-up bar. How far does he jog?

2. Jan rides her bike from the sit-up area to the pull-up bar. How far does she ride?

3. Joy reaches the pull-up bar. What part of the course has she completed? How far is it to the end of the course?

4. Joe reaches the rope climb. He realizes that he left his glasses at the sit-up area. How much of the course has he completed? How far back must he travel?

CUMULATIVE REVIEW

Write the letter of the correct answer.

1. What digit is in the thousands place in 543,192?

   a. 3 b. 4
   c. 5 d. not given

2. $602 - 412 = \blacksquare$

   a. 190 b. 210
   c. 1,014 d. not given

3. $4 \times 2 \times 3 = \blacksquare$

   a. 8 b. 14
   c. 28 d. not given

4. $\$12.93 + 1.88$

   a. $11.05
   b. $11.15
   c. $13.81
   d. not given

5. Compare. $7.01 ● $6.99

   a. < b. >
   c. = d. not given

6. Which sentence completes the fact family?

   $20 \div 4 = 5$
   $20 \div 5 = 4$
   $4 \times 5 = 20$

   a. $4 + 5 = 9$
   b. $5 \times 4 = 20$
   c. $20 - 5 = 15$
   d. not given

7. $30 \div 5 = \blacksquare$

   a. 6 b. 7
   c. 25 d. not given

8. $36 \div 6 = \blacksquare$

   a. 6 b. 9
   c. 30 d. not given

9. $8\overline{)72}$

   a. 7
   b. 8
   c. 9
   d. not given

10. Ricky has 232 chestnuts. He gets 848 more chestnuts. Estimate the total number of chestnuts Ricky has now.

    a. 900 chestnuts
    b. 1,000 chestnuts
    c. 10,000 chestnuts
    d. 848, 232 chestnuts

11. Erica had 18 school photos. She divided them equally among her 3 friends. How many photos did each friend receive?

    a. 3 photos b. 6 photos
    c. 54 photos d. not given

9 MEASUREMENT

Your school is having a field day. Your job is to plan an obstacle course for a race. Decide which obstacles to use. Make a drawing that someone could use to set up your course for you.

Measuring in Centimeters

A. The **centimeter (cm)** is a metric unit. You can use centimeters to measure short lengths.

How long is the key?

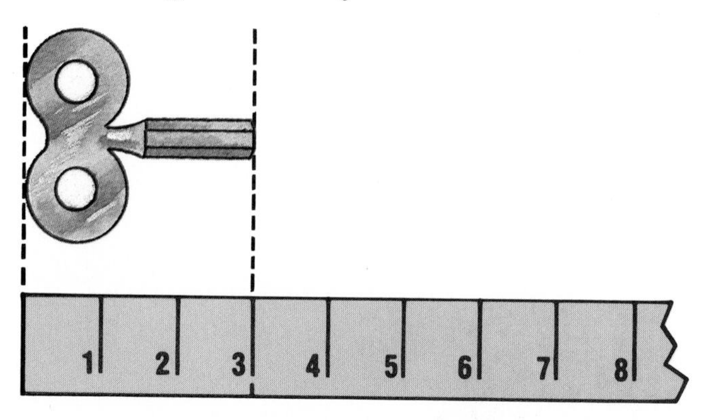

The key is 3 centimeters long.

B. You can measure to the nearest centimeter.

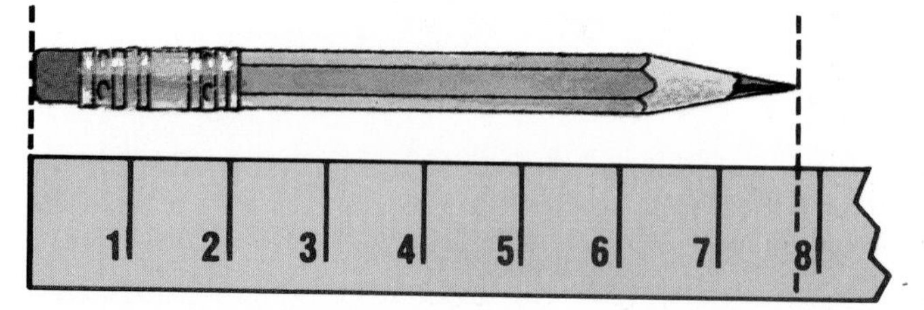

This pencil is between 7 cm and 8 cm long, but it is closer to 8 cm. The pencil is 8 cm long to the nearest centimeter.

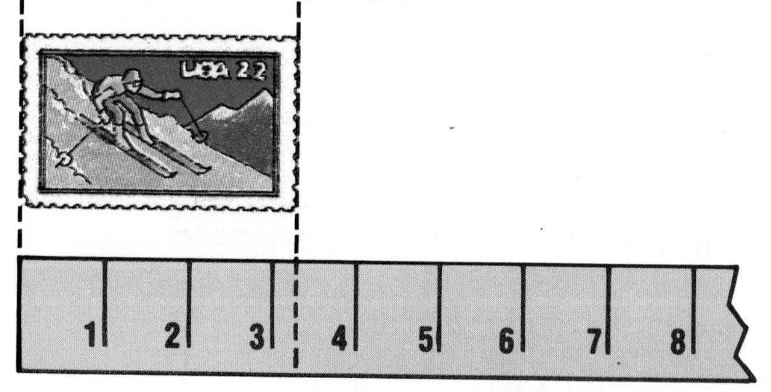

This stamp is between 3 cm and 4 cm long, but it is closer to 3 cm. The stamp is 3 cm long to the nearest centimeter.

Use your metric ruler to measure each object to the nearest centimeter.

1. 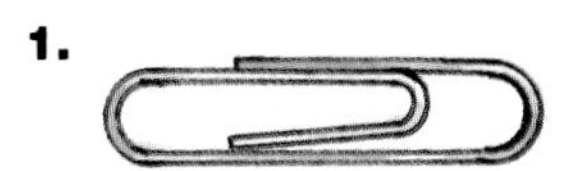

2. 

3.

4. 

5.

6. 

Use your ruler to draw each length.

7. 3 cm **8.** 15 cm **9.** 1 cm **10.** 5 cm **11.** 2 cm

12. 9 cm **13.** 4 cm **14.** 6 cm **★15.** 5.5 cm **★16.** 3.5 cm

FOCUS: ESTIMATION

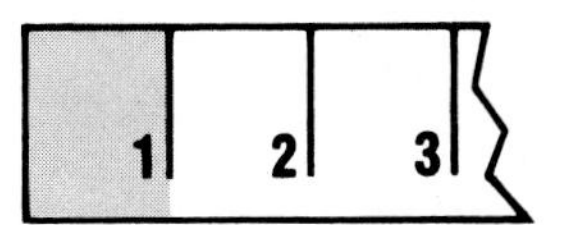

1. Name two things that are about 1 cm long.
2. Find two pictures in this book that are about 10 cm long.
3. Name two things in your classroom that are about 20 cm long.
4. Measure to check your answers.

Metric Units of Length

A. The **meter (m)** and the **kilometer (km)** are other metric units of length.

You use meters to measure length.

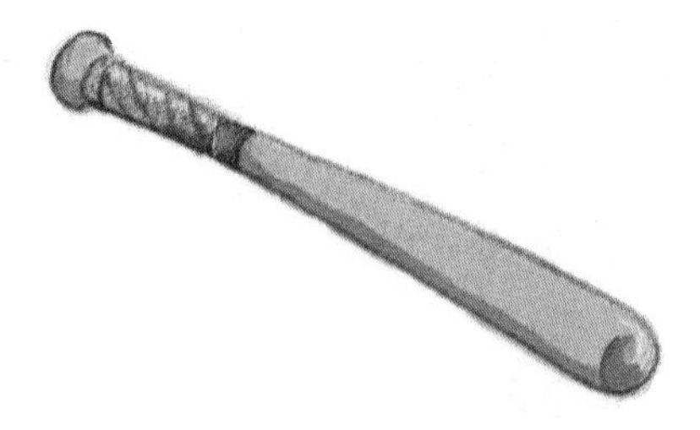

A baseball bat is about 1 meter long.

Other examples:

A door is about 1 meter wide.

You use kilometers to measure long distances.

A map can show distances in kilometers.

The length of 12 city blocks is about 1 kilometer.

B. You can rename measurements by using a pattern.

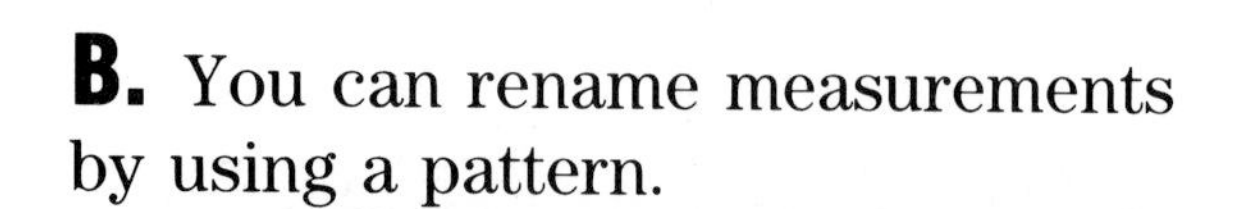

1 meter (m) = 100 centimeters (cm)
1 kilometer (km) = 1,000 meters (m)

You can rename a larger unit with a smaller unit.

m	1	2	3	4	5
cm	100	200	300	400	500

4 m = ▒ cm
4 m = 400 cm

1 m = 100 cm
So, 4 m = 400 cm.

You can rename a smaller unit with a larger unit.

m	1,000	2,000	3,000	4,000	5,000
km	1	2	3	4	5

3,000 m = ▒ km
3,000 m = 3 km

1,000 m = 1 km
So, 3,000 m = 3 km.

Choose the unit that you would use to measure each. Write *cm*, *m*, or *km*.

1. the length of a bowling lane ____
2. the distance a school bus travels ____
3. the length of your footprint ____
4. the edge of a book ____
5. the length of a soccer field ____

Choose the better estimate. Write the letter of the correct answer.

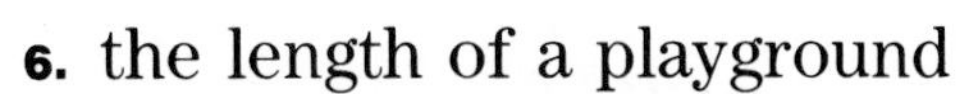

6. the length of a playground
   a. 200 cm b. 200 m
7. the length of a sled
   a. 1 m b. 1 km
8. the distance of a canoe trip
   a. 5 m b. 5 km
9. the length of a basketball court
   a. 40 m b. 40 km
10. the width of a record cover
    a. 30 cm b. 30 km
11. the height of a basketball hoop
    a. 3 m b. 3 km

Copy and complete.

12. 3 km = ____ m
13. 3 m = ____ cm
14. 5 km = ____ m
15. 1,000 m = ____ km
16. 200 cm = ____ m
17. 5 m = ____ cm
18. 6 km = ____ m
19. 4 m = ____ cm
20. 4,000 m = ____ km

FOCUS: ESTIMATION

Estimate the length of each.
Write *more than 1 m* or *less than 1 m*.

1. a classroom
2. a dollhouse room
3. a textbook
4. a library shelf
5. a pencil
6. a car

Perimeter

Perimeter is the distance around a figure.

To find the perimeter of this figure, measure each side. Then add the measures.

$3 + 5 + 3 + 5 = 16$

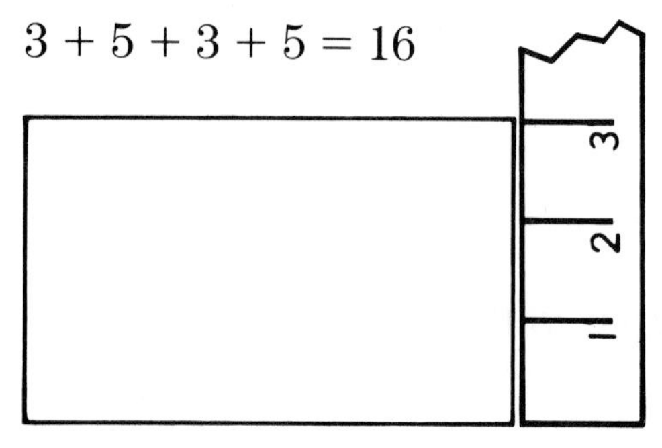

The perimeter is 16 cm.

Other examples:

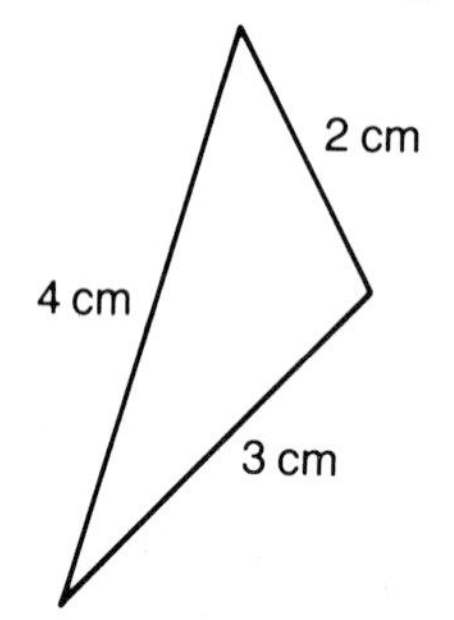

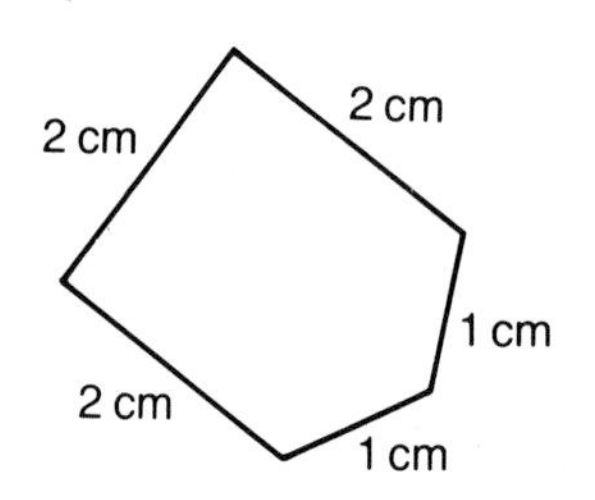

$4 + 2 + 3 = 9$
The perimeter is 9 cm.

$2 + 2 + 2 + 1 + 1 = 8$
The perimeter is 8 cm.

Use your metric ruler to measure each side.
Then find the perimeter.

1.

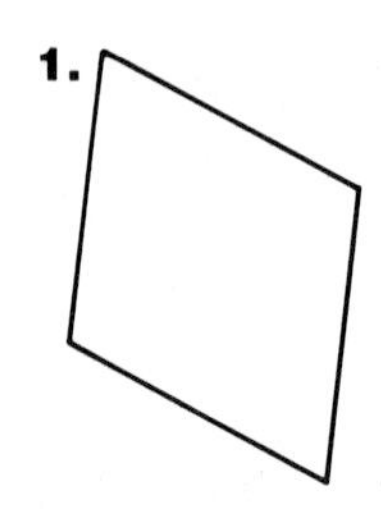

2.

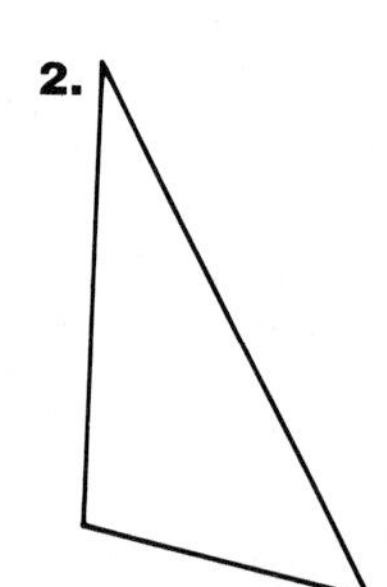

3.

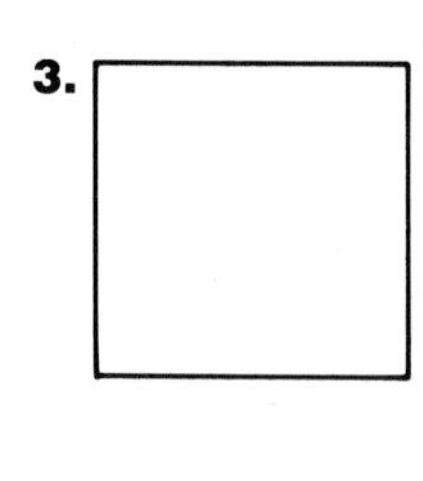

Use your metric ruler to measure each side.
Then find the perimeter.

4. 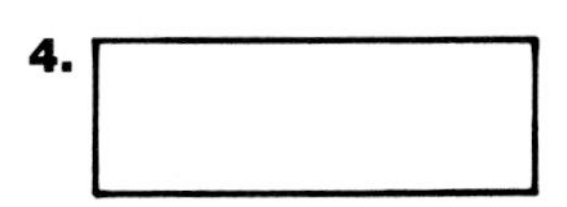

5. 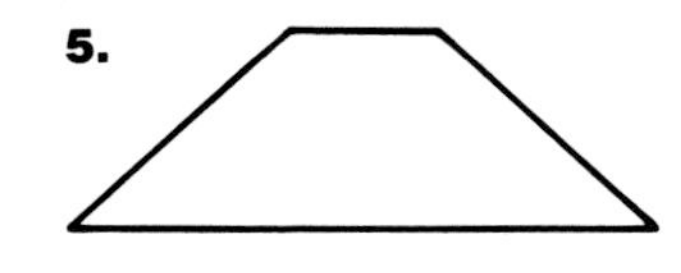

6. 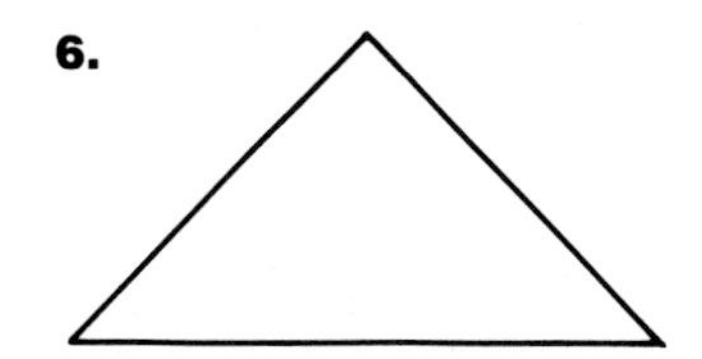

7. 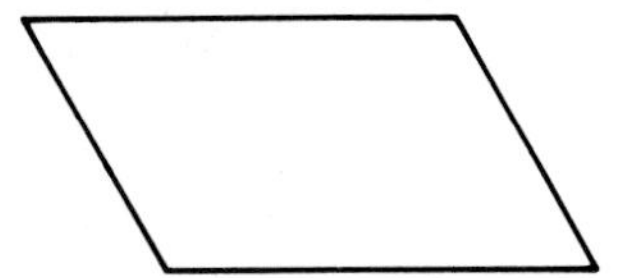

8.

9. 

10. How many meters is it around the soccer field?

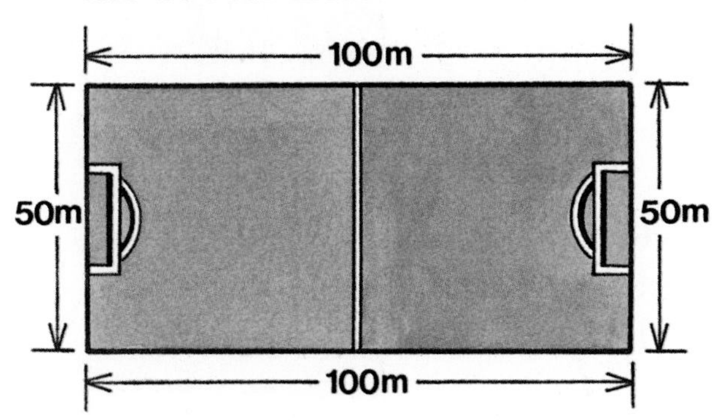

11. What is the perimeter of the sign?

Solve.

12. Eileen is making flags for field day. She trims each flag with ribbon. The sides of each flag are 26 cm, 13 cm, 26 cm, and 13 cm. How much ribbon does she use for each flag?

★13. Jerry runs around the basketball court one time. The court is 16 m by 30 m. How many meters does Jerry run?

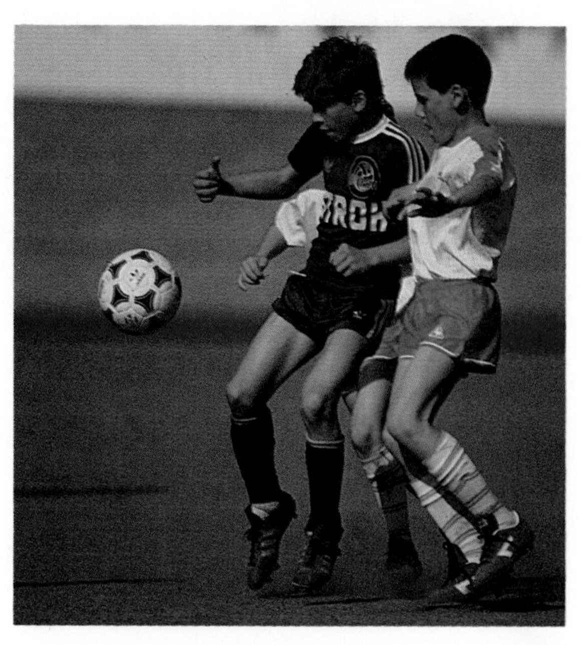

PROBLEM SOLVING

Help File Review

Remember to use the Help File on page 397 if you are having trouble solving a problem. Each part of the file answers a different question you might have.

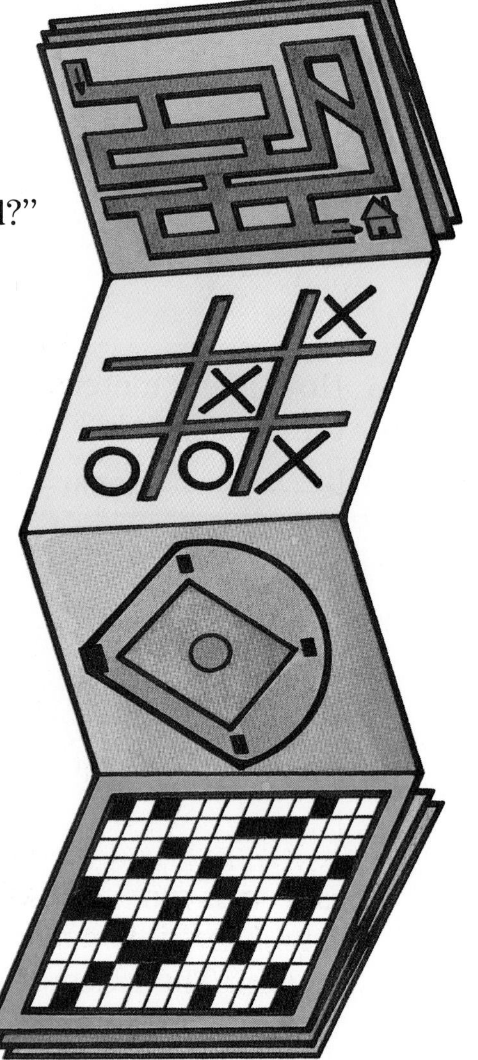

Questions: "What is the problem asking me to find?"

Tools: "What must I do to solve this problem?"

Solutions: "How can I do the math needed in this problem?"

Checks: "How can I check to see whether my answer is correct?"

Read the problem. Help each student below by deciding where in the Help File each can find ideas for solving the problem. Write the letter of the best answer.

> A book on games has instructions for how to play 372 games. For 223 of the games, teams of players are needed. The other games can be played by one or two people. How many games can be played by one or two people?

1. Kati knows she needs to subtract. But she can't remember how to regroup. Kati should look under:

   a. Questions **b.** Tools
   c. Solutions **d.** Checks

2. Ty understands the question. But he doesn't know whether to add or to subtract. Ty should look under:

   a. Questions **b.** Tools
   c. Solutions **d.** Checks

Solve each problem below. If you have trouble, go to the Help File.

3. Lucy, Lynn, Gary, and Herb want to play Dragon's Tail. They need 6 balloons for each player. How many balloons do they need?

4. Each player takes 6 balloons and ties them to a string 10 feet long. If the players have 50 feet of string, will 4 people be able to play?

5. The strings of balloons are tied to the players' waists. Each player tries to break the other players' balloons. Gary breaks 4 of Herb's balloons. Herb breaks 2 of Gary's balloons. How many balloons do Gary and Herb each have now?

6. Lynn breaks 2 of Lucy's balloons, 1 of Gary's balloons, and 1 of Herb's balloons. How many balloons does Lynn break?

7. In the next game, Lucy breaks 3 of Herb's balloons. Gary breaks 2 of Herb's balloons. Lynn breaks 1 of Herb's balloons. Any player who has no balloons left is out of the game. Is Herb out of the game?

★8. The children play Dragon's Tail 3 times. If all the balloons were broken during each game, how many balloons did the children use? How many feet of string did they use?

Milliliters and Liters

A. The **milliliter (mL)** and the **liter (L)** are metric units that are used to measure liquid.

You use milliliters to measure a few drops of a liquid.

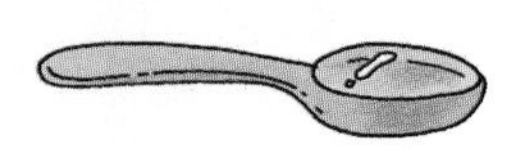

A teaspoon holds about 5 milliliters.

You use liters to measure larger amounts of a liquid.

Juice comes in 1-liter bottles.

B. You can rename liters with milliliters.

1 liter (L) = 1,000 milliliters (mL) 2 liters (L) = 2,000 milliliters (mL) 5 liters (L) = 5,000 milliliters (mL)

Choose the unit you would use to measure each. Write the letter of the correct answer.

1.

a. mL **b.** L

2.

a. mL **b.** L

3.

a. mL **b.** L

4.

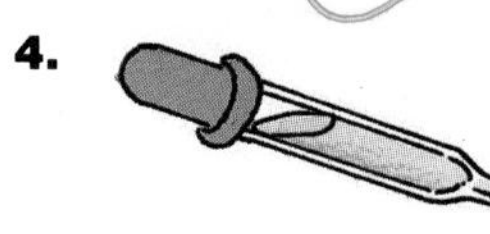

a. mL **b.** L

5.

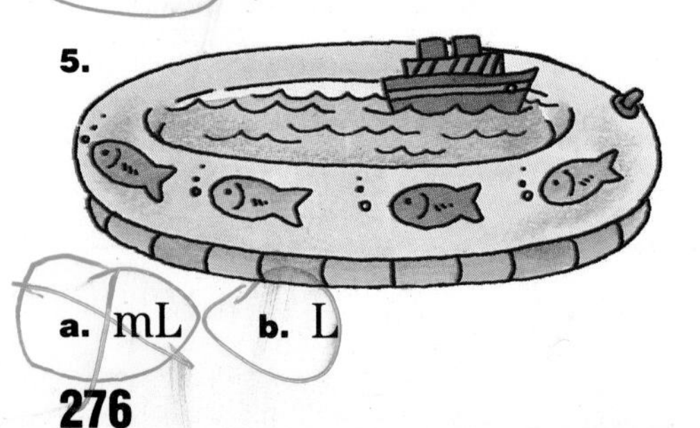

a. mL **b.** L

6.

a. mL **b.** L

Choose the better estimate. Write the letter of the correct answer.

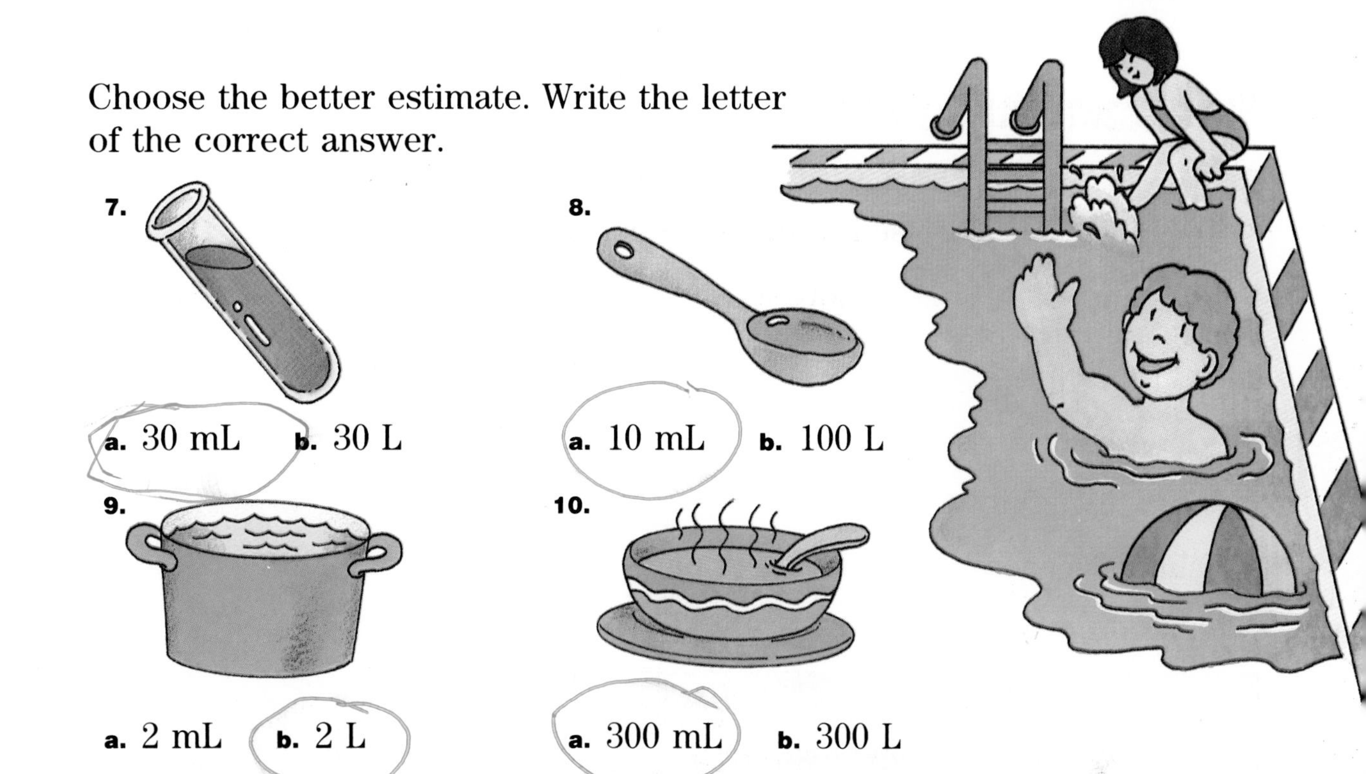

7. **a.** 30 mL **b.** 30 L

8. **a.** 10 mL **b.** 100 L

9. **a.** 2 mL **b.** 2 L

10. **a.** 300 mL **b.** 300 L

Find the pattern. Copy and complete the chart to rename liters with milliliters.

11.

milliliters	1,000	2,000	■	■	■
liters	1	2	3	4	5

Copy and complete.

12. 2,000 mL = ____ L **13.** 5,000 mL = ____ L **14.** 4,000 mL = ____ L

15. 5 L = ____ mL **16.** 3 L = ____ mL **17.** 1 L = ____ mL

★**18.** 10,000 mL = ____ L ★**19.** 12 L = ____ mL ★**20.** 35,000 mL = ____ L

FOCUS: REASONING

You know that in the metric system the prefix *centi-* means $\frac{1}{100}$. You need 100 centimeters to equal 1 meter. The prefix *milli-* means $\frac{1}{1,000}$. You need 1,000 milliliters to equal 1 liter. How many millimeters would equal 1 meter?

Grams and Kilograms

A. The **gram (g)** and the **kilogram (kg)** are metric units you can use to measure how heavy something is.

You use grams to measure light objects.

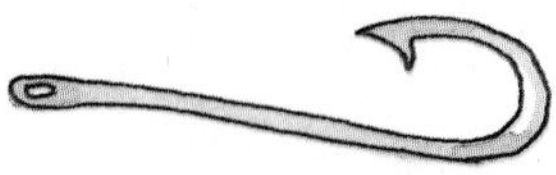
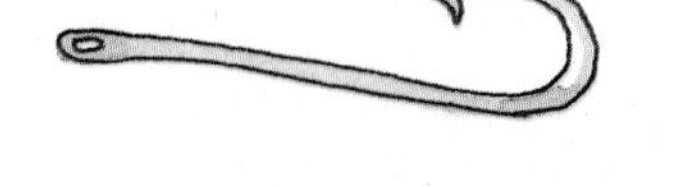

A fishhook is about 1 gram.

You can use kilograms to measure heavier objects.

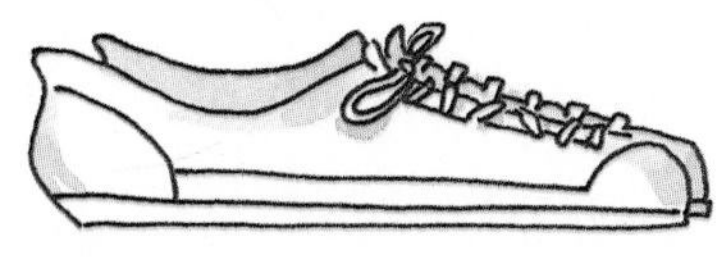

A pair of tennis shoes is about 1 kilogram.

B. You can rename kilograms with grams.

1 kilogram (kg) = 1,000 grams (g)

Choose the unit you would use to measure each of these. Write *g* or *kg*.

1. 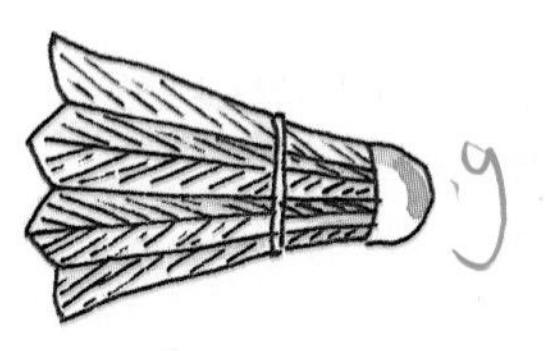

2. 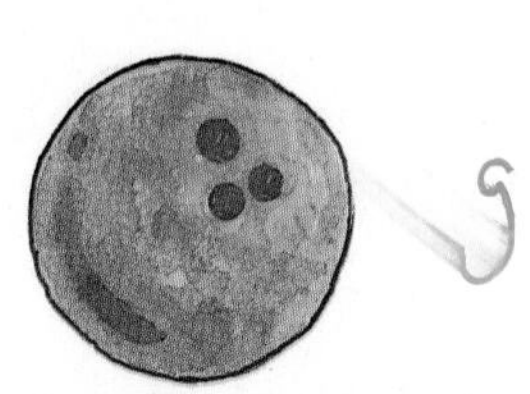

3. 

How heavy is each one? Choose the better estimate. Write the letter of the correct answer.

4. 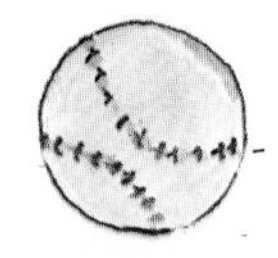

a. 300 g **b.** 300 kg

5. **a.** 250 g **b.** 250 kg

6. 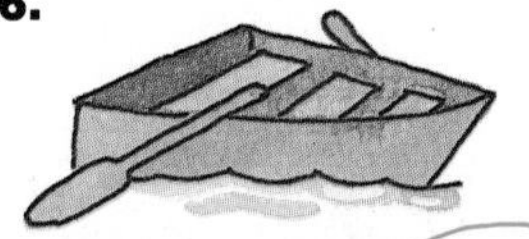

a. 75 g **b.** 75 kg

Find the pattern. Copy and complete the chart to rename kilograms with grams.

7.

grams	1,000	2,000	___	___	___
kilograms	1	2	3	4	5

More Practice, page 413

Celsius

This is a **Celsius thermometer.**
It is used to measure temperature.

Each line shows two degrees Celsius.

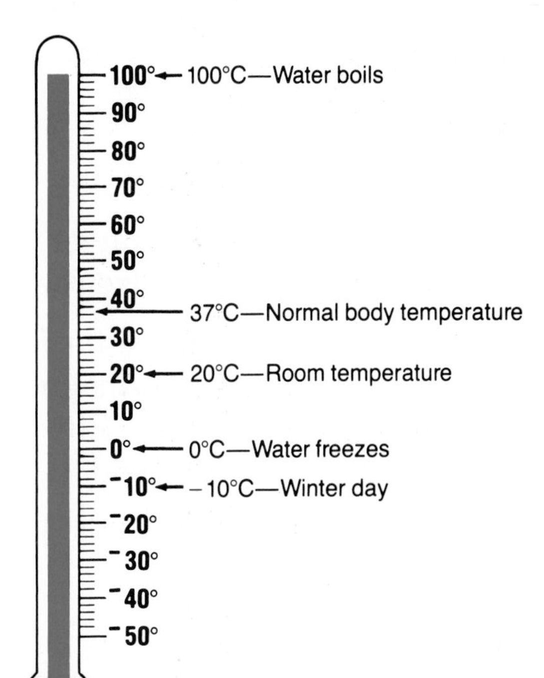

Read: one hundred degrees Celsius.
Write: 100°C.

Choose the most likely temperature for each.
Write the letter of the correct answer.

1. shoveling snow
a. −5°C
b. 5°C

2. playing baseball
a. 23°C
b. 73°C

3. swimming
a. −15°C
b. 25°C

4. having a picnic
a. 18°C
b. 6°C

5. wearing a coat
a. 40°C
b. 4°C

6. wearing shorts
a. 35°C
b. 8°C

MIDCHAPTER REVIEW

Choose the better estimate.
Write the letter of the correct answer.

1. a dog
a. 5 g **b.** 5 kg

2. a jug of cider
a. 2 mL **b.** 2 L

3. a shoelace
a. 45 cm **b.** 45 m

4. a belt
a. 1 cm **b.** 1 m

5. a teardrop
a. 2 mL **b.** 2 L

6. a penny
a. 2 g **b.** 2 kg

PROBLEM SOLVING
Practice

Write the letter of the number sentence that you could use to solve the problem.

1. The city pool has room for 8 swimmers to race. At one meet, 48 swimmers will race. How many races must be held?

a. $48 - 8 =$ 40
b. $48 \div 8 =$ 6

2. When Bob swims, he takes a deep breath every 4 strokes. He took 8 deep breaths per lap. How many strokes did he take to swim the lap?

a. $8 + 4 =$ 12
b. $4 \times 8 =$ 32

Estimate.

3. Hale School sent 26 swimmers to a meet. Webb School sent 42 swimmers. Truman School sent 89 swimmers. About how many swimmers did all three schools send?

4. Deanna swam 1,352 m during practice one day, 1,287 m the next day, and 1,561 m the next day. About how many meters did she swim during the 3 days of practice?

Without finding the exact answer, write the letter of the most reasonable answer.

5. Swimmer Mark Spitz set 26 world speed records. Kornelia Ender set 23 records. What is their combined number of records?

a. 3 records **b.** 30 records
c. 50 records

6. Walter Poenish swam 205 km at one time. Ricardo Hoffman swam 478 km. About how much farther did Hoffman swim?

a. 20 km **b.** 200 km
c. 2,000 km

Which sentence answers the question? Write the letter of the correct answer.

7. A hockey rink's perimeter is 182 m. Look at the swimming pool in the picture. Is the pool's perimeter smaller?

   a. yes b. no c. 40 m

8. A swimmer swam 150 m of a 200-m race. How many laps were there left to swim?

   a. 50 m b. 1 lap c. yes

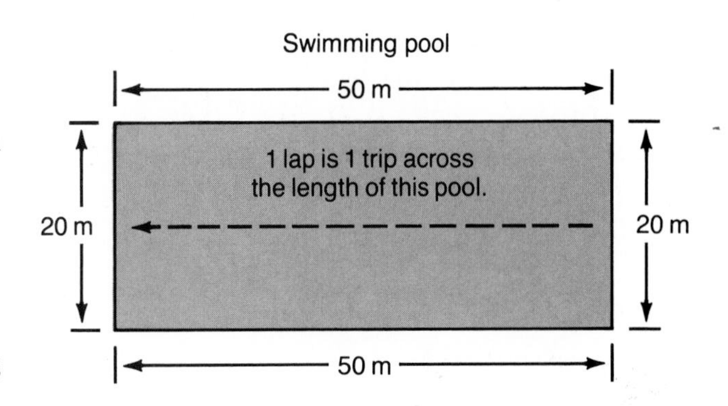

Solve.

9. A roller skate has 4 wheels. How many new wheels will Ivan need for 2 *pairs* of skates?

10. One shop sells 42 pairs of figure skates, 51 pairs of racing skates, and 67 pairs of hockey skates. How many pairs of skates did the shop sell?

11. One rink in Jo's county is 121 km away from her home. Another rink is 139 km away. How much farther must Jo travel to reach the second rink?

12. Last month, Fay skated in a 500-m race, a 3,000-m race, a 1,500-m race, and a 1,000-m race. How many *kilometers* did she skate?

13. In a relay race, each team member skated twice around the rink. The race took 12 trips around the rink. How many skaters were there on a team?

14. Roller skates were invented in 1760. They were first used in the United States 103 years later. How long did American skaters have to wait to skate across the Brooklyn Bridge, which opened in 1883?

Measuring in Inches

A. The **inch (in.)** is a customary unit of length.

1 inch

You can use inches to measure short lengths.

How long is the leaf?

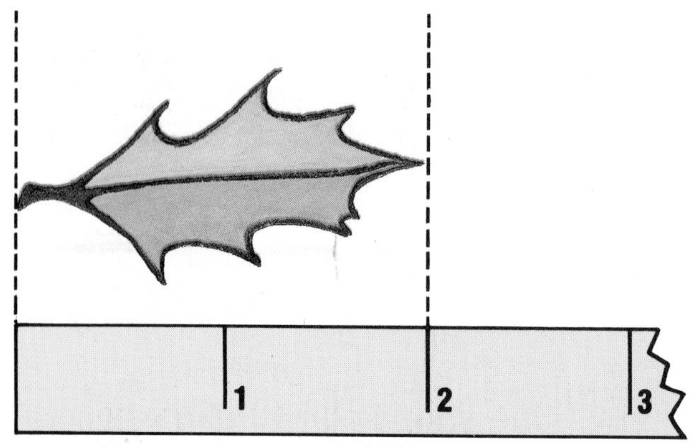

The leaf is 2 inches long.

B. You can measure to the nearest inch.

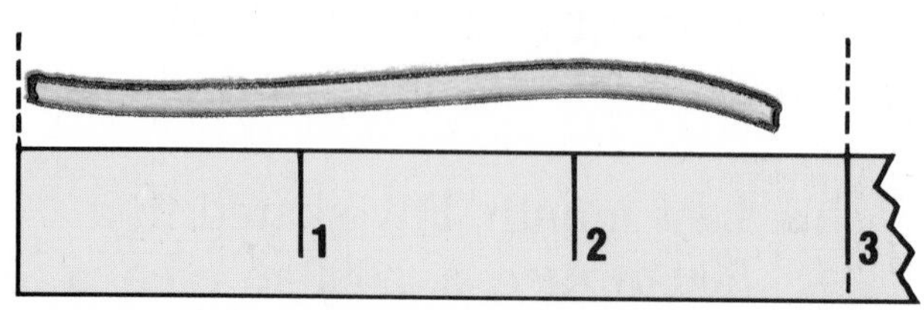

The string is between 2 inches and 3 inches in length, but it is closer to 3 inches. The string is about 3 inches long to the nearest inch.

C. You can also measure half inches and quarter inches.

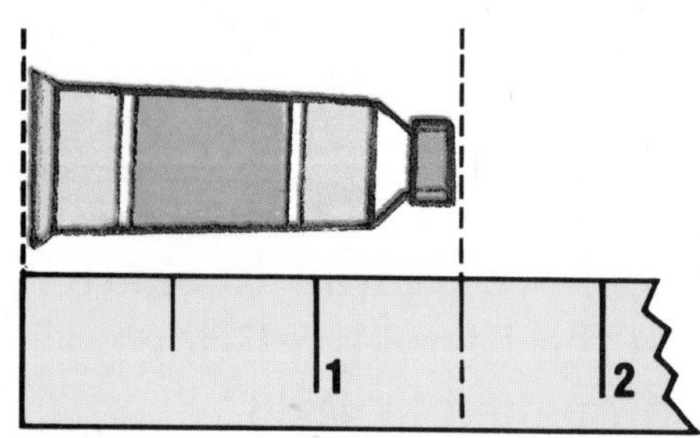

The tube of paint is $1\frac{1}{2}$ inches long.

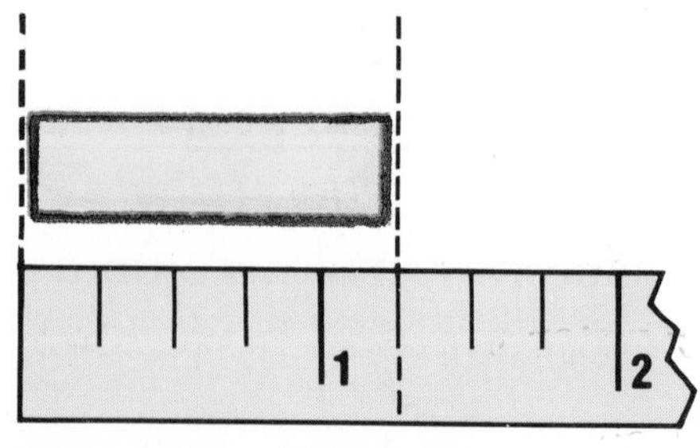

The piece of chalk is $1\frac{1}{4}$ inches long.

Use your customary ruler to measure each length to the nearest inch.

1. 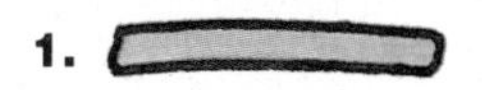

2. 

3. 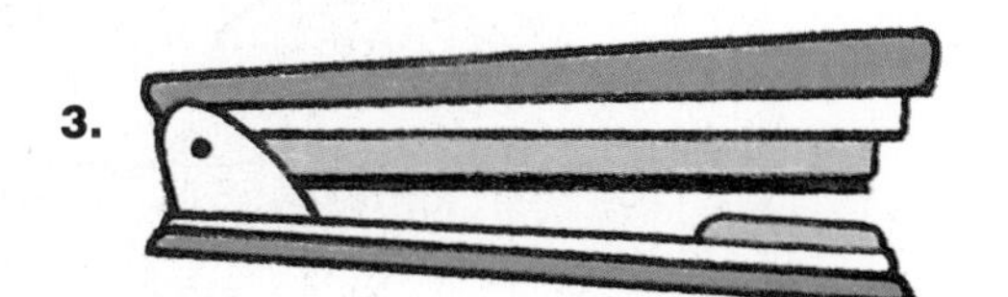

4.

Use your ruler to measure each length to the nearest $\frac{1}{2}$ inch or $\frac{1}{4}$ inch.

5. 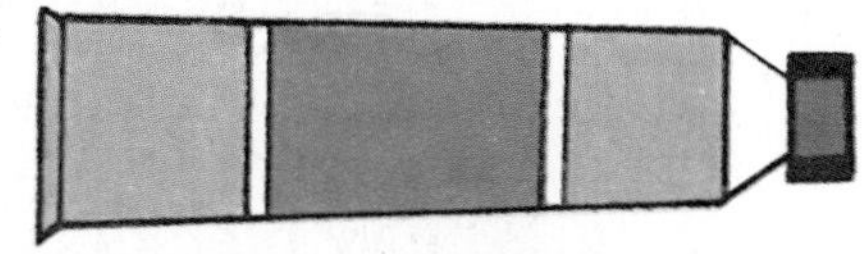

6. 

7.

8.

Use your ruler to help you choose the best measurement. Write the letter of the correct answer.

9. 

a. 2 in.

b. $2\frac{1}{4}$ in.

c. $2\frac{1}{2}$ in.

10. 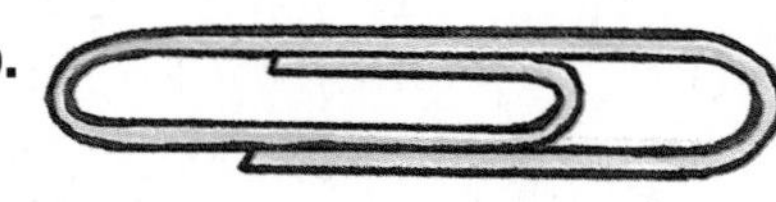

a. $1\frac{1}{2}$ in.

b. 2 in.

c. $2\frac{1}{2}$ in.

Use your ruler to draw each length.

11. 3 inches

12. 1 inch

13. 2 inches

14. $5\frac{1}{2}$ inches

15. $1\frac{1}{2}$ inches

16. $3\frac{1}{4}$ inches

17. $2\frac{3}{4}$ inches

18. $4\frac{1}{2}$ inches

Customary Units of Length

A. The **foot (ft),** the **yard (yd),** and the **mile (mi)** are other customary units that are used to measure length.

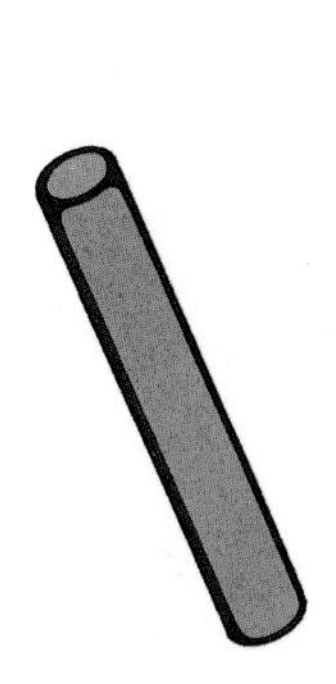

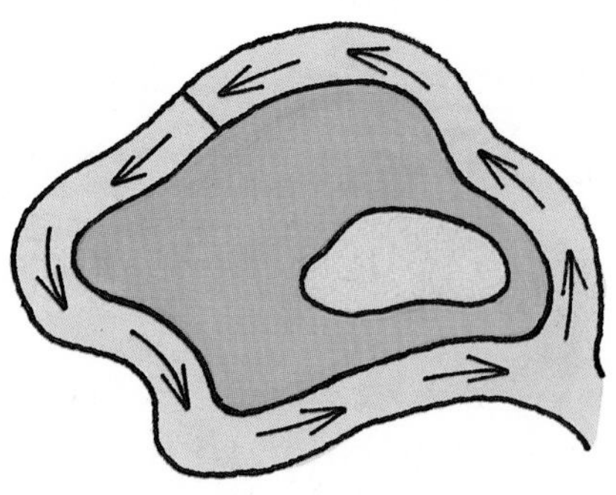

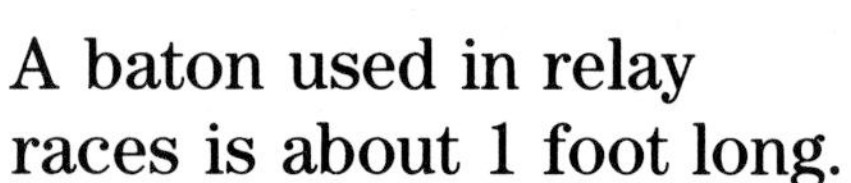

A baton used in relay races is about 1 foot long.

A baseball bat is about 1 yard long.

A map can show distance in miles.

Other examples:

A football is about 1 foot long.

A bench in a rowboat is about 1 yard long.

20 city blocks is about 1 mile.

B. You can rename measurements by using a pattern.

12 inches (in.) = 1 foot (ft)
36 inches (in.) = 1 yard (yd)
3 feet (ft) = 1 yard (yd)
5,280 feet (ft) = 1 mile (mi)
1,760 yards (yd) = 1 mile (mi)

You can rename a larger unit with a smaller unit.

ft	1	2	3	4	5
in.	12	24	36	48	60

2 ft = ___ in.
2 ft = 24 in.

1 ft = 12 in.
So, 2 ft = 24 in.

You can rename a smaller unit with a larger unit.

ft	3	6	9	12	15
yd	1	2	3	4	5

15 ft = ___ yd
15 ft = 5 yd

3 ft = 1 yd
So, 15 ft = 5 yd

Choose the unit that you would use to measure each length. Write *in.*, *ft*, *yd*, or *mi*.

1. the length of a sneaker ▀

2. the length of a hockey stick ▀

3. the length of a short race ▀

4. the length of a table-tennis paddle ▀

5. the distance across the United States ▀

Choose the better estimate. Write the letter of the correct answer.

6. the length of a football field
   a. 100 yd b. 100 mi

7. the length of a roller skate
   a. 9 in. b. 9 ft

8. the height of a bicycle
   a. 2 ft b. 2 yd

9. the length of a golf club
   a. 3 in. b. 3 ft

10. the length of a swimming pool
   a. 16 ft b. 16 yd

11. the length of a diving board
   a. 10 ft b. 10 mi

Copy and complete.

12. 3 ft = ▀ yd

13. 36 in. = ▀ ft

14. 2 yd = ▀ ft

15. 4 ft = ▀ in.

16. 18 ft = ▀ yd

17. 1 mi = ▀ ft

18. 1 mi = ▀ yd

19. 60 in. = ▀ ft

20. 15 ft = ▀ yd

Solve. Use the Infobank on page 403.

21. Jerome ran from the start of the track to the tower. How many yards did he run?

22. Irene ran from the forest to the pond. How many yards did she run?

23. Warren ran from the bench, past the tower, to the start of the track. How many yards did he run?

★24. Mary ran from the start of the track to the bench. José ran from the tower to the start of the forest and back. Who ran farther?

More Practice, page 413

Perimeter

You can use customary units to measure perimeter.

To find the perimeter, measure each side. Then add the measures.

$1 + 2 + 2 + 1 = 6$

The perimeter is 6 in.

Other examples:

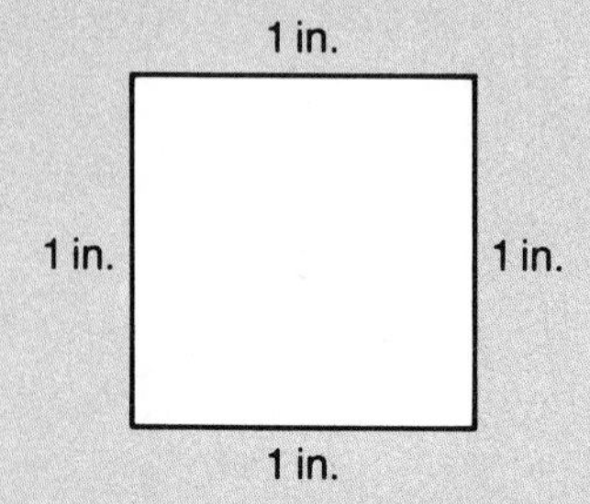

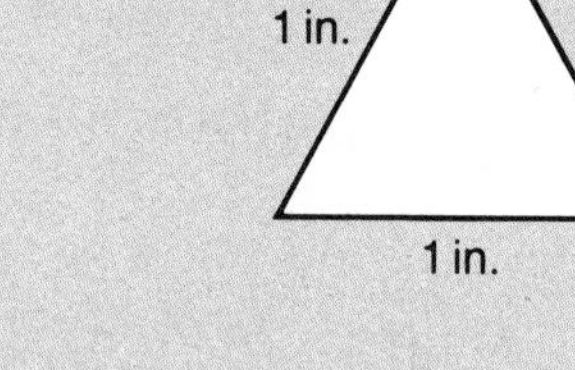

$1 + 1 + 1 + 1 = 4$
The perimeter is 4 in.

$1 + 1 + 1 = 3$
The perimeter is 3 in.

Use your customary ruler to measure each side. Then find the perimeter.

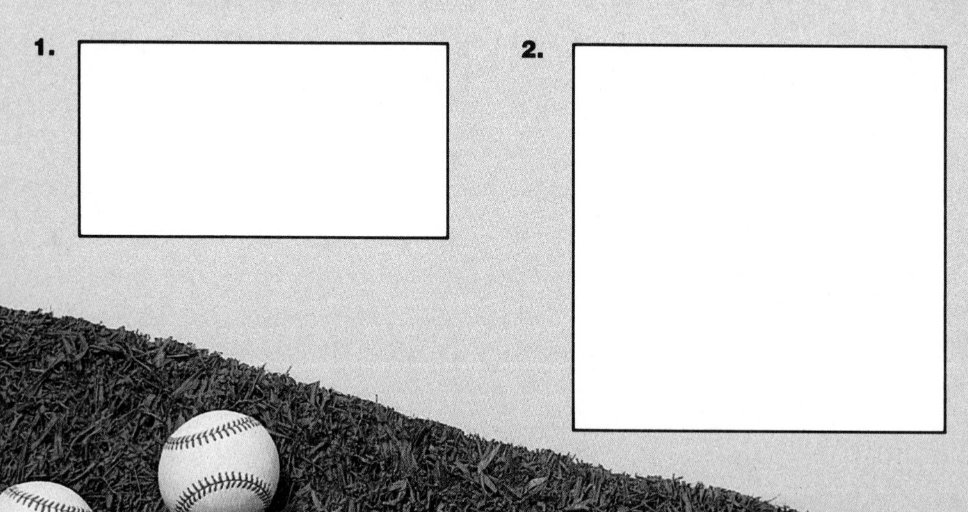

Use your customary ruler to measure each side.
Then find the perimeter.

3.

4.

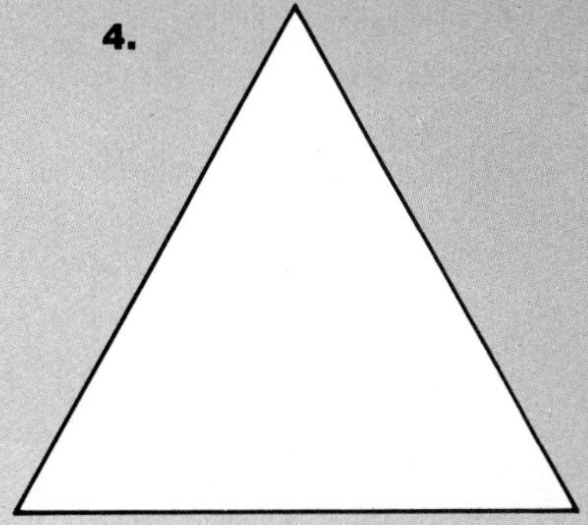

5. What is the perimeter of the sandbox?

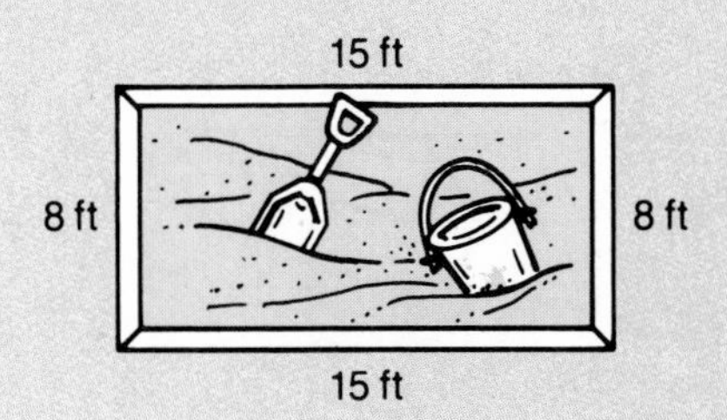

6. How many inches is it around home plate?

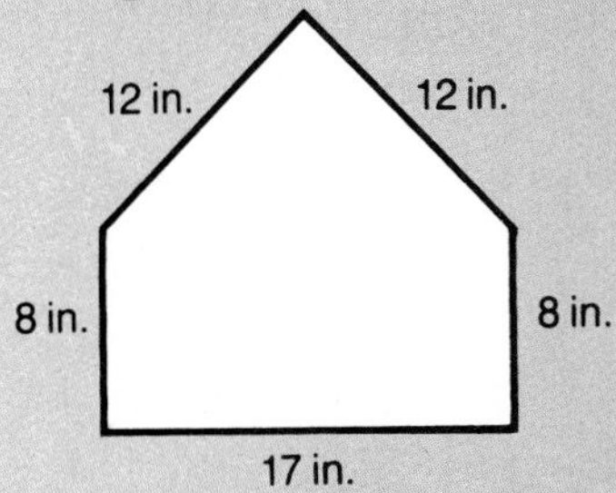

7. Katy's gym class uses a mat for the broad jump. The sides of the mat are 12 ft, 4 ft, 12 ft, and 4 ft. What is the perimeter of the mat?

8. Sam wants to put a frame around a sign for field day. The sides of the sign are 24 in., 24 in., 18 in., and 18 in. How much wood does he need to frame the sign?

CHALLENGE

Use the perimeter to find the missing lengths.

1.

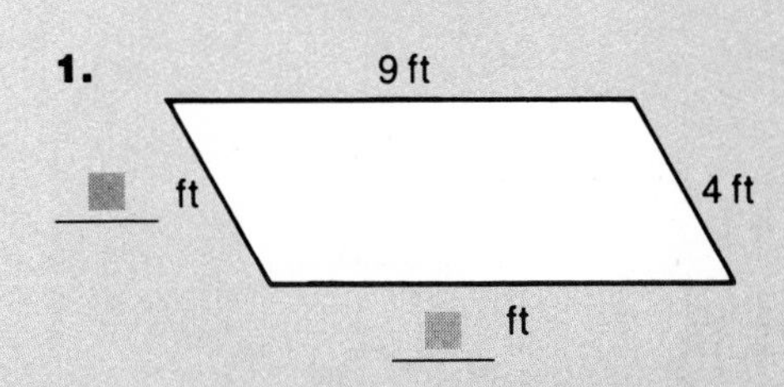

The perimeter is 26 feet.

2.

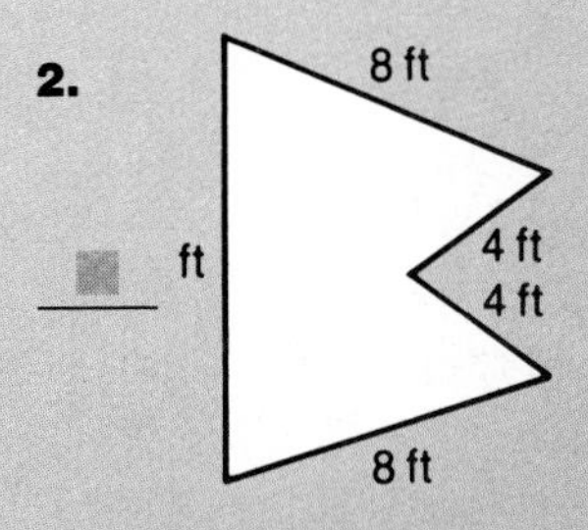

The perimeter is 34 feet.

Cups, Pints, Quarts, and Gallons

A. The **cup (c), pint (pt), quart (qt),** and **gallon (gal)** are customary units that are used to measure liquid.

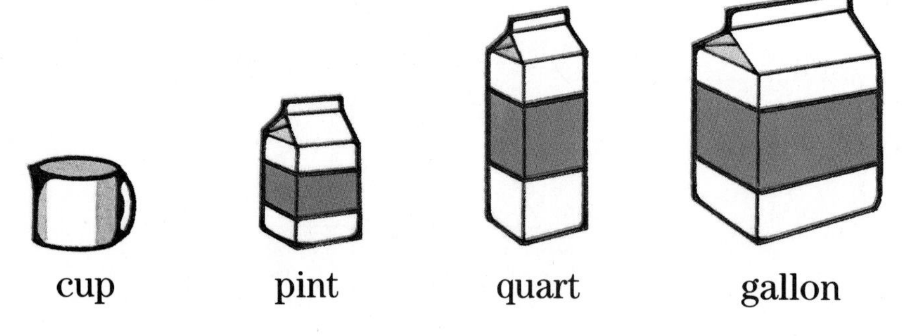

B. You can rename one measurement of liquid with another.

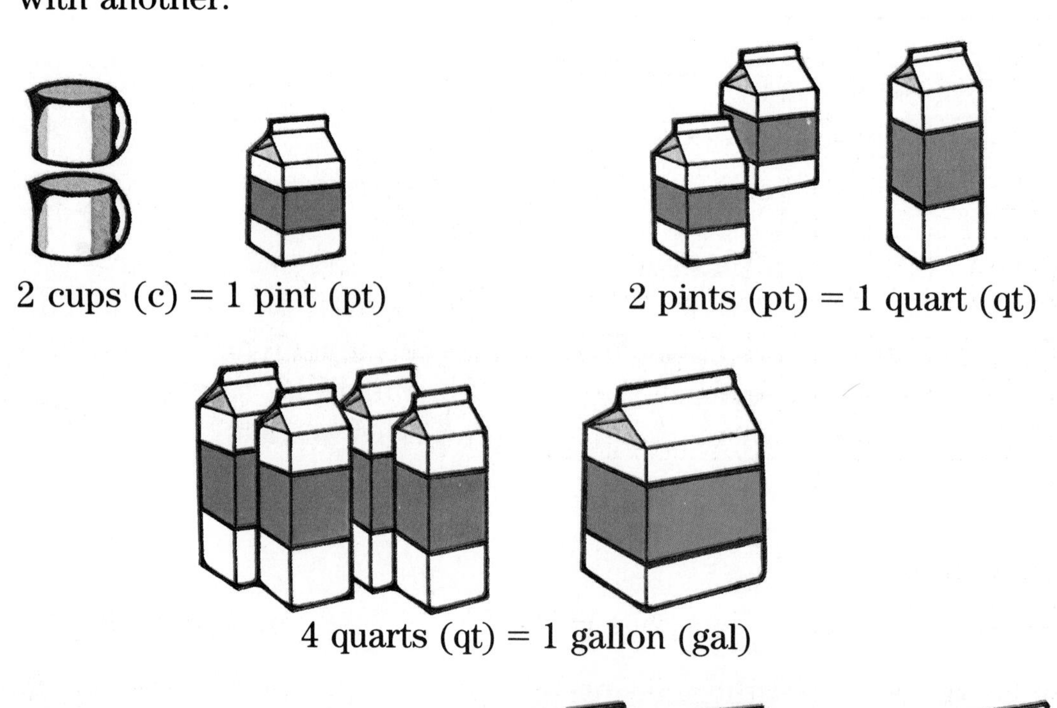

2 cups (c) = 1 pint (pt)

2 pints (pt) = 1 quart (qt)

4 quarts (qt) = 1 gallon (gal)

4 cups (c) = 2 pints (pt)

8 quarts (qt) = 2 gallons (gal)

How much will each hold? Choose the better estimate.
Write the letter of the correct answer.

1. 

a. 3 gallons **b.** 3 pints

2. 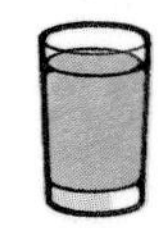

a. 1 pint **b.** 1 quart

3. 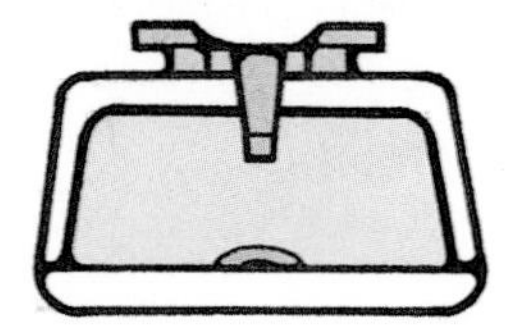

a. 20 pints **b.** 20 gallons

★4. 

a. 2 pints **b.** 2 quarts

Find each pattern. Copy and complete each chart.

5.

cups	2	4	___	___	___
pints	1	2	3	4	5

6.

pints	2	4	___	___	___
quarts	1	2	3	4	5

7.

quarts	4	8	___	___	___
gallons	1	2	3	4	5

Copy and complete.

8. 6 c = ___ pt **9.** 2 qt = ___ pt **10.** 4 pt = ___ c

11. 2 gal = ___ qt **12.** 6 pt = ___ qt **13.** 12 qt = ___ gal

ANOTHER LOOK

Multiply.

1. $2 \times 5 =$ ___ **2.** $3 \times 9 =$ ___ **3.** $4 \times 7 =$ ___ **4.** $8 \times 1 =$ ___

5. $6 \times 4 =$ ___ **6.** $8 \times 2 =$ ___ **7.** $9 \times 5 =$ ___ **8.** $0 \times 7 =$ ___

Ounces and Pounds

A. The **ounce (oz)** and the **pound (lb)** are customary units that are used to measure weight.

You use ounces to measure light objects.

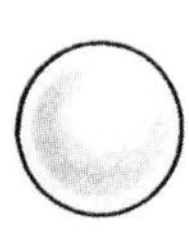

A table-tennis ball weighs about 1 ounce.

You use pounds to measure heavier objects.

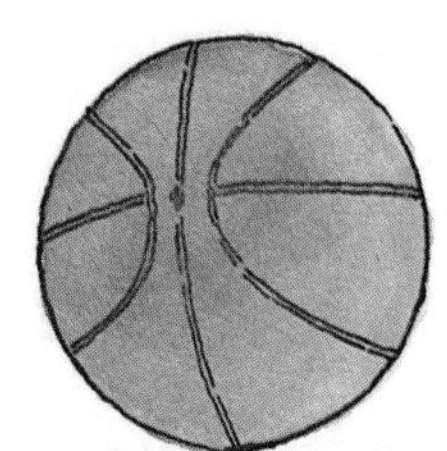

A basketball weighs about 1 pound.

B. You can rename ounces as pounds.

16 ounces (oz) = 1 pound (lb)

Choose the unit you would use to measure each. Write *oz* or *lb*.

1. 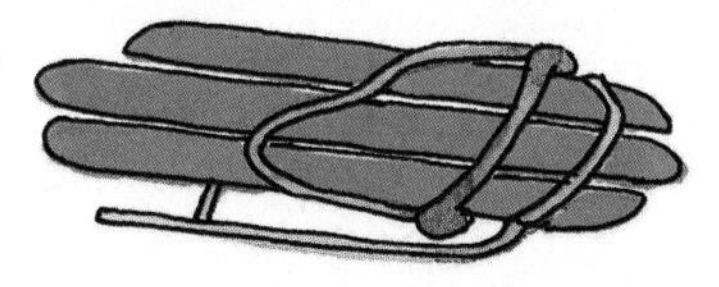

2. 

3. 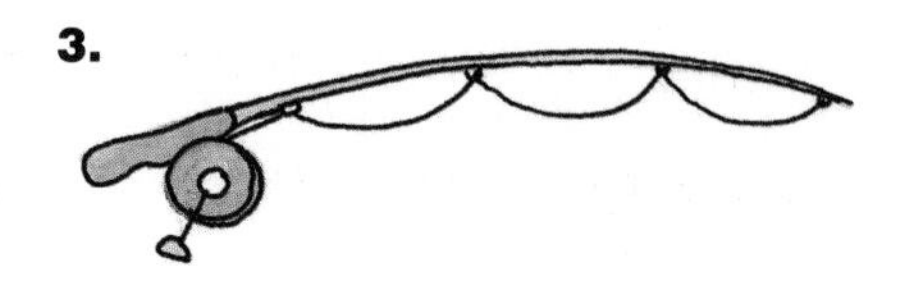

Find the pattern. Copy and complete the chart to rename ounces with pounds.

4.

ounces	16	32	48	64	80	96
pounds	1	2	___	___	___	___

Copy and complete.

5. 16 oz = ___ lb

6. 5 lb = ___ oz

7. 32 oz = ___ lb

8. 4 lb = ___ oz

9. 48 oz = ___ lb

10. 6 lb = ___ oz

More Practice, page 413

Fahrenheit

This is a **Fahrenheit thermometer.** It is used to measure temperature.

230°
220°
210°
212°F—Water boils
140°
130°
120°
110°
100°
90°
85°F—Warm summer day
80°
70°
68°F—Room temperature
60°
50°
40°
32°F—Water freezes
30°
20°
10°
$^{-}5$°F—Very cold day
0°
$^{-}10$°
$^{-}20$°

Each line shows two degrees Fahrenheit.

Read: eighty-five degrees Fahrenheit.
Write: 85°F.

Choose the most likely temperature for each. Write the letter of the correct answer.

1. feeling hot
a. 97°F
b. 28°F

2. feeling cold
a. 85°F
b. 30°F

3. raking leaves
a. 91°F
b. 60°F

4. planting flowers
a. 70°F
b. $^{-}10$°F

5. making a snowman
a. 27°F
b. 60°F

6. wearing mittens
a. 70°F
b. 40°F

FOCUS: REASONING

Tim has a pet dog that weighs 6 pounds less than Anna's dog. Together, the two dogs weigh 26 pounds. What are their weights?

PROBLEM SOLVING
Using a Map

A map can be used to show the distance between two or more places.

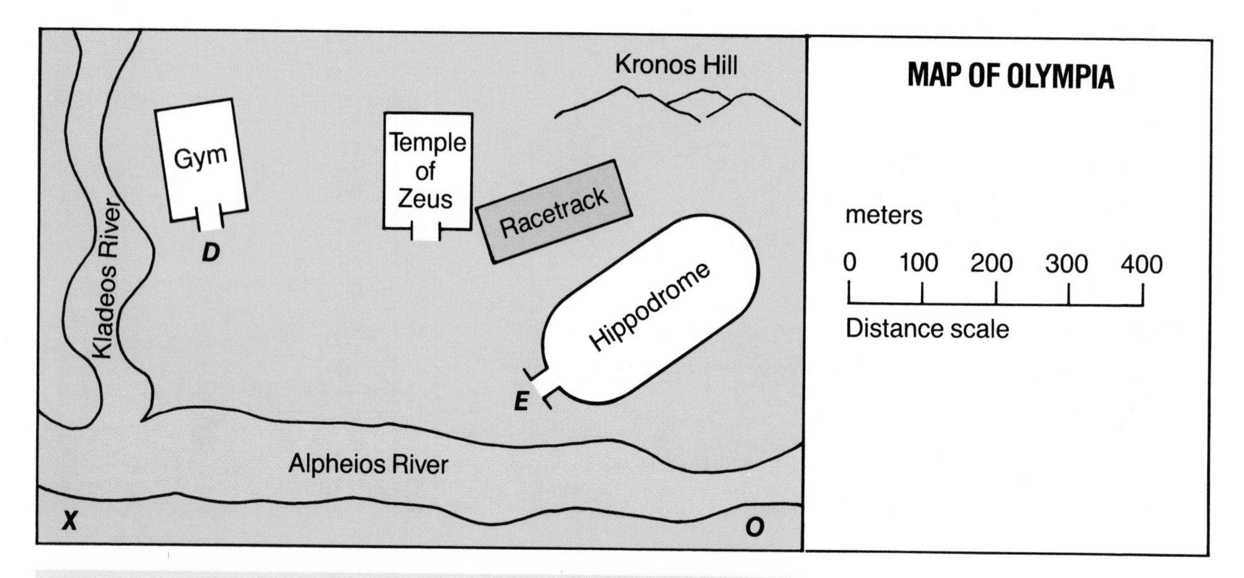

This is a map of Olympia. The first Olympic Games were held there. Only one footrace was held in the first Olympics. The runners ran from one end of the racetrack to the other. About how many meters did they run?

You can use this distance scale to measure the racetrack.

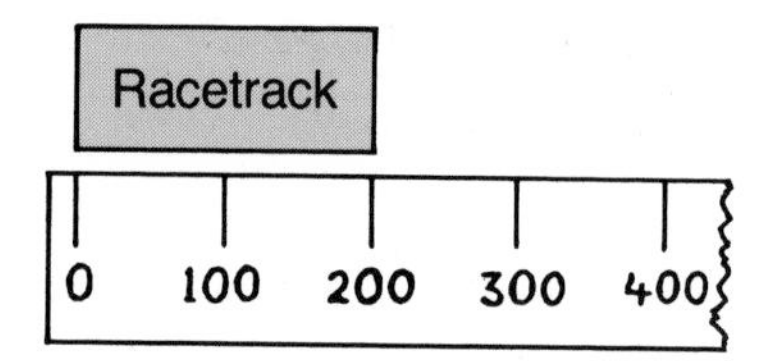

- Copy the distance scale on another piece of paper.
- Put your copy next to the racetrack. Line up the beginning of the track with the 0.
- The end of the racetrack lines up with the 200-meter mark on the distance scale.

The racetrack is 200 meters long. The runners ran 200 meters.

Use the map to solve.

1. Helen walked along the river from the place marked *X* to the place marked *0.* How far did Helen walk? Give your answer first in meters and then in kilometers.

2. What is the distance from the Gym to the Hippodrome? Measure from *D* to *E.*

3. Long ago, the Alpheios River flooded and washed away the Hippodrome. How far did the river have to spread to do this?

4. The chariot races were held in the Hippodrome. About how many meters did the chariots travel to go around the Hippodrome once? **Hint:** You can use a string to measure a curve.

5. The Kladeos River used to flow outside the Gym. What does this map tell you about the river?

6. If you were standing on Kronos after the races and wanted to go to the Hippodrome, what would be the shortest way?

7. George walked from the front of the Temple of Zeus to the Alpheios River. How far did he walk?

★8. The Hippodrome is North of the river. In what direction did George walk when he went from the Temple to the river?

9. There was no town at Olympia. People camped out. Pick a camping spot. Explain where it is and why you picked it.

LOGICAL REASONING

On Saturday, June went for a swimming lesson and a tennis lesson. Here is what happened.

June went home without her tennis racket.

She left home at 1:00

She finished her tennis lesson at 3:30.

She left her tennis racket in a locker while she was swimming.

She had her swimming lesson.

She left her tennis racket in the locker after swimming.

Did June have her swimming lesson before her tennis lesson?

Sometimes you will need to put events in order to solve a problem. Make a list that shows the order of the events above.

June still had her tennis racket.

1. June left home (1:00)
2. finished tennis lesson (3:30)
3. put tennis racket in locker
4. had swimming lesson
5. left tennis racket in the locker
6. went home without tennis racket

So, June had her swimming lesson after her tennis lesson.

Solve.

Clint bought some baseball cards at the store.
He checked a book out at the library at 8:30 A.M.
Clint left his baseball cards at Joe's house.
The store opened at 9:00 A.M.

1. Did Clint have his baseball cards in the library?
2. Did Clint have his book at Joe's house?
3. Did Clint go to Joe's house before or after 8:30?

GROUP PROJECT

Taking a Survey

The problem: Sometimes we need to know how everyone in a community would answer one question. To find out, we can take a survey. Newspapers take surveys. They ask many people a question such as "Do you agree with the Mayor?" Then they count the results for their readers.

You can take your own survey. Find which sport is more popular, baseball or soccer. Count how many of your classmates like baseball better. How many prefer soccer? Set up a chart like this one.

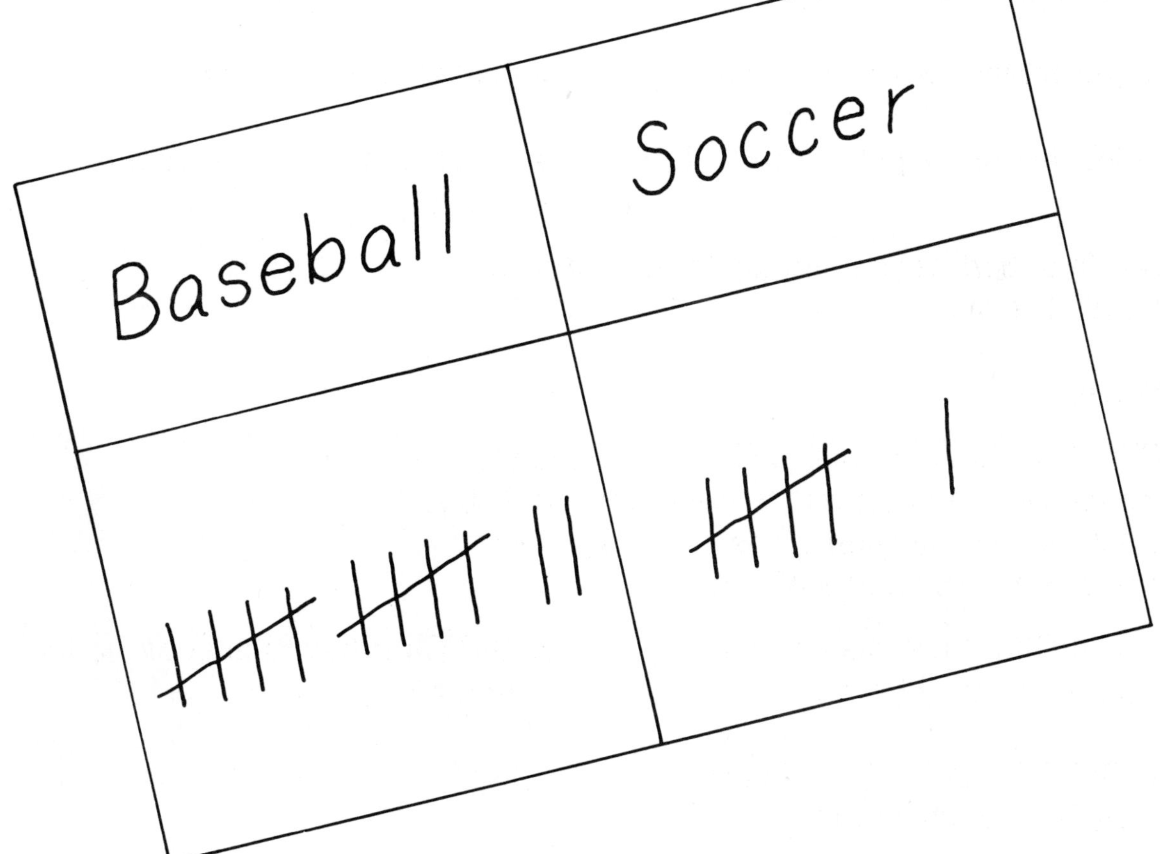

CHAPTER TEST

Use your centimeter ruler to measure each length to the nearest centimeter. (page 268)

1.

2.

Choose the unit that you would use to measure each. Write *cm, m,* or *km.* (page 270)

3. the length of your father's necktie ___

4. the width of your bedroom wall ___

Choose the unit that you would use to measure each. Write *mL* or *L.* (page 276)

5. a bathtub full of water ___

6. a few drops of rain ___

Find the perimeter. (pages 272 and 286)

7.

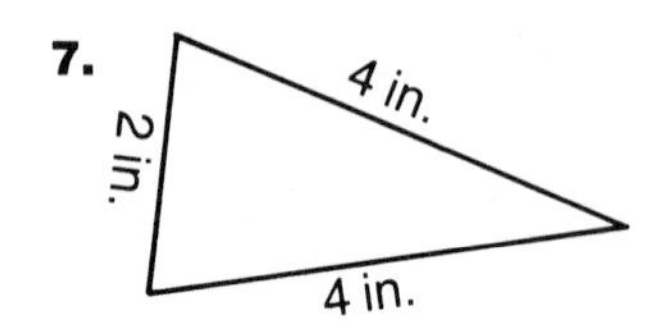

8.

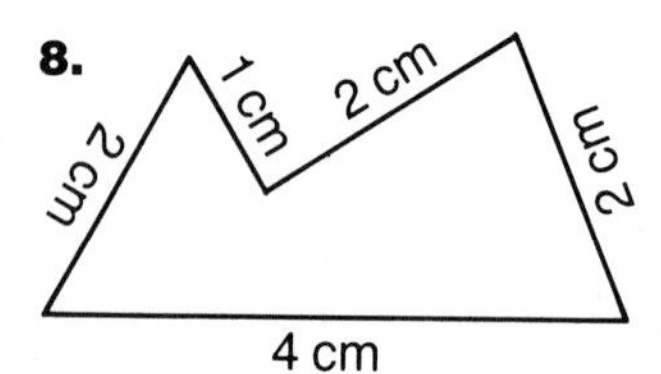

Choose the unit that you would use to measure each. Write *g* or *kg.* (page 278)

9. a lion ___

10. an egg ___

Use the map to solve. (page 292)

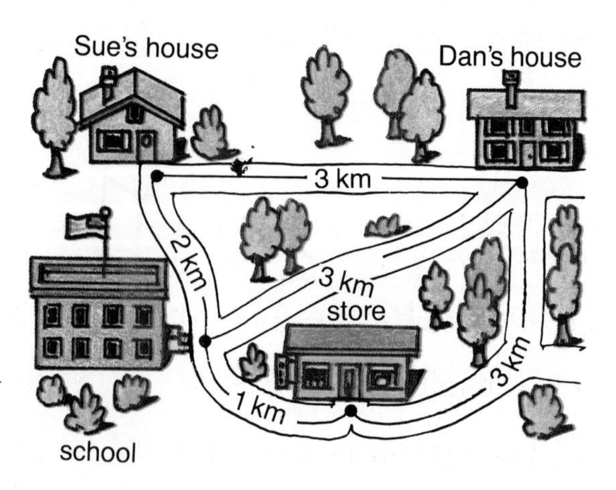

11. After school, Dan rides his bicycle to the store. Then he rides home. How many kilometers does he ride?

12. Dan chooses the shortest way from his house to the school. How many kilometers does he ride?

Use your customary ruler to measure each length to the nearest $\frac{1}{4}$ inch or $\frac{1}{2}$ inch. (page 282)

13. •————————•

14. •————————•

Give the temperature in degrees Celsius. (page 279)

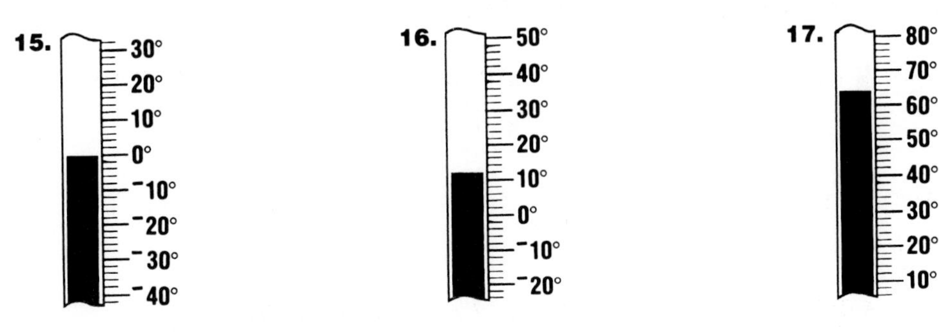

Choose the unit that you would use to measure each. Write *in.*, *ft*, *yd*, or *mi.* (page 284)

18. the length of a pencil ____

19. the height of a flagpole ____

20. the length of a highway ____

21. the length of your leg ____

Complete. (page 288)

22. 2 cups = ____ pint

23. 8 quarts = ____ gallons

Choose the unit that you would use to measure each. Write *oz* or *lb.* (page 290)

24. a bag of apples ____

25. a letter ____

BONUS

Choose the best metric unit of measure.

1. McNight Avenue is 79 ____ long.

2. Joe's father is 2.5 ____ tall.

3. The spoon holds 3 ____ of juice.

4. The flower weighs 2 ____ .

RETEACHING

You can use **cup (c), pint (pt), quart (qt),** and **gallon (gal)** to measure liquid. These are called **customary units.**

You can rename one measurement of liquid with another.

2 cups (c) = 1 pint (pt)

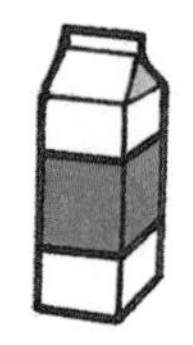

2 pints (pt) = 1 quart (qt)

4 quarts (qt) = 1 gallon (gal)

Complete.

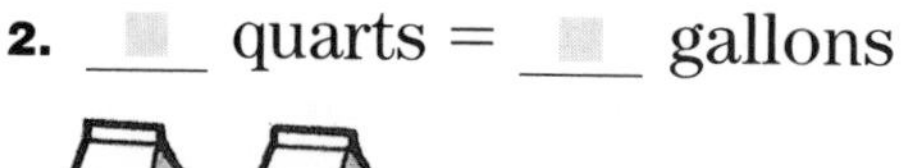

1. ____ cups = ____ pints
2. ____ quarts = ____ gallons

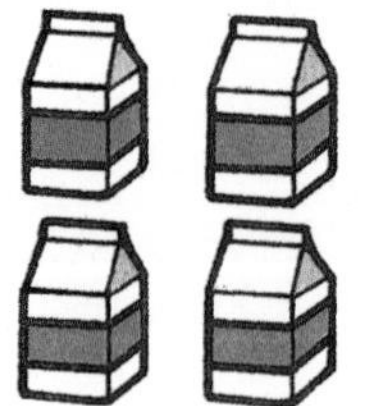

3. ____ quarts = ____ gallons
4. ____ pints = ____ quarts
5. ____ cups = 4 pints
6. ____ pints = 4 quarts
7. ____ quarts = 4 gallons
8. ____ quarts = 5 gallons

ENRICHMENT

Optical Illusions

1. Which is farther, the distance from A to B or the distance from C to D?

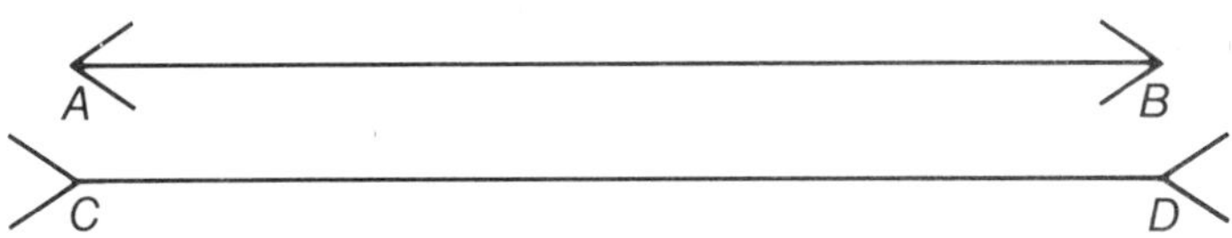

2. Will the long lines ever meet?

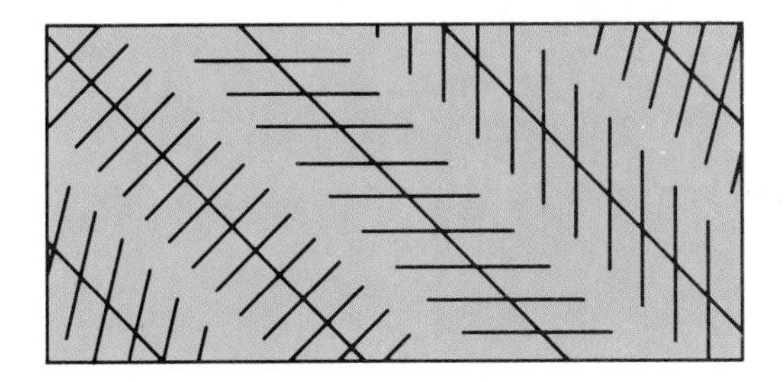

3. Do the sides of the square curve inward?

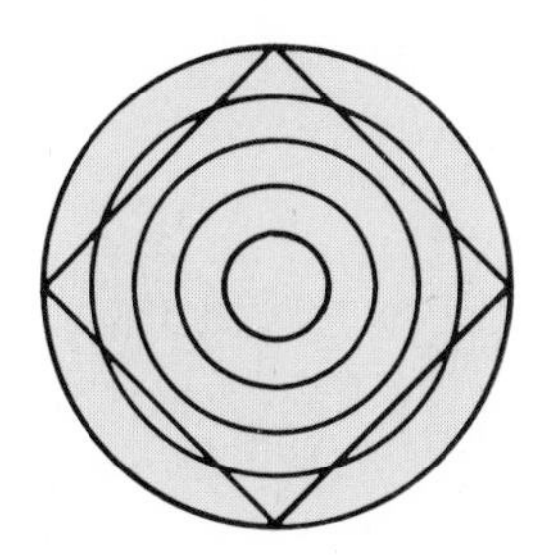

4. Which circle is larger, A or B?

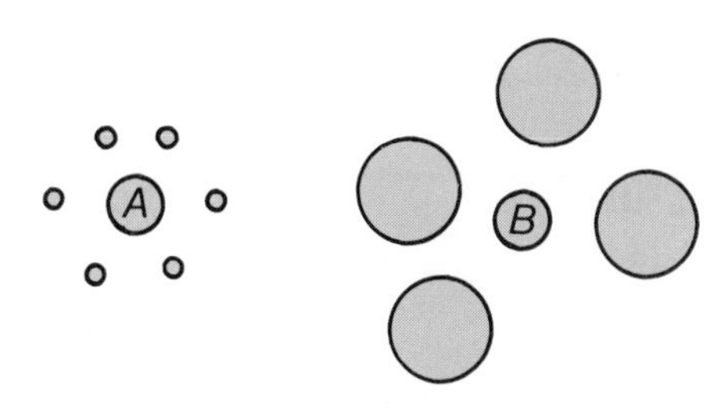

5. Which block is the largest, A, B, or C?

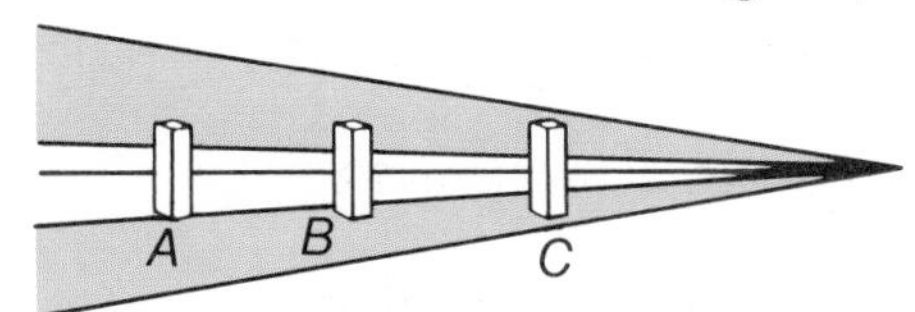

TECHNOLOGY

In BASIC, you can use the PRINT statement to have the computer print on the screen.

Statement	What the Computer prints
PRINT "4 − 3 IS"	4 − 3 IS
PRINT 4 − 3	1

When you finish each line, press the RETURN or the ENTER key. The computer will print its answer.

If you want the computer to remember several statements, you can write a program. Then you can have the computer carry out several statements in order.

A program for the statements above might look like this.

```
10 PRINT "4 - 3 IS"
20 PRINT 4 - 3
30 END
```

In front of each PRINT statement is a line number. The line numbers tell the computer the order in which to carry out the instructions. You can RUN the program by typing the letters *R U N* and then pressing the RETURN key. The computer will do the statement with the lowest line number first.

If you run the program, you will see this on the screen.

```
4 - 3 IS
1
```

If you typed in the program in this order:

```
20 PRINT 4 – 3
10 PRINT “4 – 3 IS”
30 END
```

and then typed RUN, the computer would run the program with the statements in the correct order.

The symbol * is used to mean multiplication in BASIC. Here is how it works.

Statement	What the computer prints
PRINT 2 * 4	8

1. What do you type to make the computer begin to do a program?

2. Here is a program in which the lines are out of order. When you run the program, what will the computer print?

```
50 PRINT “LAKE”
10 PRINT “JIM”
40 PRINT “THE”
20 PRINT “JUMPED”
30 PRINT “INTO”
```

3. Finish the program. It prints the product of 4 and 3. Write what the computer prints when the program is RUN.

```
10 PRINT “ ”
20
30 END
```

CUMULATIVE REVIEW

Write the letter of the correct answer.

1. Round 26 to the nearest ten.

   a. 20 b. 25
   c. 30 d. not given

2. 1,692 + 3,457 = ___

   a. 1,765 b. 4,049
   c. 5,149 d. not given

3. Estimate.

   $8.93
   − 3.56

   a. $5.00
   b. $5.47
   c. $6.00
   d. $13.00

4. What part is shaded?

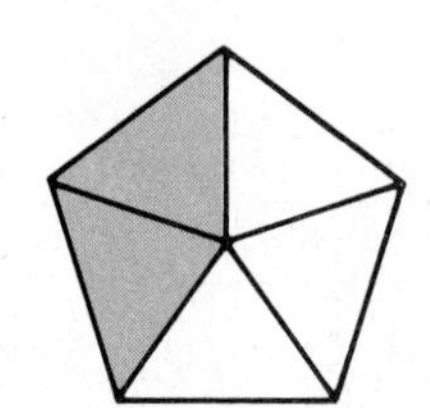

   a. $\frac{2}{5}$ b. $\frac{3}{5}$
   c. $\frac{2}{3}$ d. not given

5. Name the equivalent fractions.

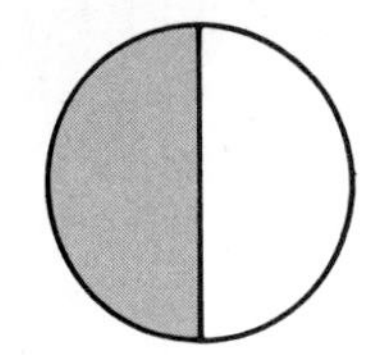

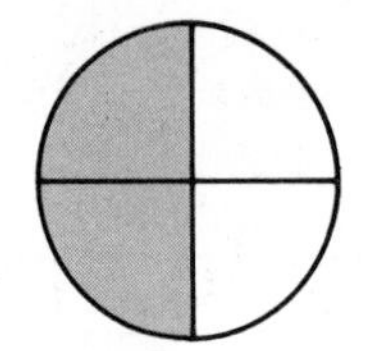

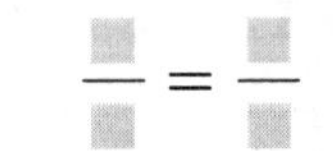

   a. $\frac{1}{2} = \frac{2}{4}$ b. $\frac{1}{2} = \frac{3}{4}$
   c. $\frac{1}{1} = \frac{2}{1}$ d. not given

6. What part is shaded?

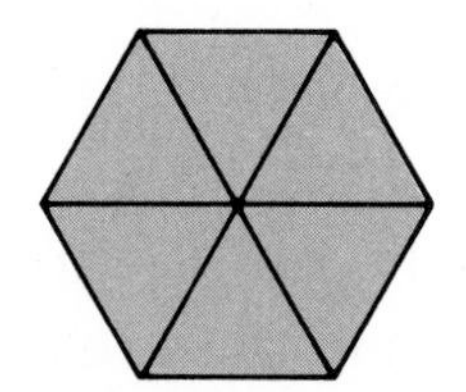

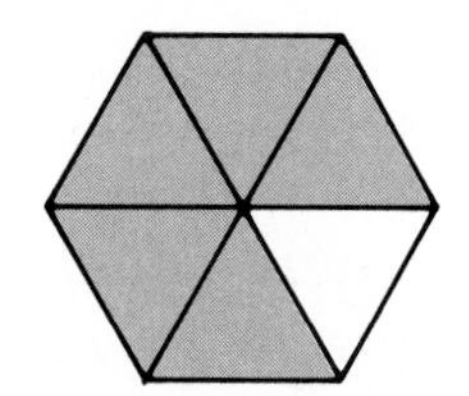

   a. $\frac{1}{11}$ b. 11 c. $1\frac{5}{6}$ d. not given

7. Write the decimal: twenty-one hundredths.

   a. 0.21 b. 21.21
   c. 2,100 d. not given

8. 3.2 + 5.9 = ___

   a. 8.1 b. 9.1
   c. 9.11 d. not given

9. Pete has 7 bags of marbles with 8 marbles in each bag. Which number sentence shows how many marbles Pete has?

   a. 8 − 7 = 1 b. 8 + 7 = 15
   c. 8 × 7 = 48 d. not given

10. Tom jogged 1.45 kilometers on Thursday and 3.02 kilometers on Friday. How many more kilometers did he jog on Friday?

   a. 1.57 km b. 1.68 km
   c. 4.48 km d. not given

10 GEOMETRY AND GRAPHING

Have your ever thought what the future will be like? Plan a bedroom for someone your age in the year 2025. Your bedroom can be up to 200 square feet in size.

Basic Ideas of Geometry

A. Charlie drew a plan for a house of the future.

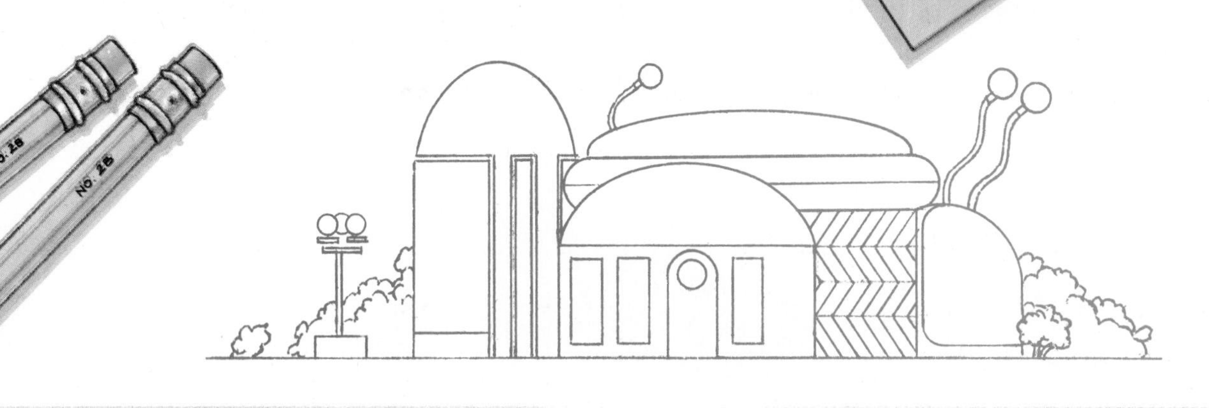

Charlie drew line segments to make the shapes.
A **line segment** is straight. It has two **endpoints.**

Read: line segment *ST* or line segment *TS*.
Write: $\overline{ST}$ or $\overline{TS}$.

B. A line segment is part of a line. A **line** is straight. It goes on forever in both directions.

G H

The arrows mean that the line continues.

Read: line *GH* or line *HG*.
Write: $\overleftrightarrow{GH}$ or $\overleftrightarrow{HG}$.

Is this a line segment? Write *yes* or *no*.

1.

2.

3.

4.

5.

6.

Name each line segment in two ways.

7. J K

8. S T

9. H I

10. V W

11. E D

12. P Q

Write *line* or *line segment* for each.

13.

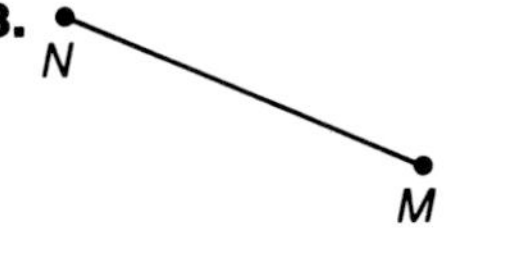

14.

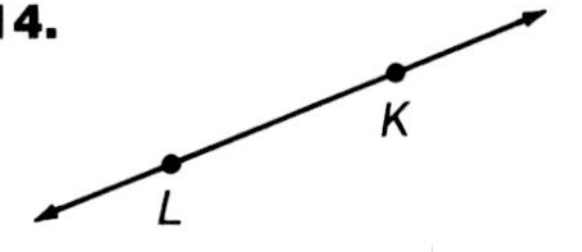

15.

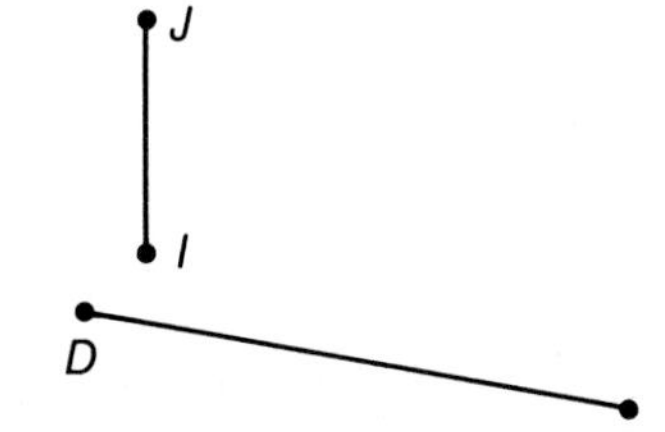

16. H G

17. E F

18. D C

Name each line in two ways.

19. F G

20.

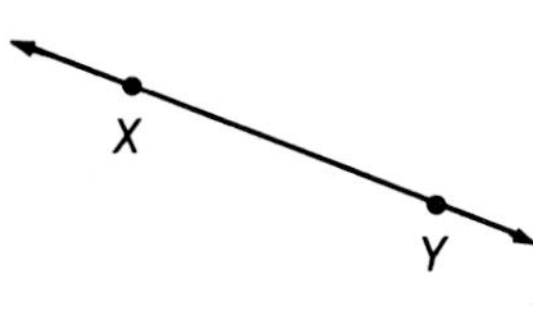

21.

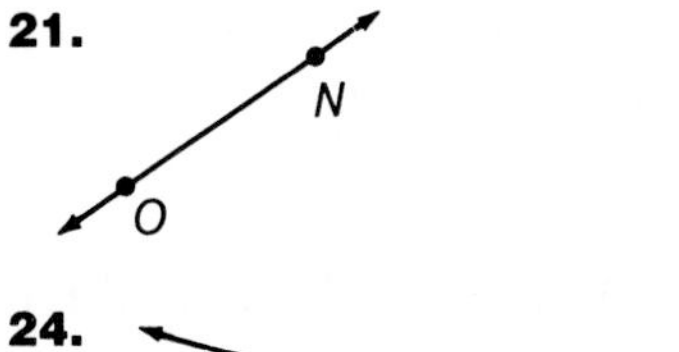

22. O P

23.

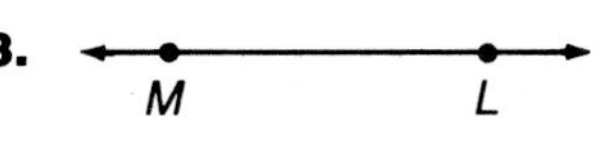

24. X W

FOCUS: ESTIMATION

Without counting, find which group has the most dots. Which group has the fewest dots?

1. 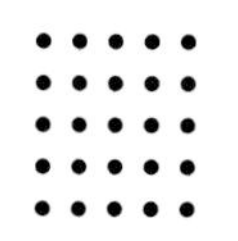

2. 

3. 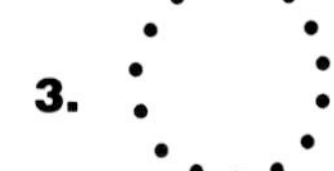

4. 

Angles

A. The wings and the body of this airplane meet to form an **angle.**

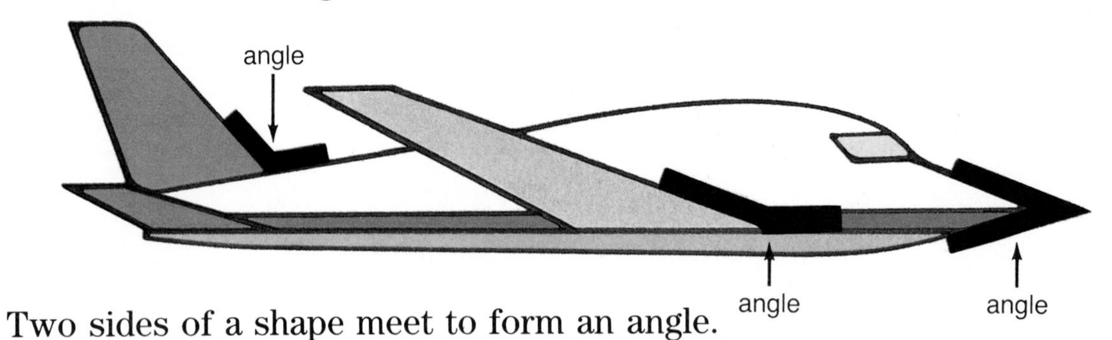

Two sides of a shape meet to form an angle.

Other examples:

B. An angle that forms a square corner is called a **right angle.**

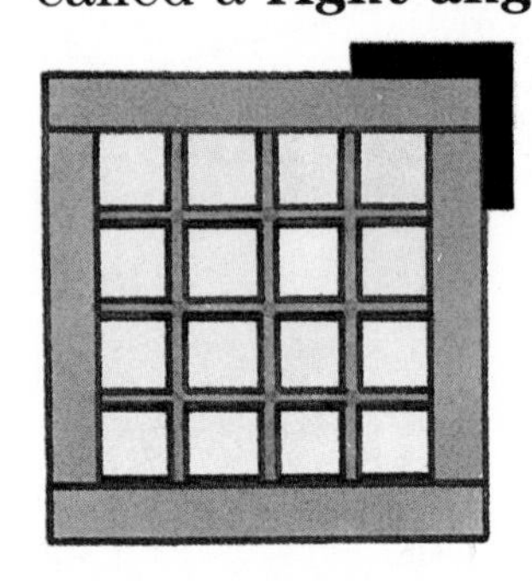

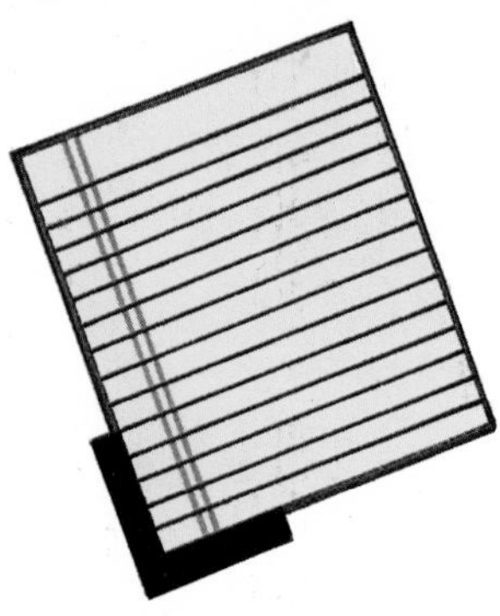

Is this an angle? Write *yes* or *no.*

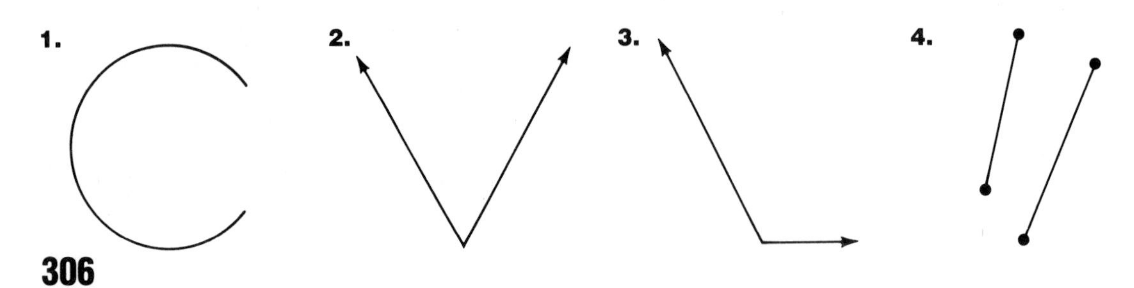

Is this a right angle? Write *yes* or *no.*

5. 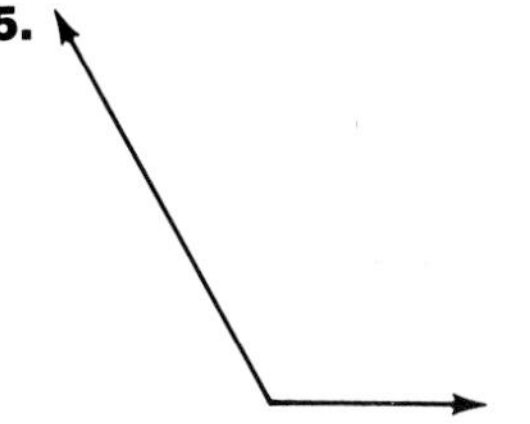

6. 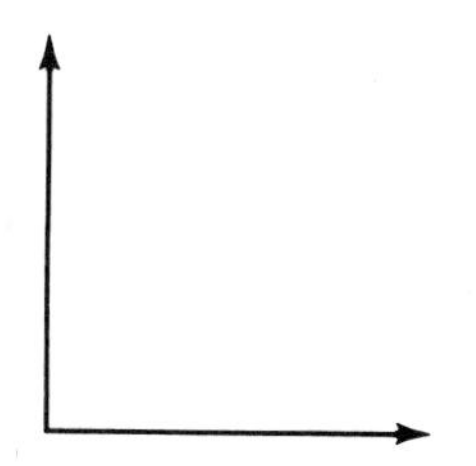

7. 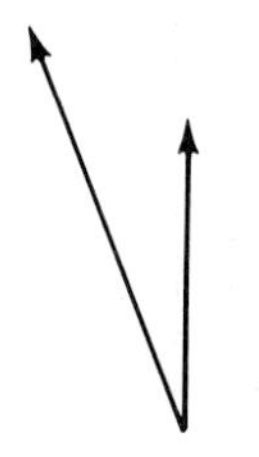

8. 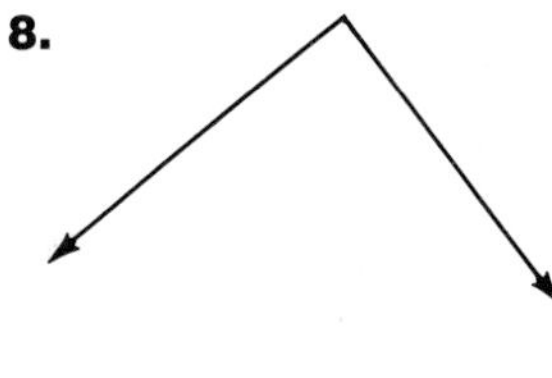

9. 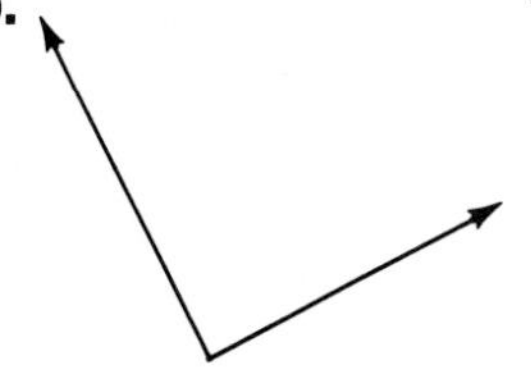

10. 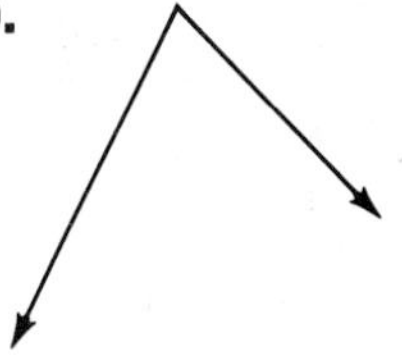

11. 

12. 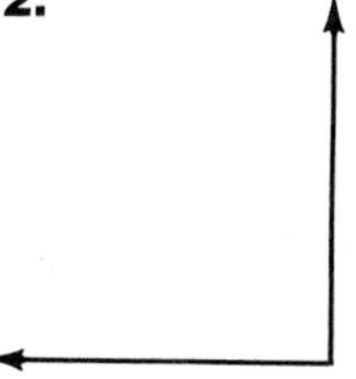

Write how many angles there are in each figure.

13. 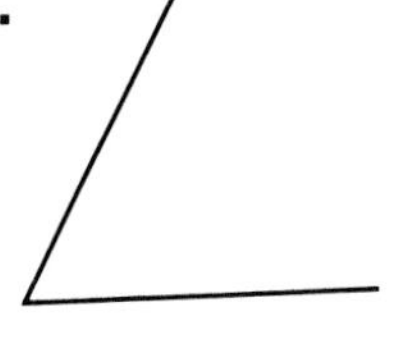

14. 

15. 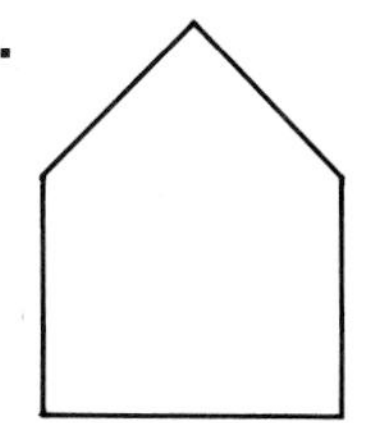

16. 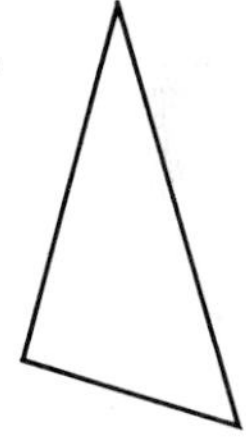

17. 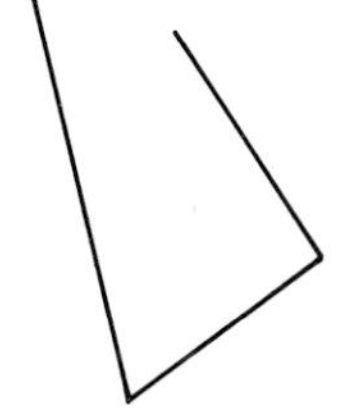

18. 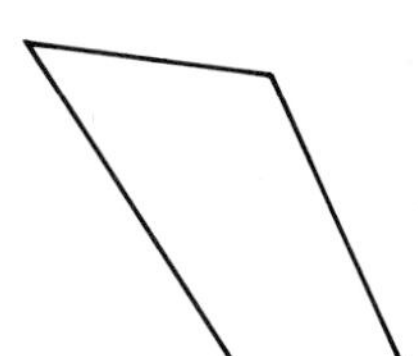

19. 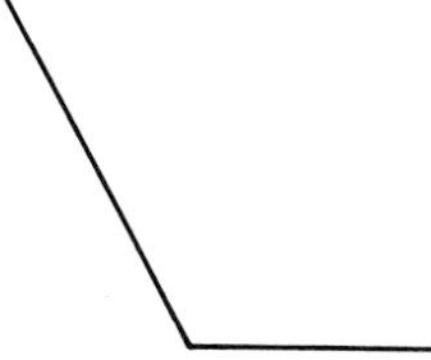

20. 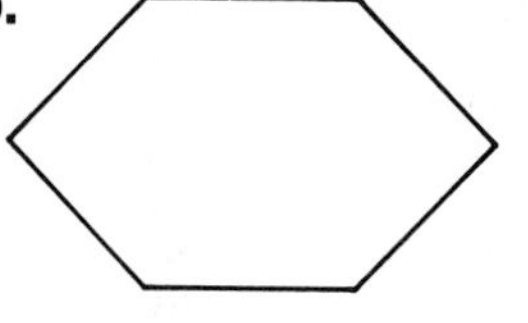

CHALLENGE

Write how many angles there are on the inside of each figure.

1. 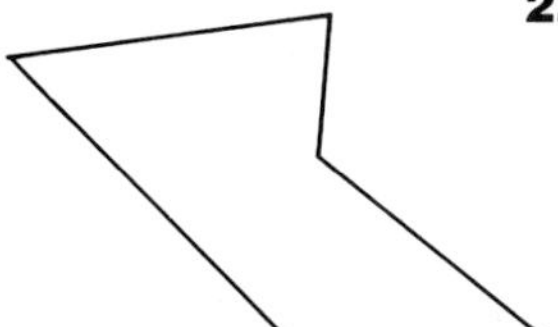

2. 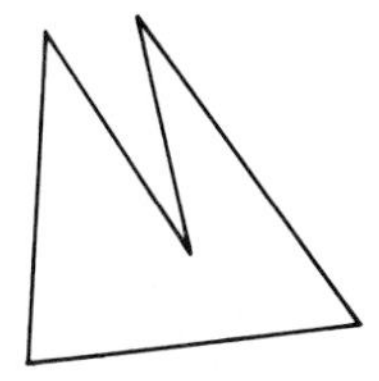

3. 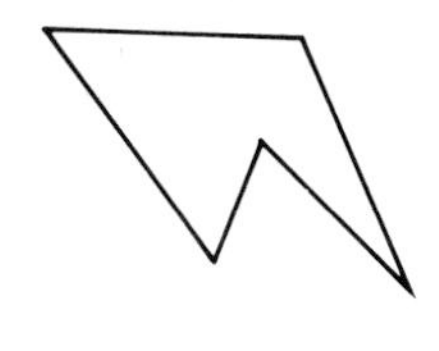

4. 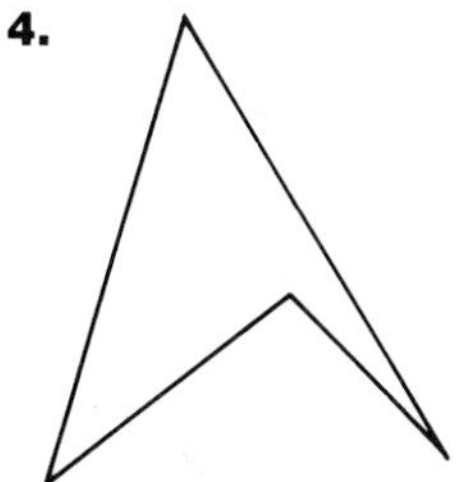

Plane Figures

Raoul's class cut out shapes for a picture of what their town might look like in the year 2050. They cut out shapes for houses and spaceships.

Here are some of the shapes they cut. Each side of a shape is a line segment. Line segments meet at a **vertex.**

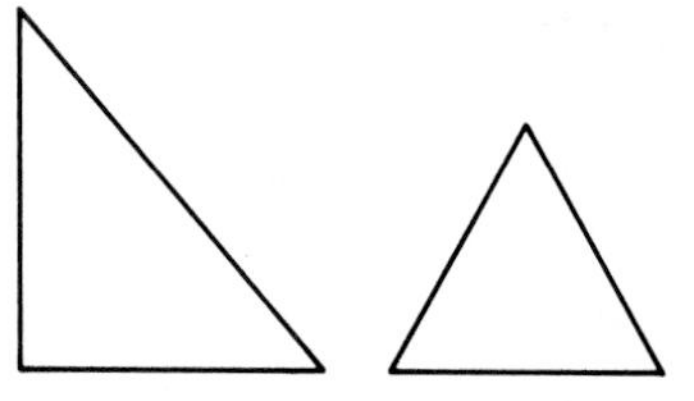

Triangles have 3 sides and 3 vertices.

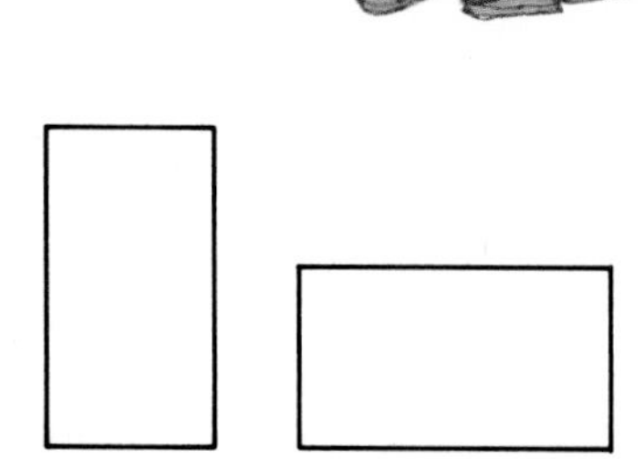

Rectangles have 4 sides, 4 right angles, and 4 vertices.

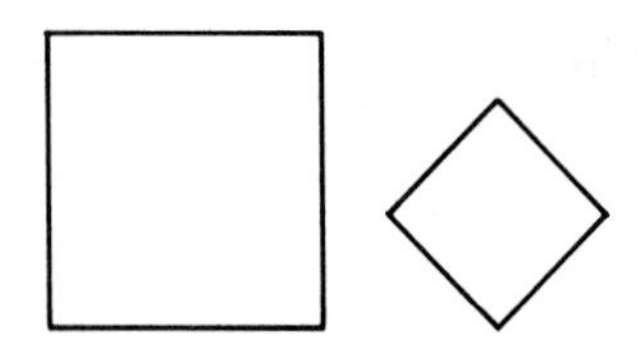

Squares have 4 sides of equal length, 4 right angles, and 4 vertices.

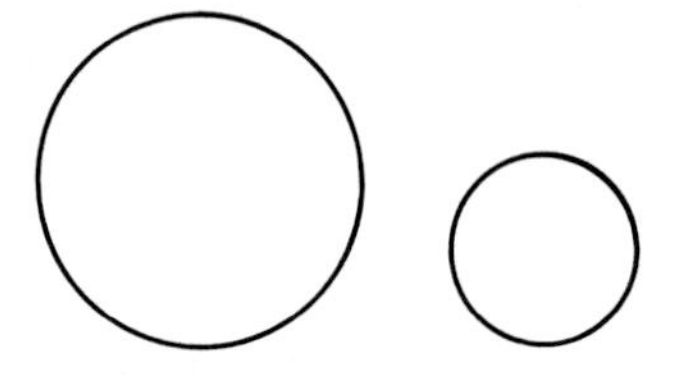

Circles have no sides and no angles.

Other examples:

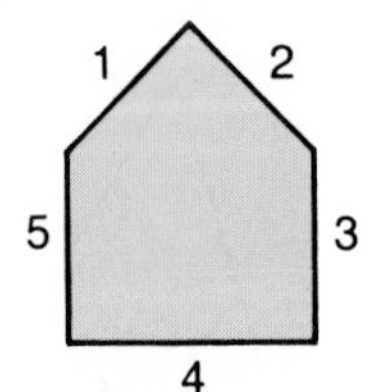

This shape has 5 sides and 5 angles.

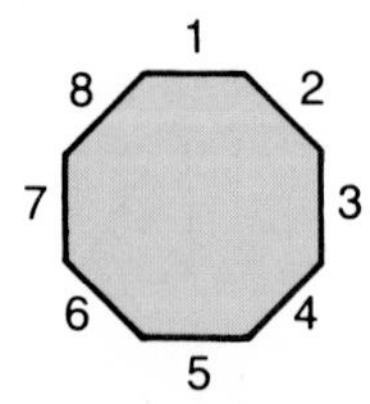

This shape has 8 sides and 8 angles.

Write the name of each shape.

1. 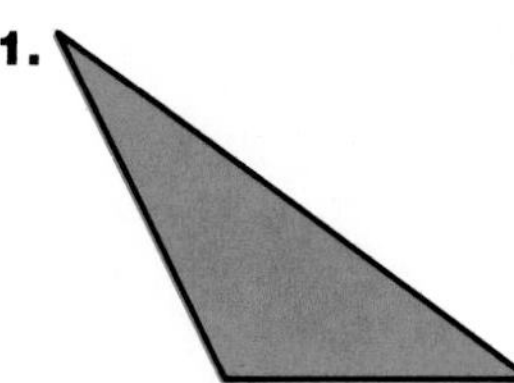

2. 

3. 

4. 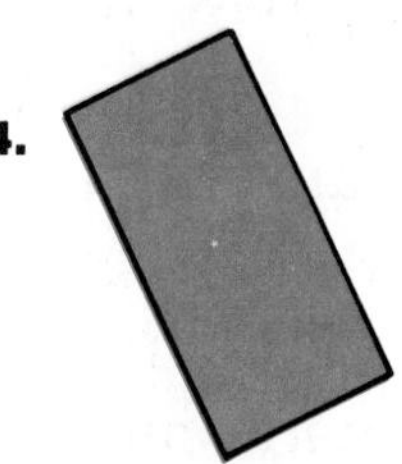

Count the sides and vertices of each shape.

5. 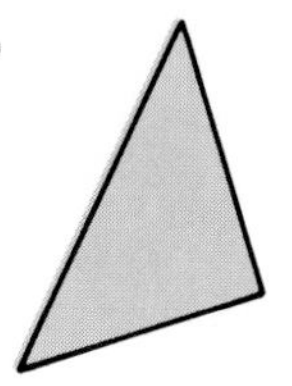
____ sides
____ vertices

6. 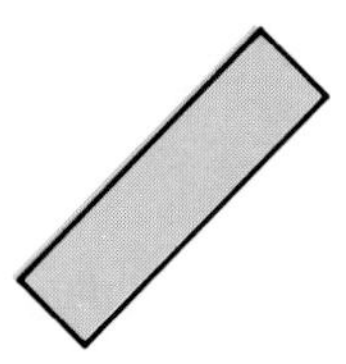
____ sides
____ vertices

7. 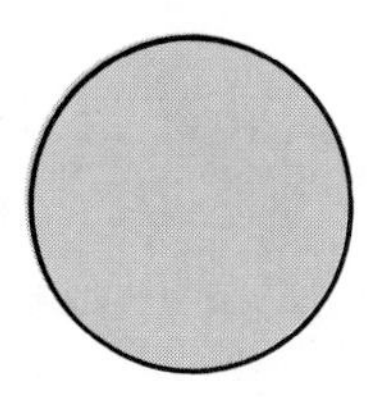
____ sides
____ vertices

8. 
____ sides
____ vertices

9. 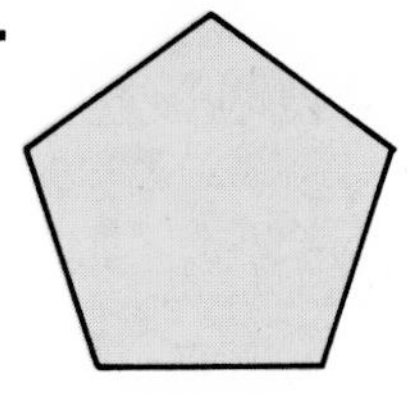
____ sides
____ vertices

10. 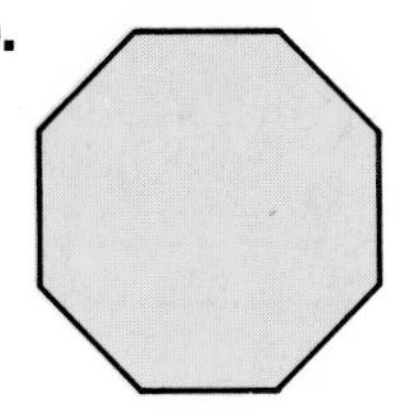
____ sides
____ vertices

11. 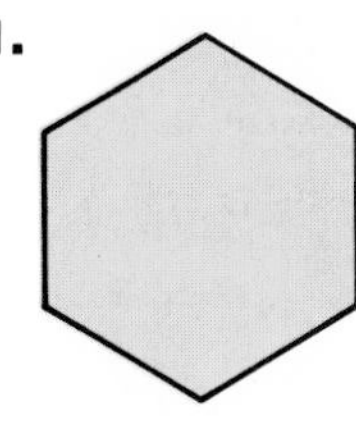
____ sides
____ vertices

12. 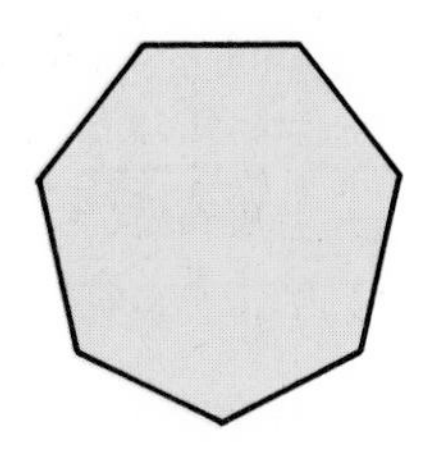
____ sides
____ vertices

Write the name of each figure.

13. It has 4 sides and 4 right angles.

14. It has 3 sides and 3 vertices.

15. It has no sides and no vertices.

16. It has 4 equal sides and 4 right angles.

ANOTHER LOOK

Complete.

1. $\frac{1}{4}$ of 32 = ____
2. $\frac{1}{3}$ of 27 = ____
3. $\frac{1}{2}$ of 14 = ____
4. $\frac{1}{2}$ of 8 = ____
5. $\frac{1}{6}$ of 36 = ____
6. $\frac{1}{8}$ of 48 = ____
7. $\frac{1}{5}$ of 35 = ____
8. $\frac{1}{8}$ of 72 = ____

More Practice, page 414

PROBLEM SOLVING
Using a Picture

Pictures contain much useful information.

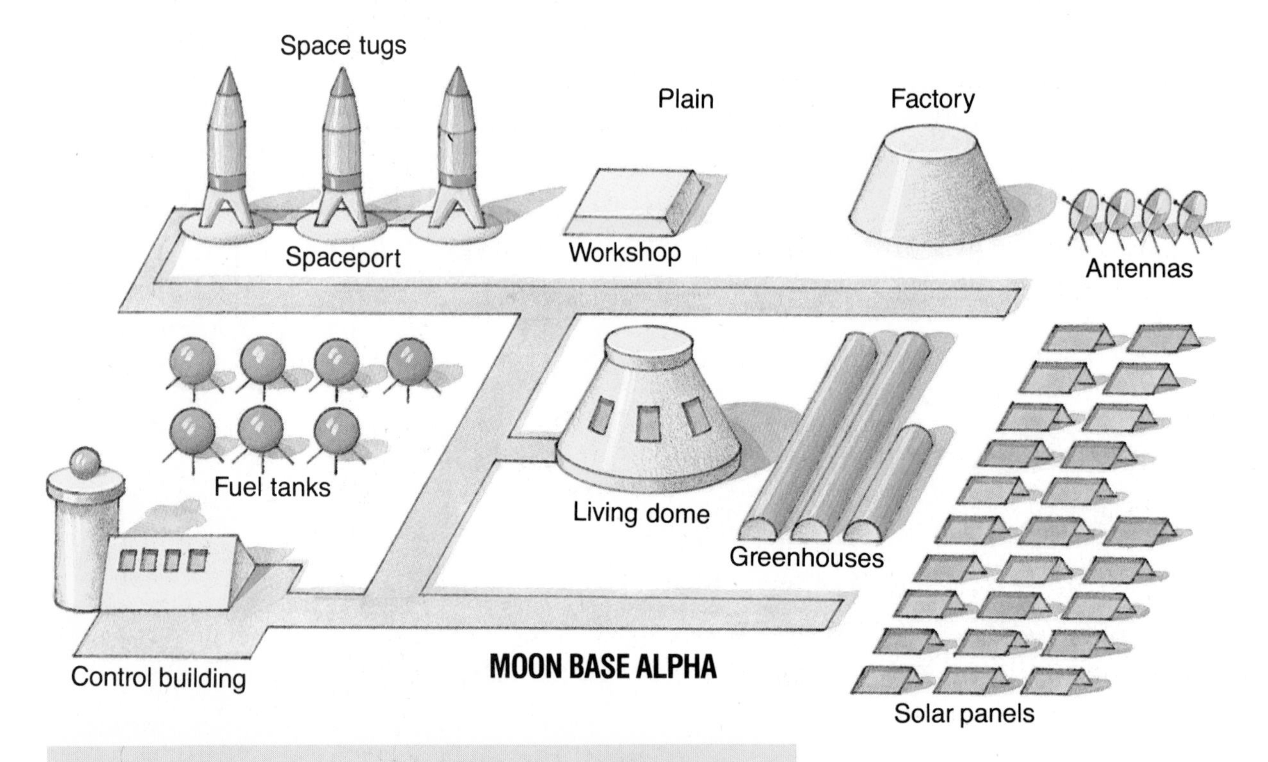

Food for Moon Base Alpha is grown in greenhouses. A large greenhouse grows enough food for 50 people. A small greenhouse grows food for 25 people. How many people can be fed on this base?

The picture shows two large greenhouses and one small one. You can add to find the total number of people who can be fed.

 50 people fed by one large greenhouse
 50 people fed by the other large greenhouse
+ 25 people fed by the small greenhouse
125 people who can be fed on the base

The moon base can feed 125 people.

Use the picture on page 310 to help you solve each problem.

1. How many windows are there in the Control building? (**Hint:** The Control building has the same number of windows on both sides.)

2. If 18 solar panels can provide all the power needed by the moon base, how many extra solar panels are there?

3. Each fuel tank holds a 3-week supply of fuel. How many weeks' supply of fuel is there?

4. Some workers are checking the antennas. There are 6 workers for each antenna. How many workers are there in all?

5. The spaceport is full. A space tug from another base is in trouble. Look at the picture. Where is the best place to land? Why do you think so?

6. You have to go from the Control building to the factory. You take the road. Which do you pass first, the living dome or the workshop?

7. The base commander wants to move 8,000 pounds of equipment at the same time. Each of the space tugs can lift 3,000 pounds. But 1 tug is broken. Can the other tugs lift all the equipment at the same time?

8. If the broken space tug was fixed, how many more pounds of equipment could the commander move at the same time?

Congruence

Patty cut out a spaceship. Blake traced it. Blake's spaceship was exactly the same size and shape as Patty's spaceship.

Figures that are the same size and shape are called **congruent figures.** Two figures are congruent if one of them can fit exactly on top of the other.

Other examples:

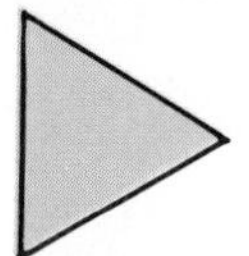
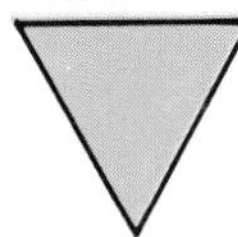
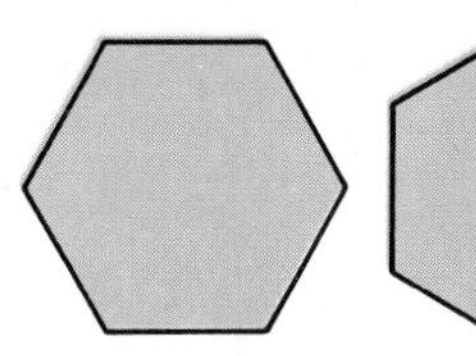

Are the figures congruent? Write *yes* or *no.*

1.

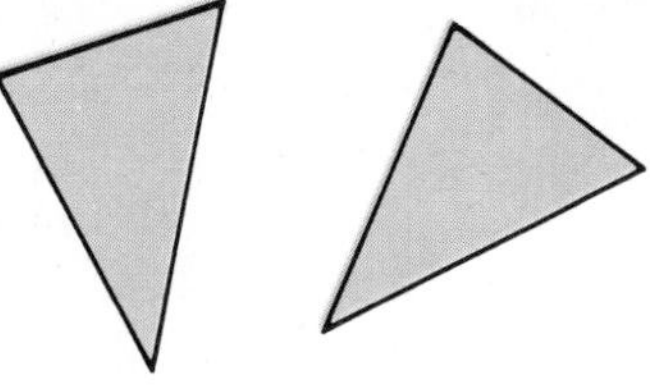

2.

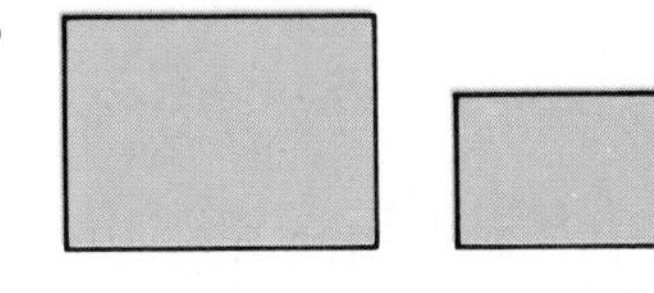

3.

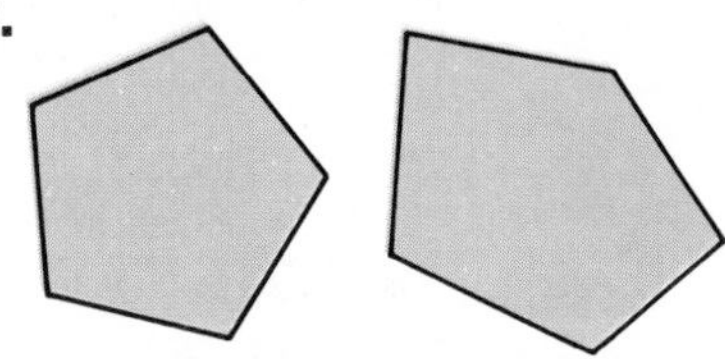

4.

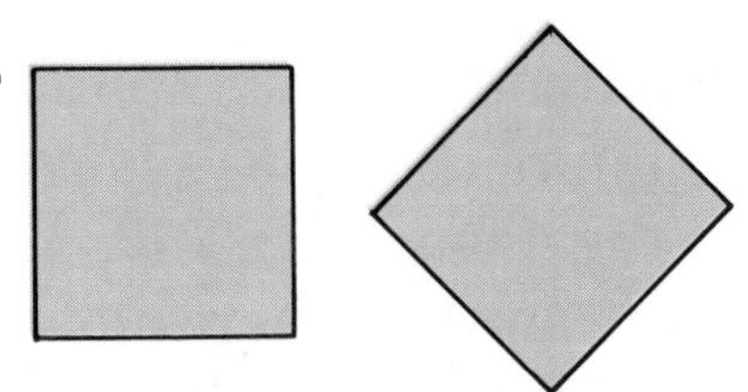

5.

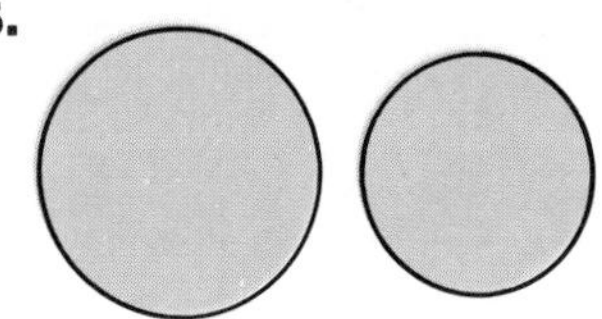

6.

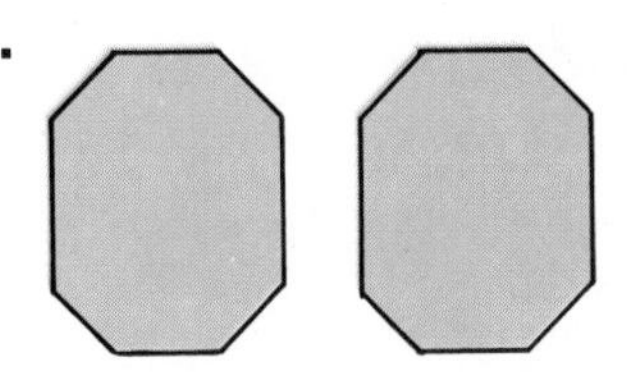

Write the letter of the figure that is congruent to the first figure shown.

7. 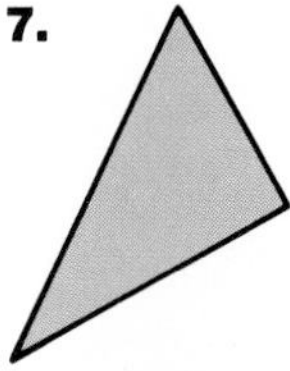 **a.** 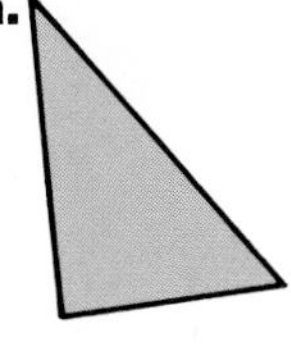 **b.** 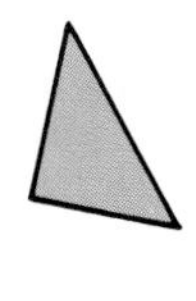 **c.** 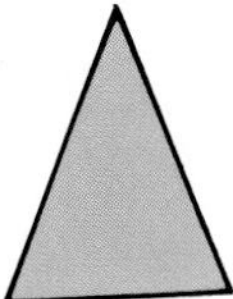

8.  **a.**  **b.**  **c.** 

9.  **a.**  **b.** 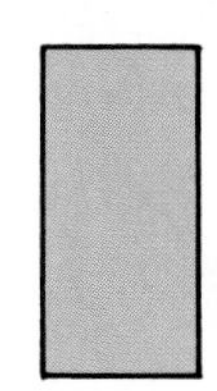 **c.** 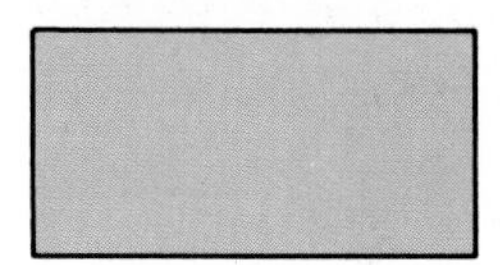

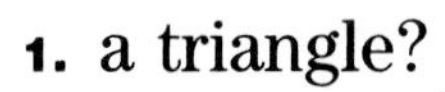

can form

Can the shapes be put together to form

1. a triangle? **2.** a rectangle? **3.** a circle?

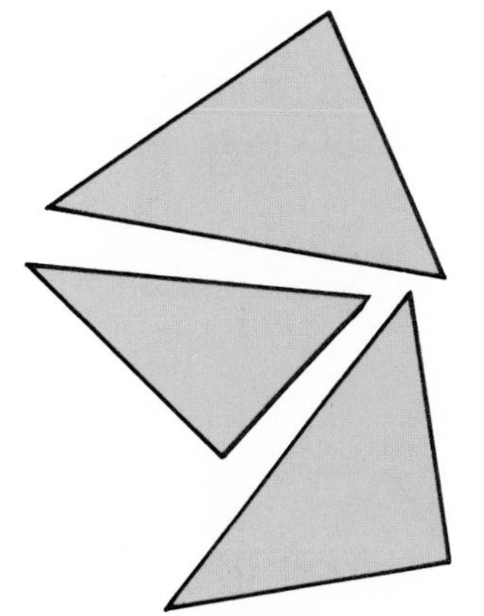

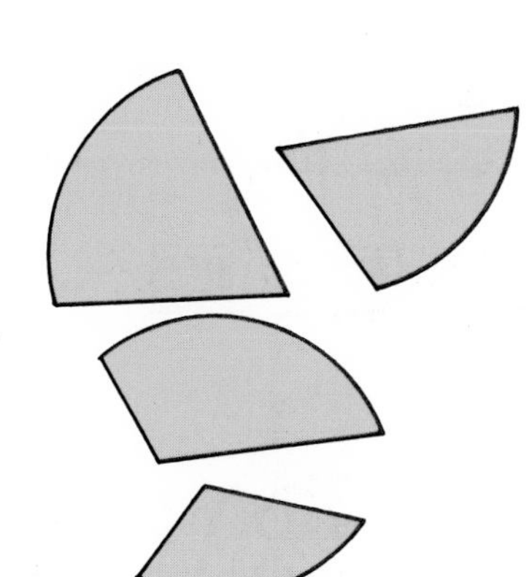

More Practice, page 414

Lines of Symmetry

Luis cut out drawings of buildings of the future. He folded a piece of paper. He cut out the shape and unfolded the paper. Both parts of the building were exactly alike.

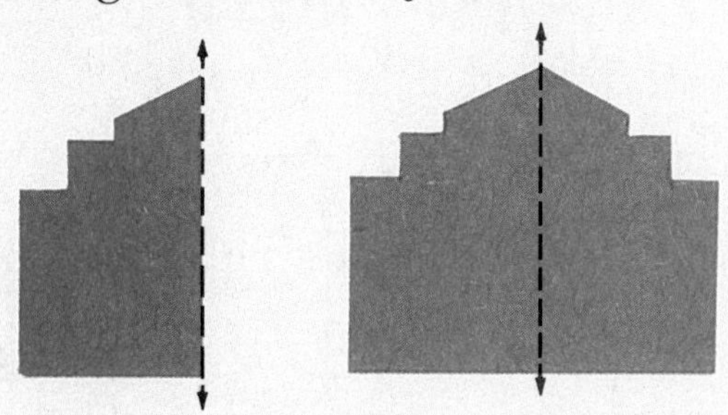

A figure has a **line of symmetry** if both parts of the figure are exactly the same size and shape.

Other examples:

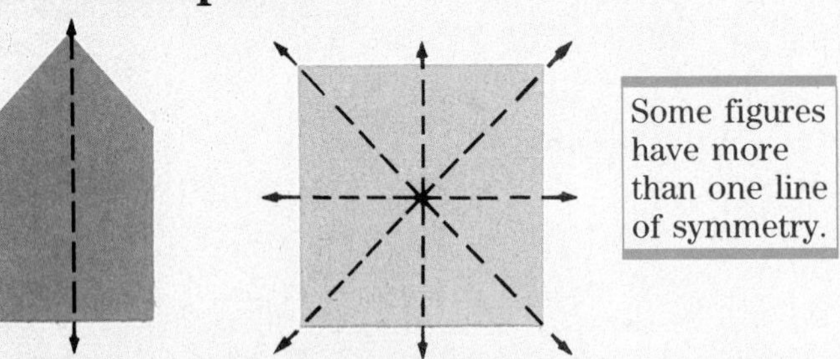

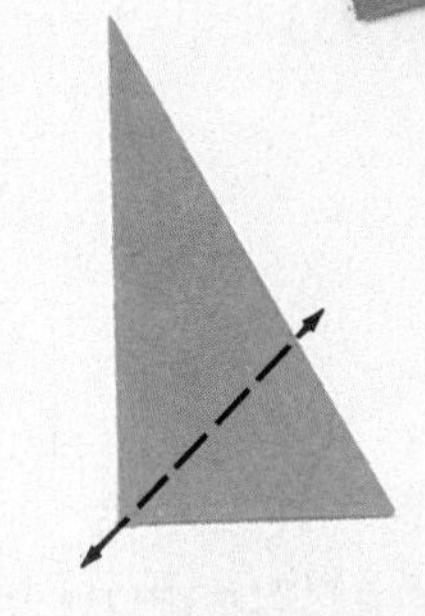

Some figures have no line of symmetry.

Does each figure have a line of symmetry? Write *yes* or *no*.

1.

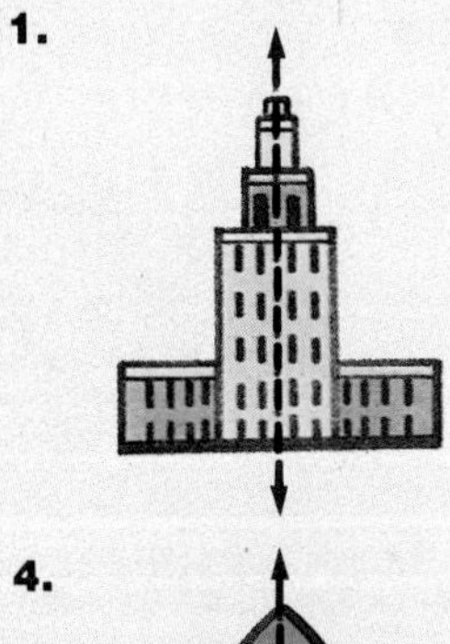

2.

3.

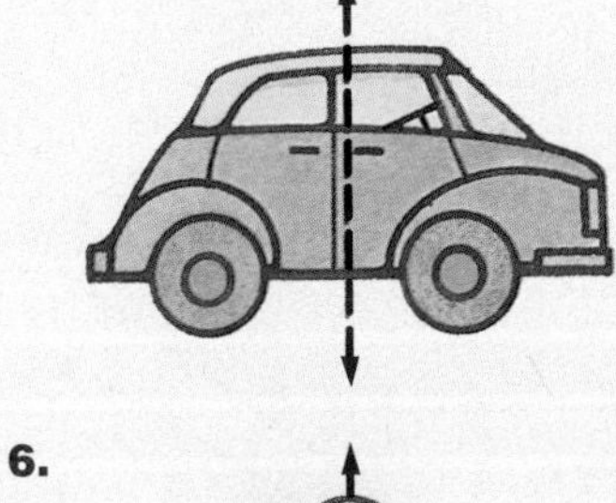

4.

5.

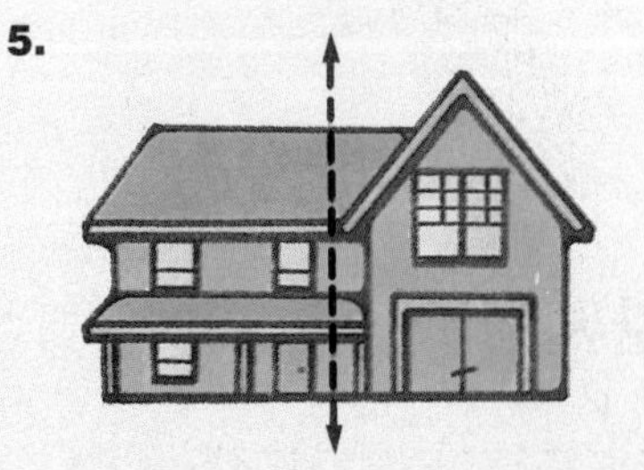

6.

Is the line a line of symmetry? Write *yes* or *no*.

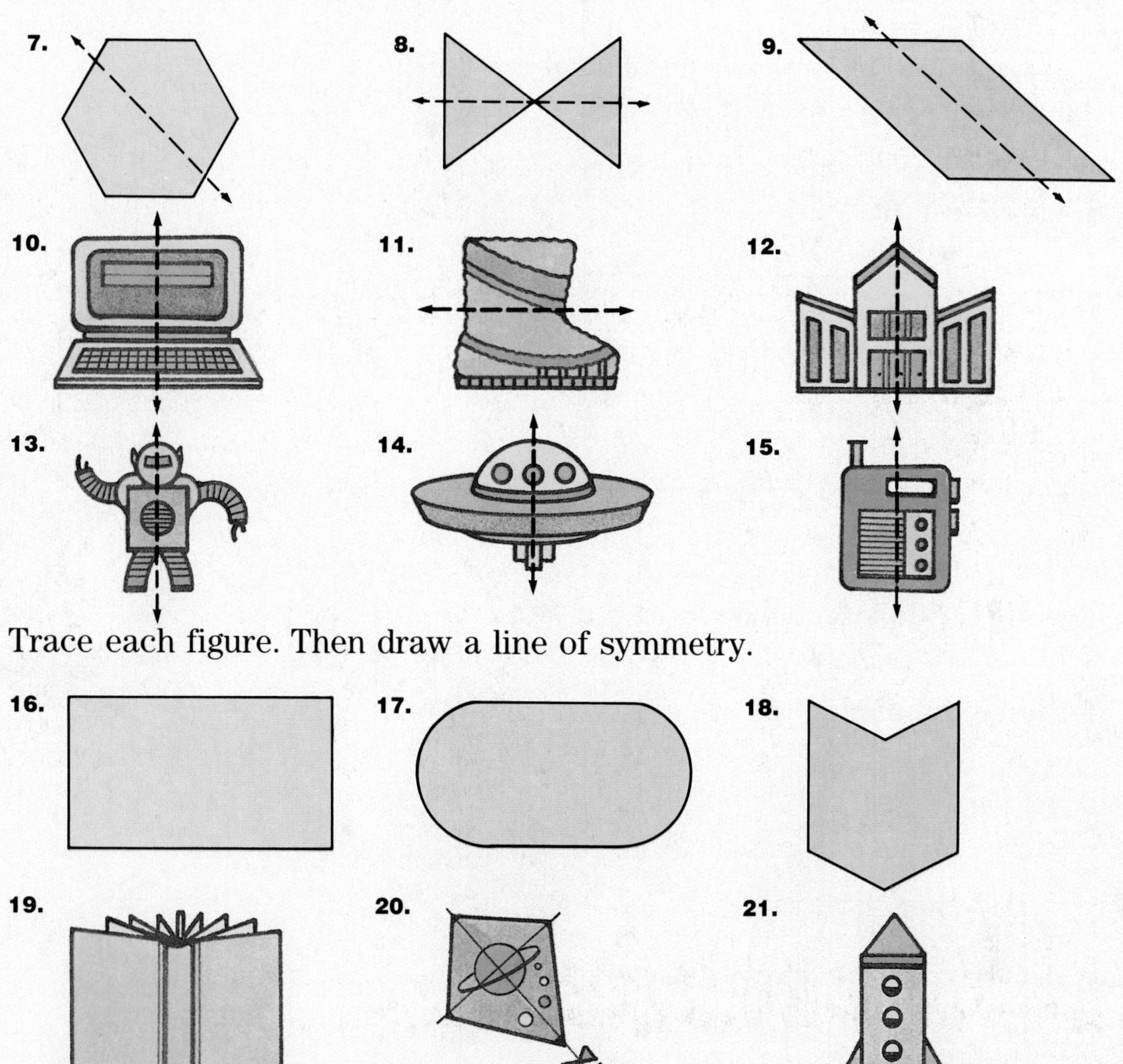

Trace each figure. Then draw a line of symmetry.

CHALLENGE

Write how many lines of symmetry each has.

1.

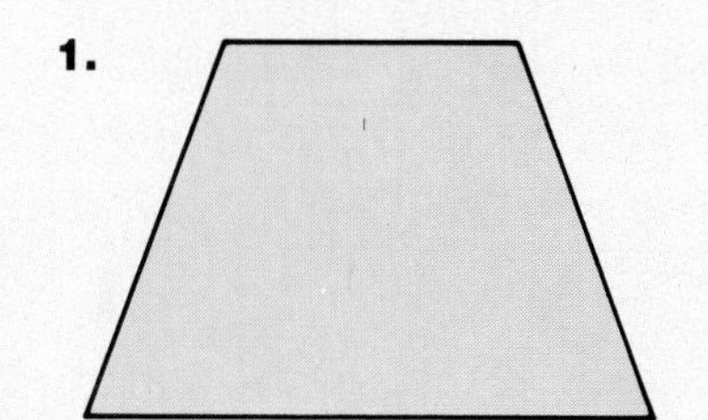

2.

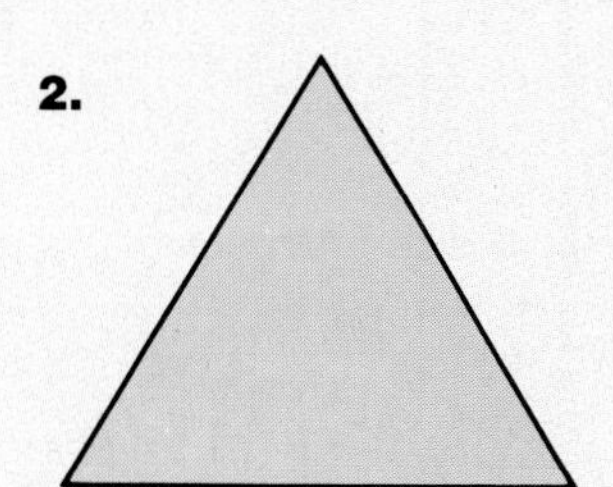

3.

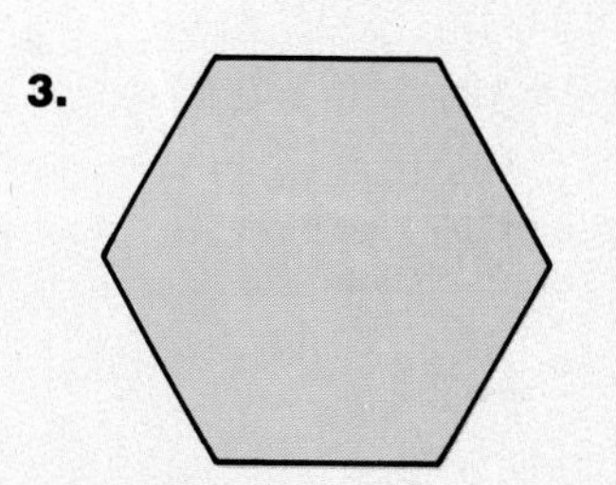

Area

The **area** of a figure is the number of square units that cover its surface.

This is a square centimeter. Each side is 1 cm long.

You can count square units to find the area of a figure.

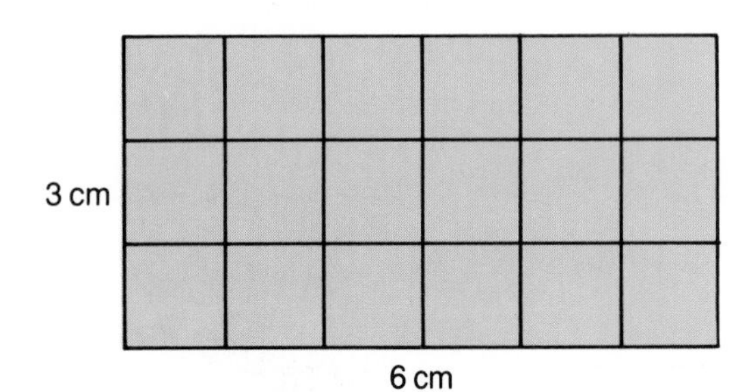

The area of this rectangle is 18 square centimeters.

Other examples:

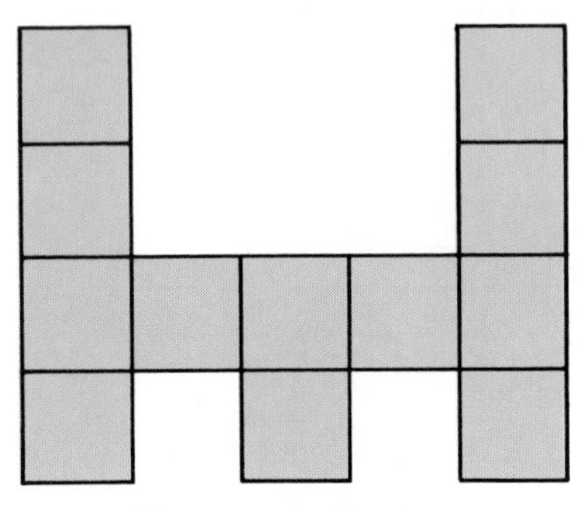

12 square centimeters

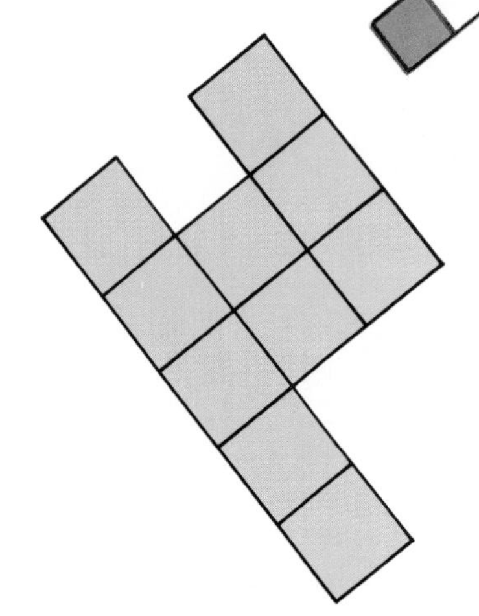

10 square centimeters

Count to find the area in square centimeters.

1.

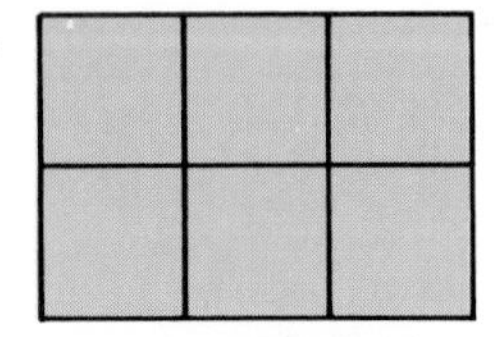

2.

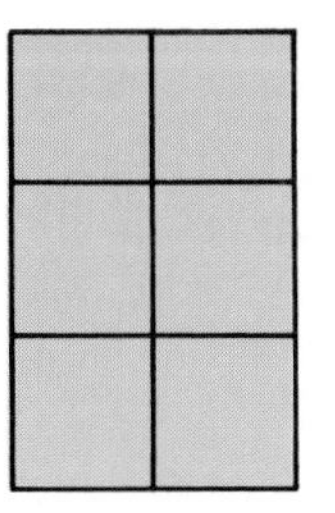

3.

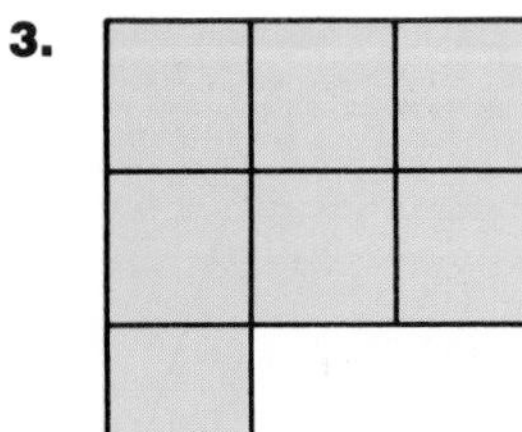

Find the area in square centimeters.

4.

5. 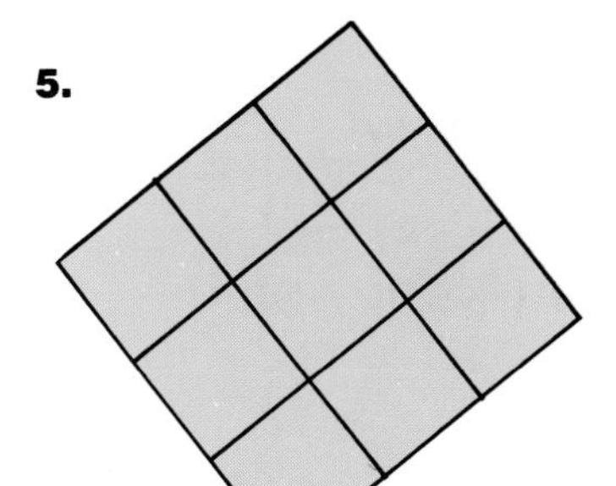

6. 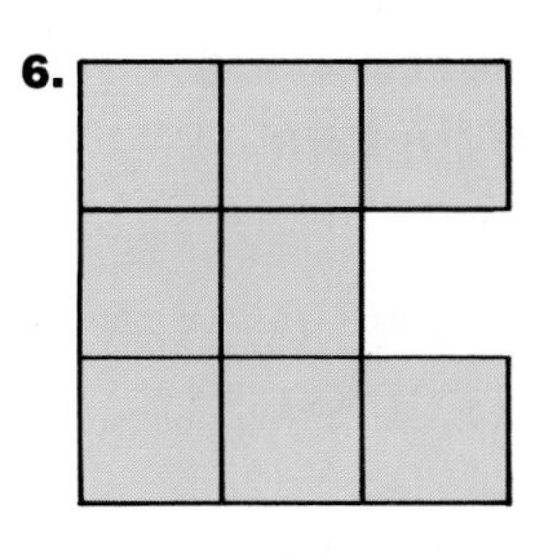

7. 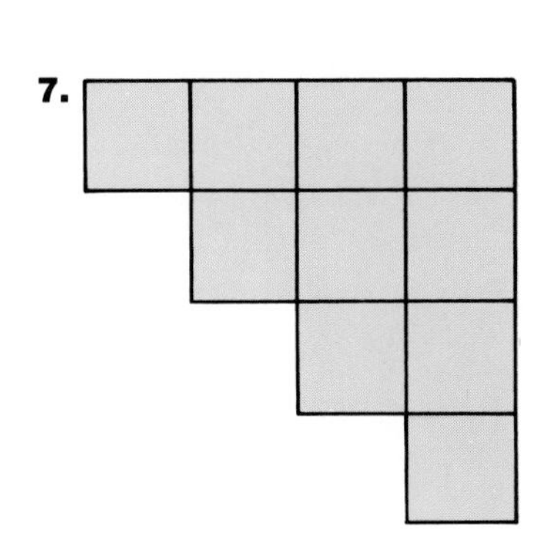

8. 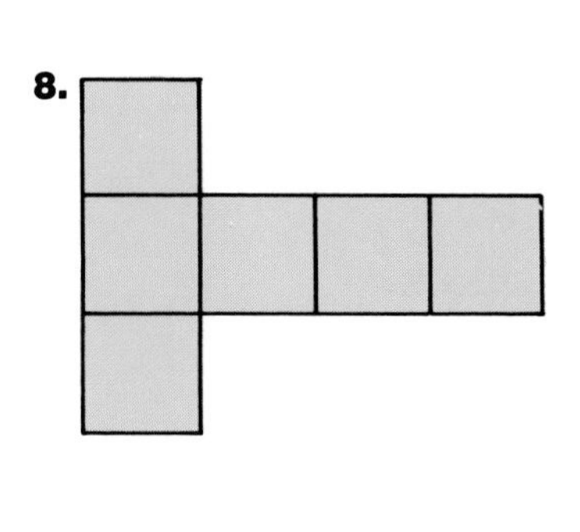

9. 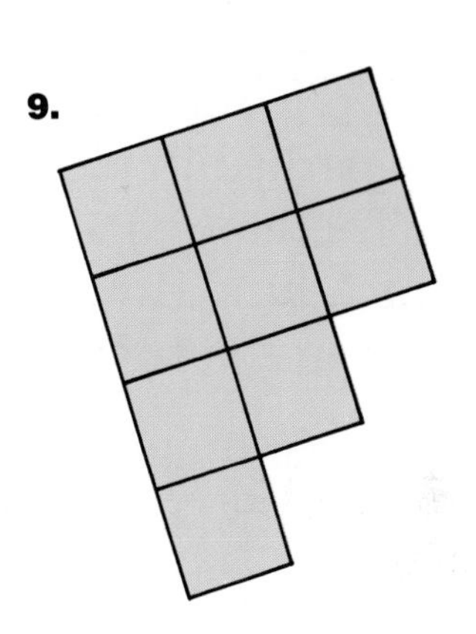

MIDCHAPTER REVIEW

How many sides are in each figure?

1.

2. 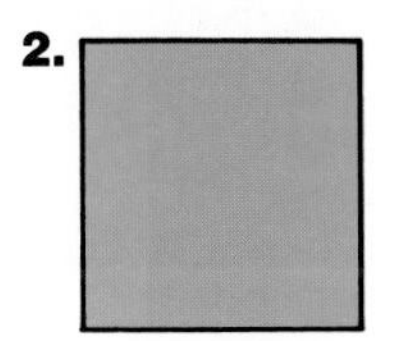

3. 

4. 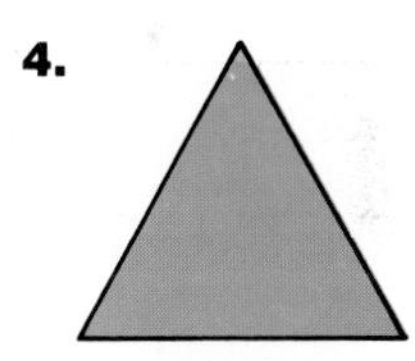

Write the letter of the figure that is congruent to the first figure shown.

5. 

a. 

b. 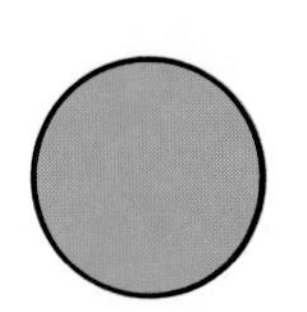

c. 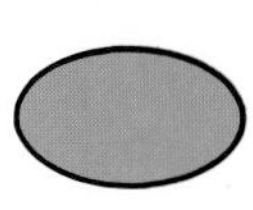

Is the line a line of symmetry?

6. 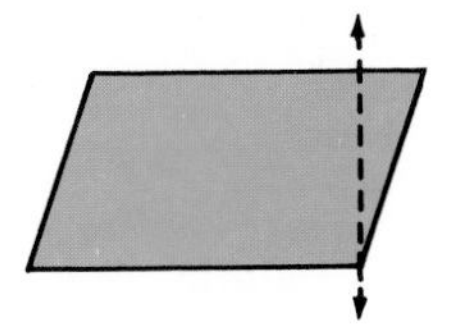

7. 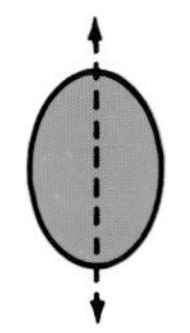

8. 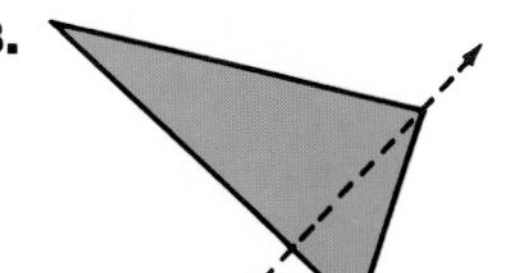

9. 

PROBLEM SOLVING
Drawing a Picture

Drawing a picture can help you solve some kinds of problems.

A. There is a row of parking spaces for airbikes in the garage. Every space is filled. Allie's airbike is parked in the middle of the row. There are 5 airbikes to the left of Allie's. How many bikes are parked in the row?

Make a simple drawing to help you count. Draw Allie's bike first. Use a circle or an x to show each bike.

left	middle bike (Allie's)	right	
● ● ● ● ●	●	● ● ● ● ●	
5 +	1 +	5	= 11

There are 11 bikes parked in the row.

You know there are 5 bikes to the left of the middle one. So, there must be 5 to the right of the middle one.

B. Each airbike rider has a locker. The lockers are numbered in order from left to right. Allie has the third locker to the right of the middle locker. There are 5 lockers between Allie's locker and Ed's locker. Ed's locker is next to the first locker. How many lockers are there in the row?

There are 9 lockers in the row.

Your drawing could look like this.

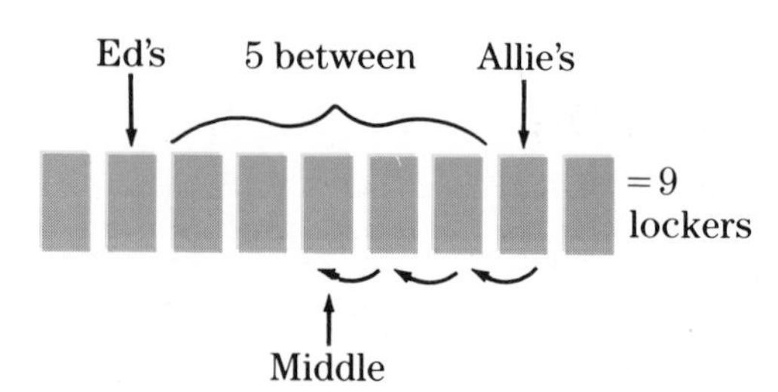

Copy and complete the picture. Solve each problem.

1. Jean lives in an apartment on the fifth floor of a building. Nathan lives four floors above her. Marie lives on the floor that is exactly between Nathan's floor and Jean's floor. Marie's floor is also the middle floor of the building. How many floors does the building have?

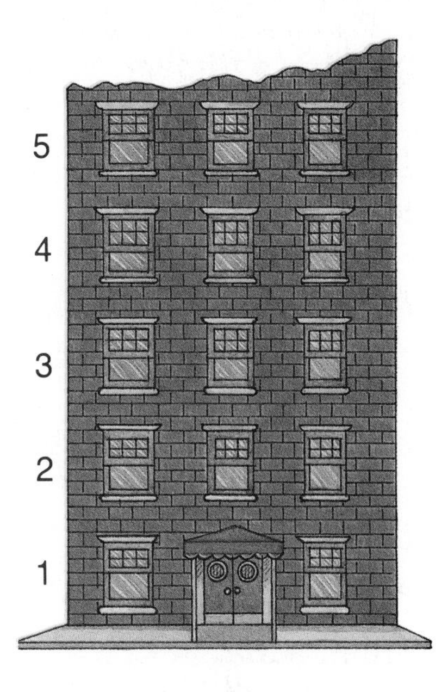

2. Elliot lives in an apartment that is three floors away from Marie and one floor away from Nathan. On which floor is Elliot's apartment?

Draw a picture to solve.

3. Mark is sitting on a step that is 2 steps above the middle step. Louise is sitting on a step that is 5 steps away from Mark. Louise is sitting on the bottom step. Mark is sitting 1 step below the top step. How many steps are there?

4. Henry is sitting on a step of the same stairway. He is 1 step away from Mark and 4 steps away from Louise. Which step is Henry on?

5. There are 9 cats sitting in a row. A black cat is sitting at each end of the row. A white cat is sitting to the right of the middle cat. A gray cat sits 3 places away from the white cat. A tiger cat sits 2 places away from the gray cat. A black cat sits next to the tiger cat. Which cat is sitting in the middle?

Hint: Show each cat as a circle. Under each cat, write *B* for a black cat, *W* for a white cat, and so on.

Space Figures

Jamie's class has built a model of their city of the future. Many of the buildings are shaped like these space figures.

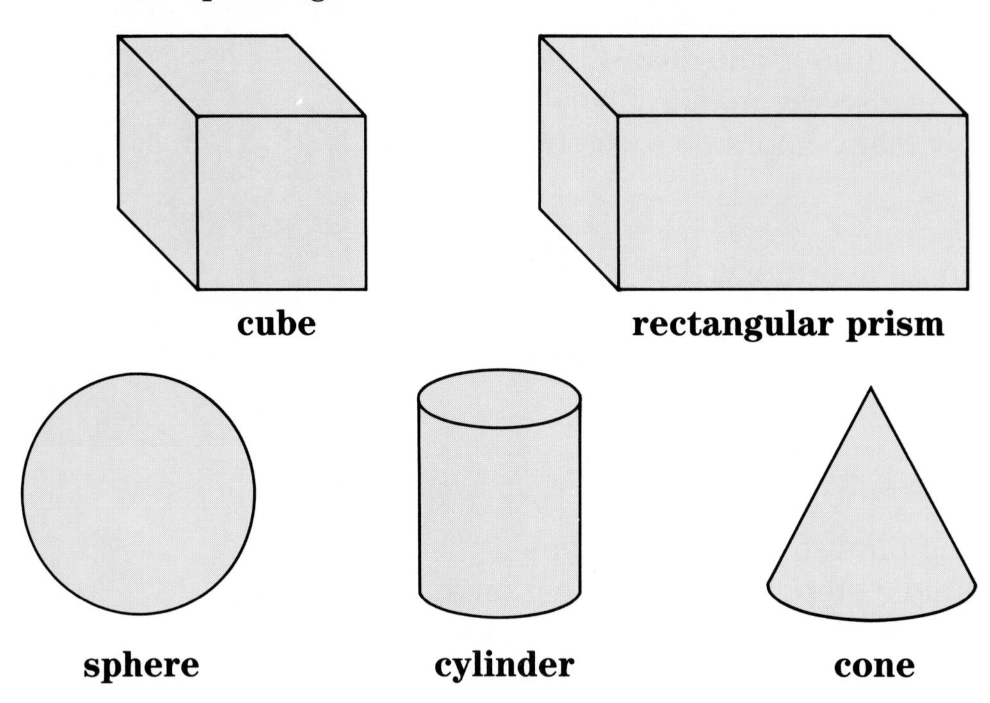

Name the space figure.

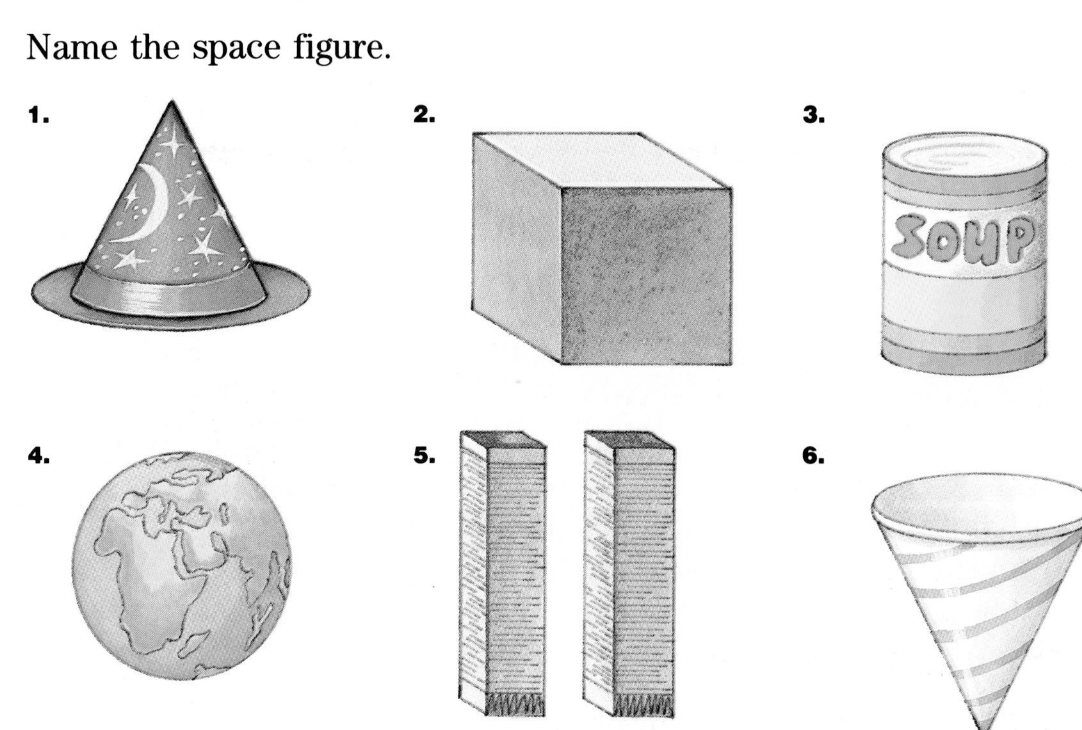

Name the space figure.

7.

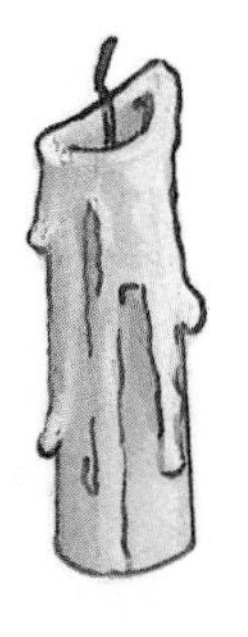

8.

9.

10.

11.

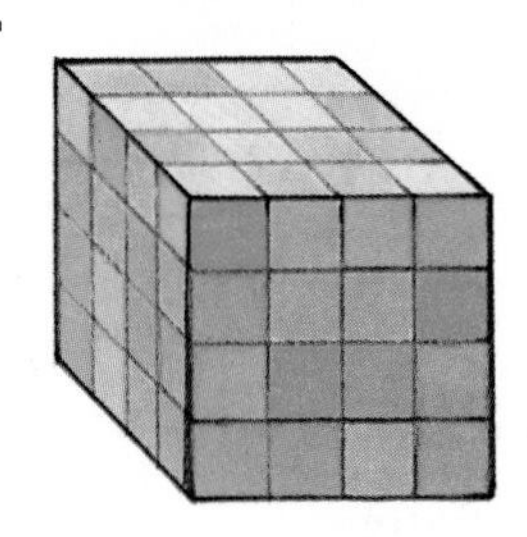

12.

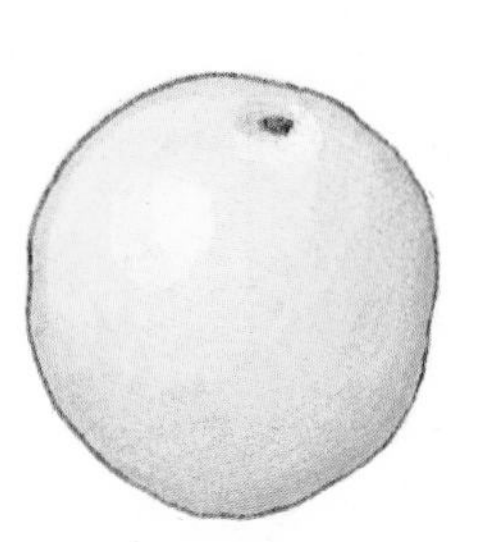

13.

14.

15.

What shape is

16. a bucket? 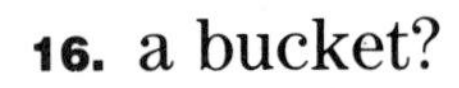

17. a book? 

18. an ice cube?

19. Write how many of each shape.

___ cubes ___ cones ___ spheres

___ rectangular prisms ___ cylinders

Faces, Edges, and Vertices

A. Wendy built a model house for the year 2050. The house had both curved and flat surfaces.

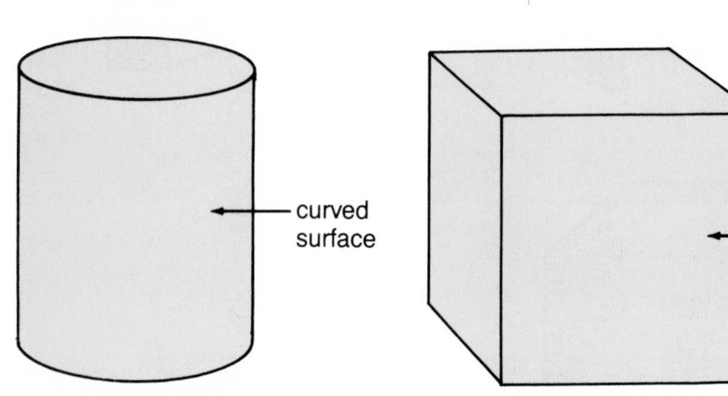

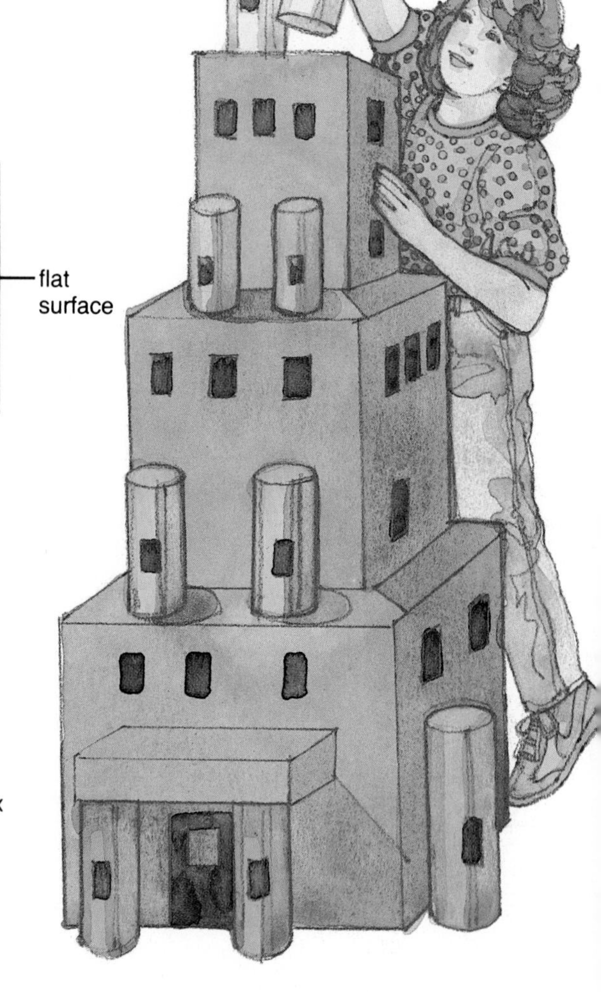

B. Any flat surface of a space figure is called a **face.**

An **edge** is where two faces meet. Edges meet at a **vertex.**

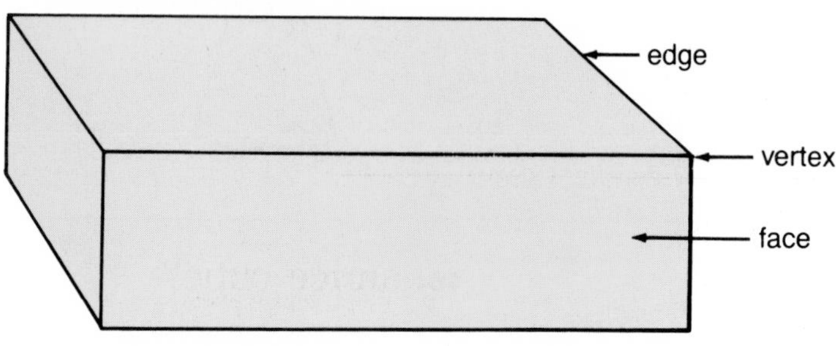

Write the number of surfaces.

1.

____ flat

____ curved

2.

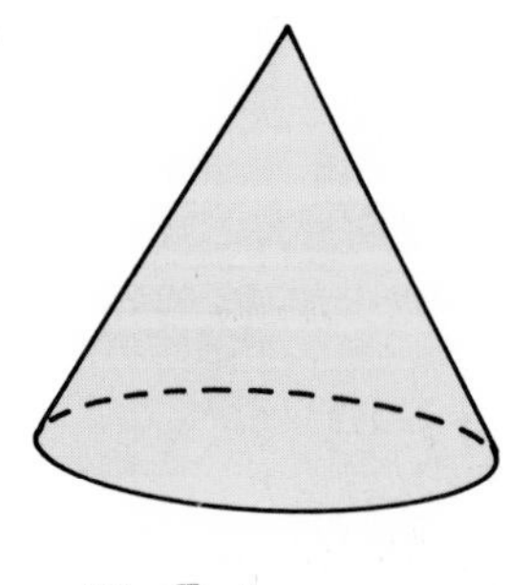

____ flat

____ curved

3.

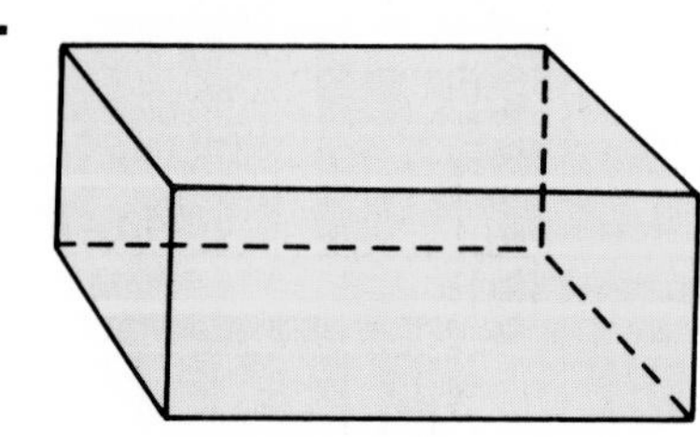

____ flat

____ curved

Write the number of surfaces.

4.

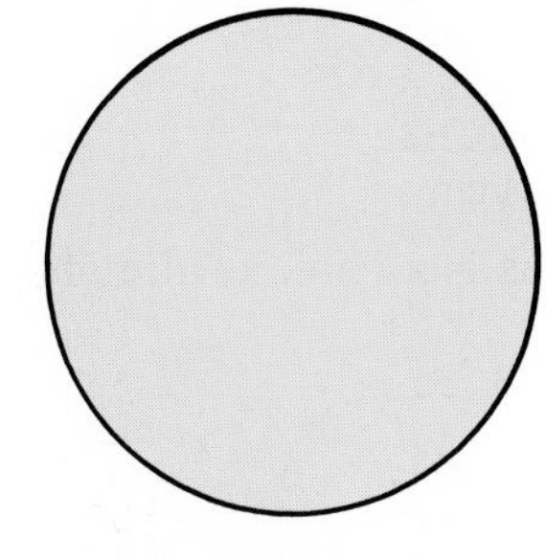

___ flat

___ curved

5.

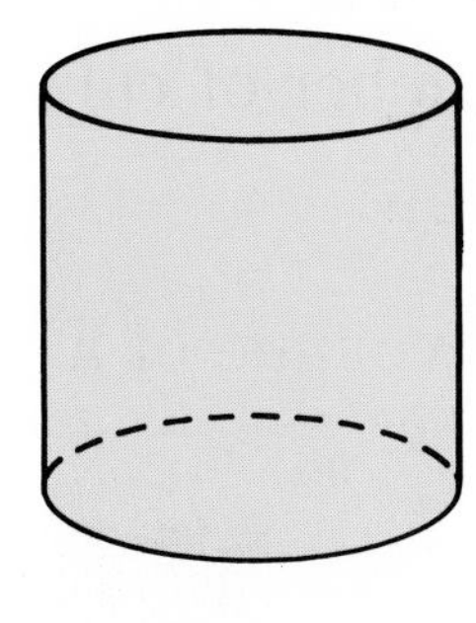

___ flat

___ curved

★6.

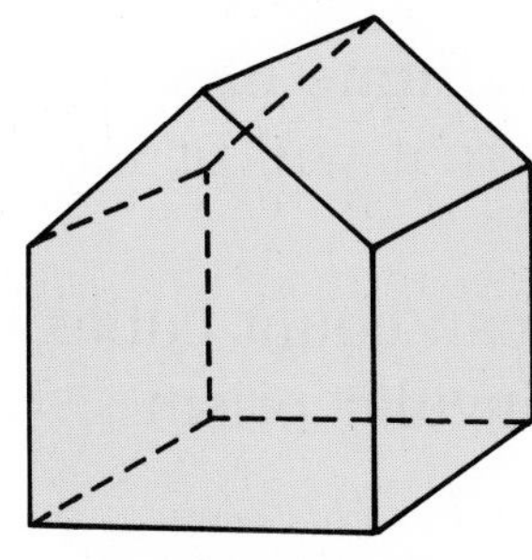

___ flat

___ curved

Write the number of faces, edges, and vertices.

7.

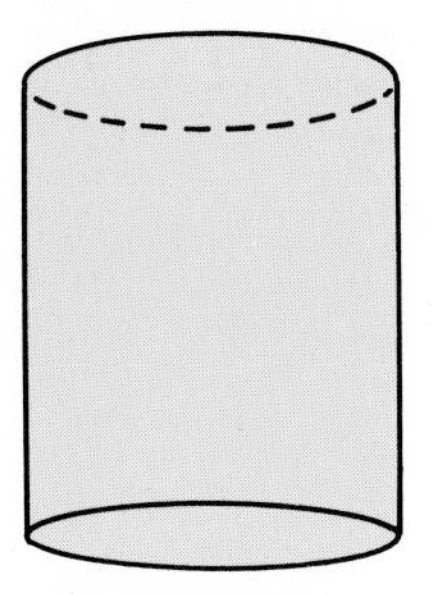

___ faces

___ edges

___ vertices

8.

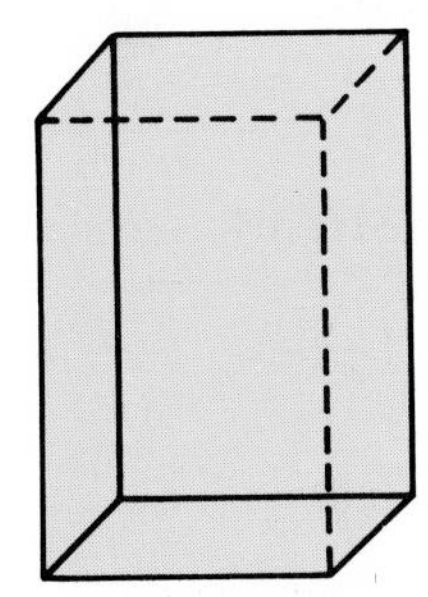

___ faces

___ edges

___ vertices

9.

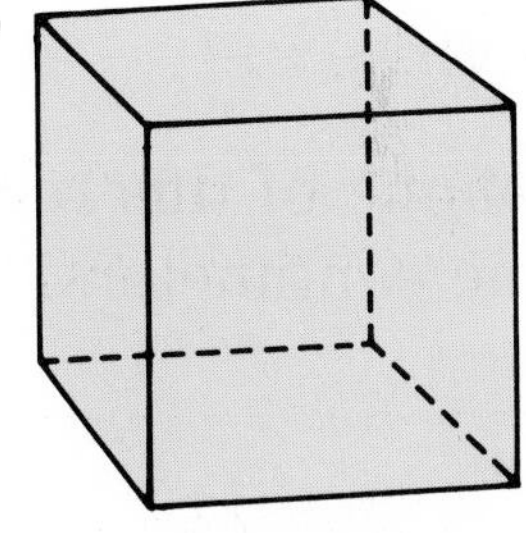

___ faces

___ edges

___ vertices

10.

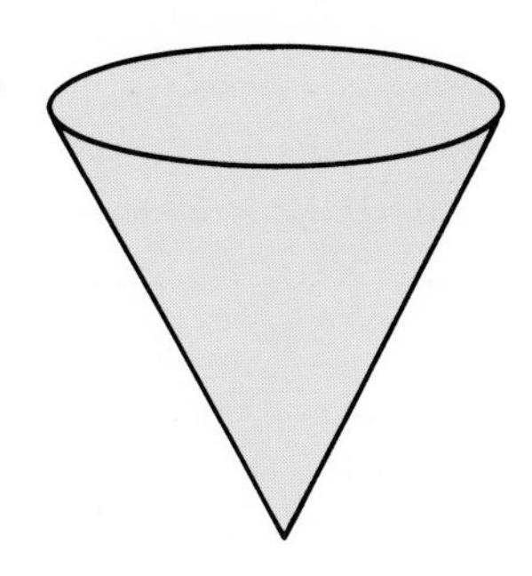

___ faces

___ edges

___ vertices

★11.

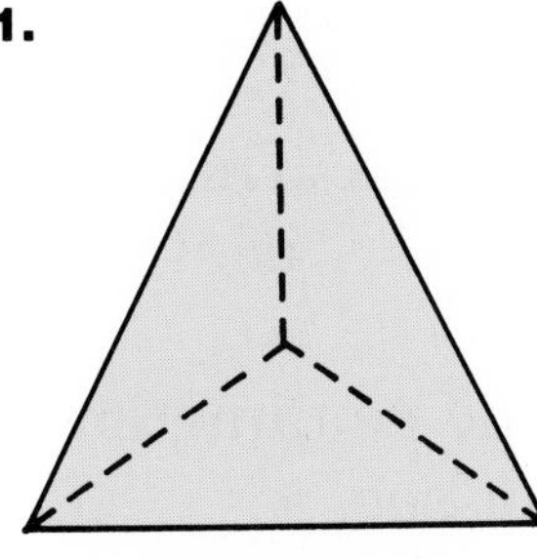

___ faces

___ edges

___ vertices

★12.

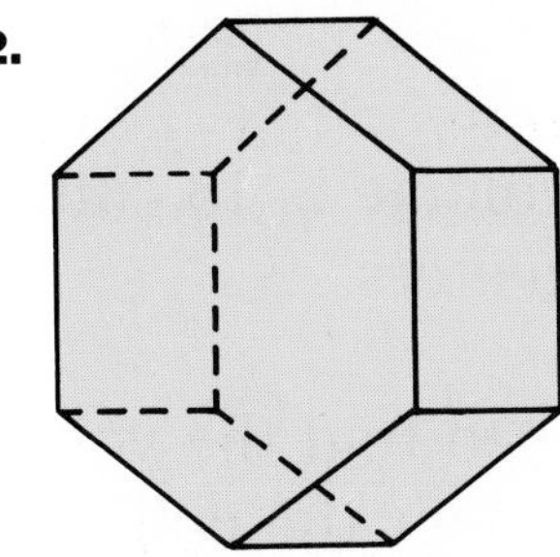

___ faces

___ edges

___ vertices

Volume

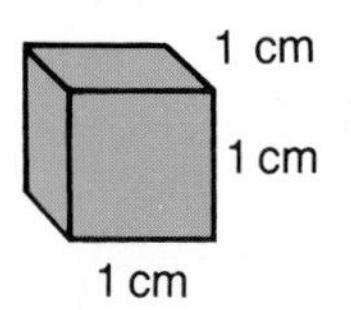

This is a cubic centimeter.

The **volume** of a figure is the number of cubic units that will fit into it.

You can count cubes to find the volume of a rectangular prism.

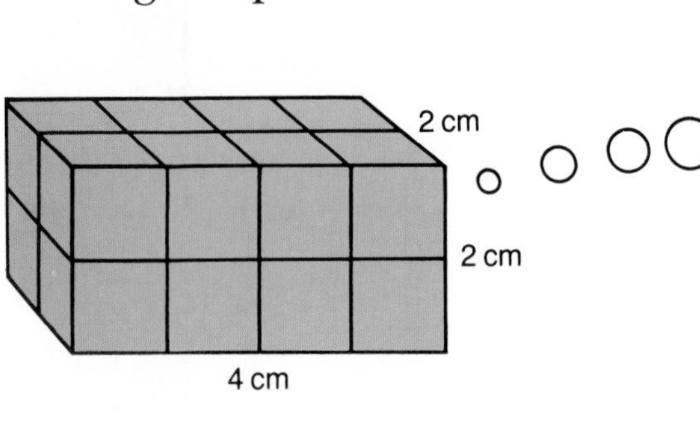

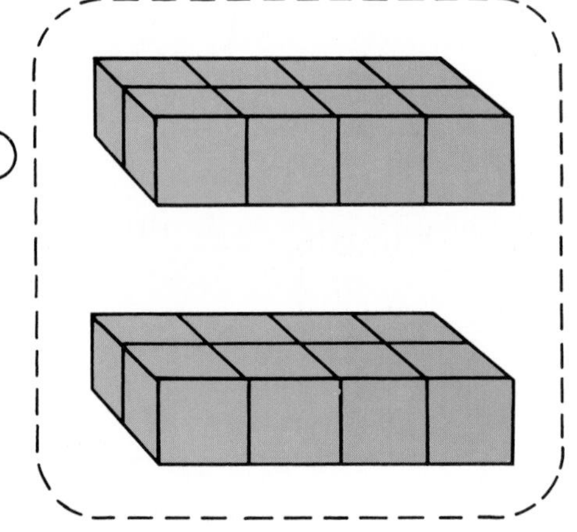

The volume of this rectangular prism is 16 cubic centimeters.

Other examples:

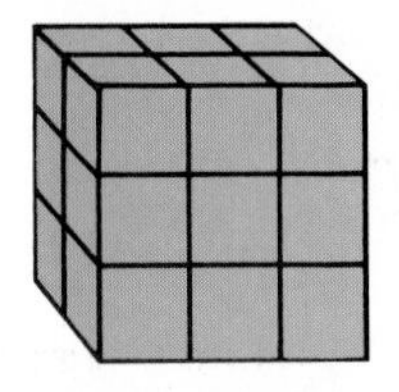

Remember the cubes you cannot see.

The volume is 18 cubic centimeters.

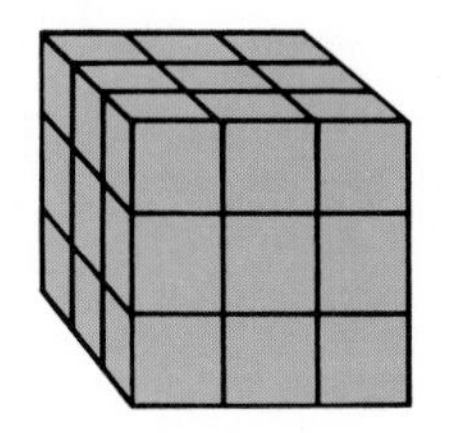

Remember the cubes you cannot see.

The volume is 27 cubic centimeters.

Count to find the volume in cubic centimeters. Remember the cubes you cannot see.

1. 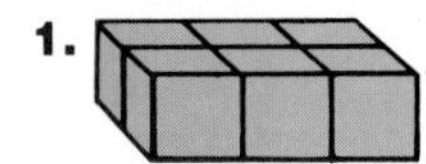

2. 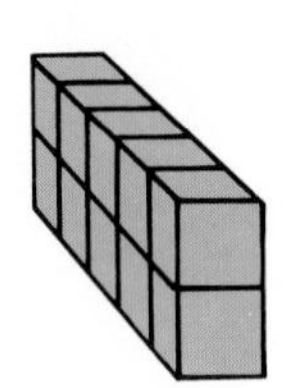

Find the volume in cubic centimeters.

3. 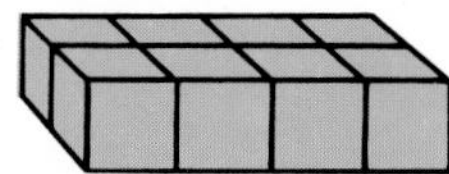

4. 

5. 

6. 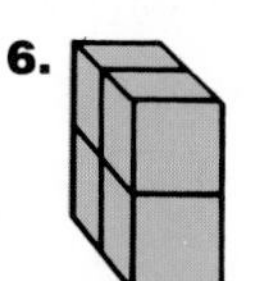

7. 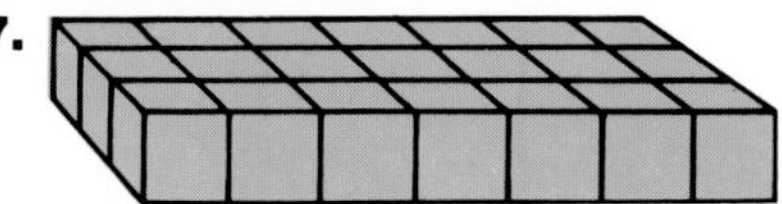

8.  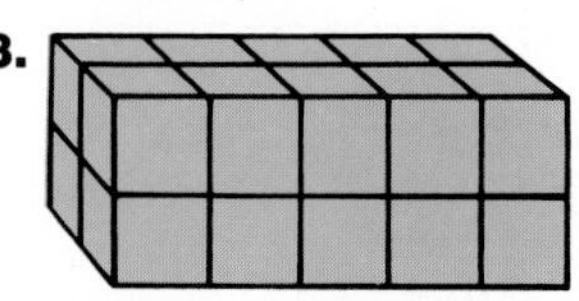

9. 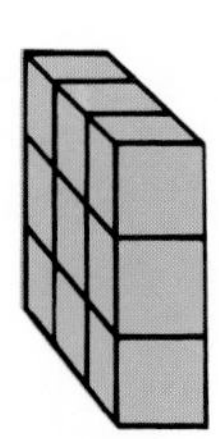

10.  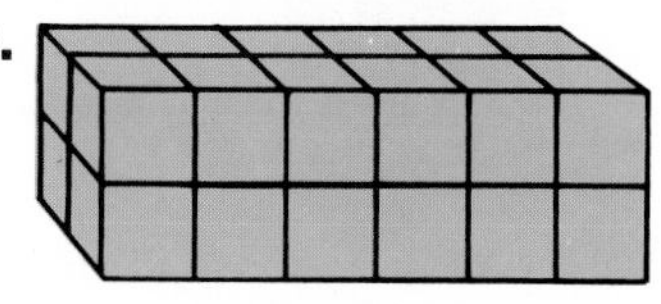

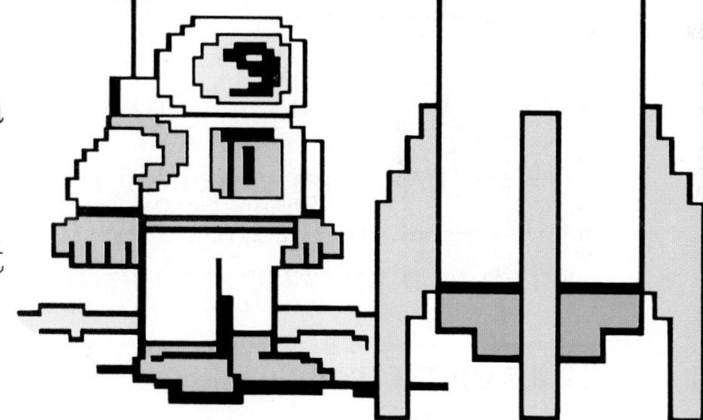

FOCUS: MENTAL MATH

You can multiply to find the volume of a cube or a rectangular prism.

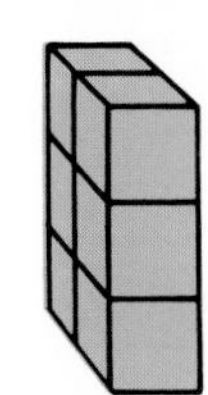

Volume = length × width × height

$V = 2 \times 1 \times 3$

$V = 6$ cubic centimeters

Multiply mentally to find the volume.

1. 

2. 

3. 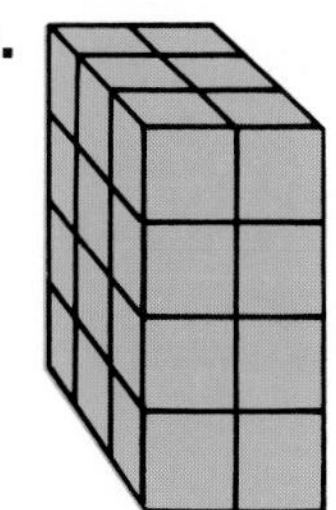

Tallies and Tables

A. The third-grade class visited a factory that made robots. Janet kept a **tally** of the different kinds of robots she saw in three hours from 9:00 to 12:00. Each time she saw one, she put a tally mark on her sheet. How many police robots did she see from 10:00 to 11:00?

Hour 1 (9:00–10:00)	
Police Robots	/
Servant Robots	/
Welder Robots	//

Hour 2 (10:00–11:00)	
Police Robots	//
Servant Robots	///
Welder Robots	//

Hour 3 (11:00–12:00)	
Police Robots	
Servant Robots	//
Welder Robots	~~////~~

~~////~~ = /////

Janet saw 2 police robots from 10:00 to 11:00.

B. Janet used her tally sheet to prepare a **table.** The table shows the number of robots Janet saw in 3 hours. How many welder robots did she see from 11:00 to 12:00?

To find how many welder robots she saw from 11:00 to 12:00, find the row for 11:00 to 12:00 and the column for welder robots.

	Police Robots	**Servant Robots**	**Welder Robots**
Hour 1 (9:00–10:00)	1	1	2
Hour 2 (10:00–11:00)	2	3	2
Hour 3 (11:00–12:00)	0	2	5

Janet saw 5 welder robots from 11:00 to 12:00.

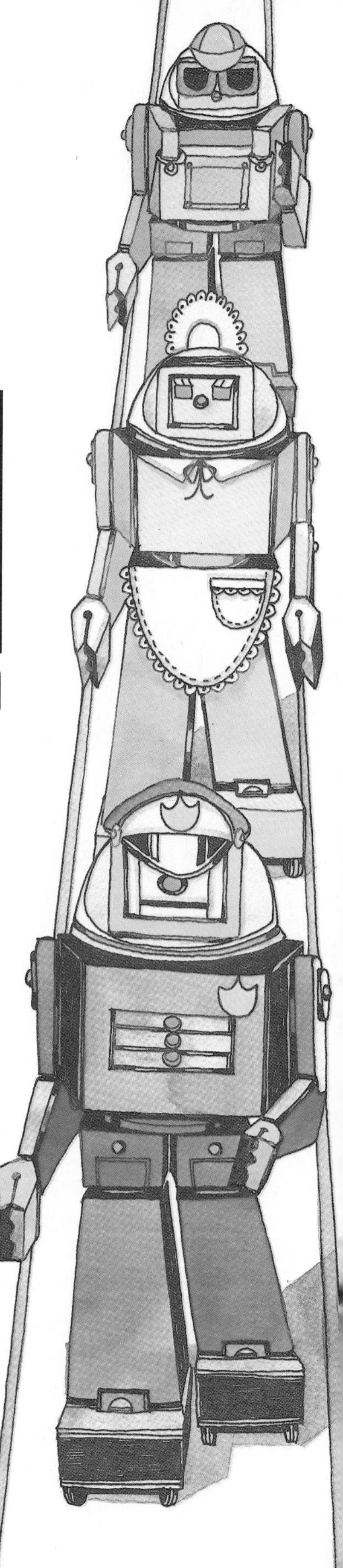

Copy the tally sheet. Use the picture to complete it.

1.

Police Robots	
Servant Robots	
Welder Robots	

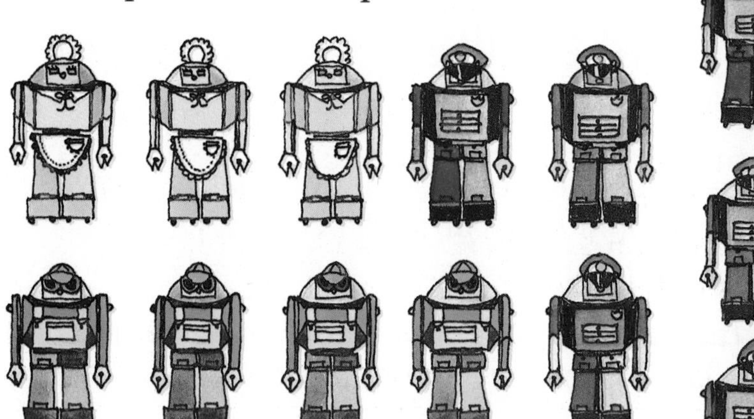

Use the information on the tally sheet above to solve.

2. How many police robots are there?

3. How many police robots and servant robots are there altogether?

Copy the table. Use the information to complete it.

4. Android, Inc., makes different types of robots. On Monday the factory made 7 waiters, 22 sweepers, and 13 doctors. On Tuesday it made 10 waiters, 15 sweepers, and 21 doctors. On Wednesday it made 17 waiters, 8 sweepers, and 14 doctors.

	Waiter Robots	Sweeper Robots	Doctor Robots
Monday			
Tuesday			
Wednesday			

Use the table above to solve.

5. On Tuesday, how many more doctor robots were made than waiter robots?

6. How many more sweeper robots were made on Monday than on Wednesday?

7. On which day were the fewest waiter robots made?

★8. On which day were the most robots made?

PROBLEM SOLVING

Using a Graph

It is easy to compare information that is shown in a bar graph. This bar graph compares the number of rooms in a space station.

Use the bar graph to answer this question. How many laboratories are there in the station?

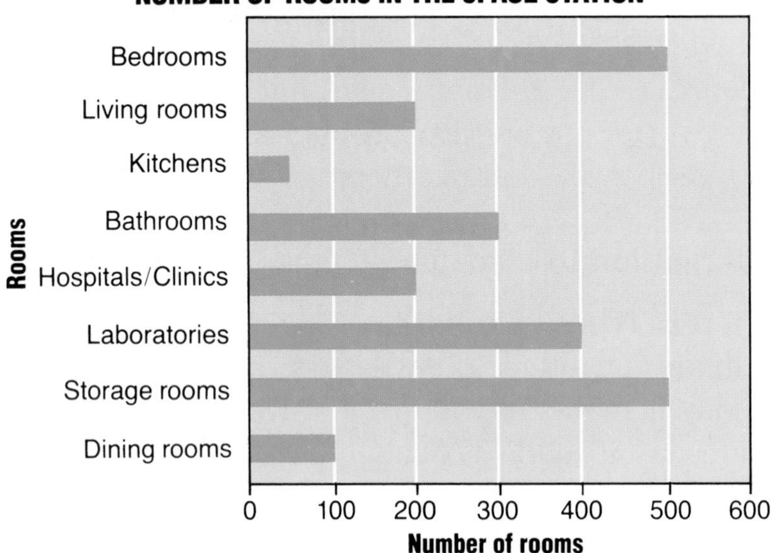

- The title tells you that this graph shows how many rooms there are in the space station.
- The labels at the left tell you what kinds of rooms there are in the space station.
- The labels at the bottom of the graph tell you how many rooms are used for each purpose.

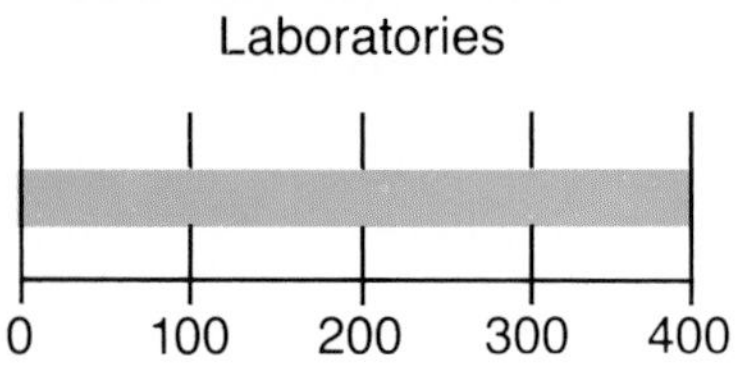

For example, the bar next to the label *Laboratories* goes to 400. This means that there are 400 laboratories in the station.

Use the bar graph on page 328 to answer each question.

1. Are there more bedrooms, living rooms, or dining rooms in the space station?

2. There are fewer than 100 of which kind of room?

3. Each living room and each dining room has a large ceiling window so people can watch the sky. How many ceiling windows are there in all the dining rooms and living rooms?

4. Are there more kitchens or bathrooms?

5. Can you tell how large the bedrooms are from the graph? Why or why not?

6. If a doctor works in each hospital or clinic, how many doctors work in the space station?

7. What is kept in the storage rooms?

8. There are 175 clinics in the space station. How many hospitals are there?

★9. Suppose there are 2 space stations in orbit. Each one has the same number of rooms as are shown on this graph. How many rooms of each kind would there be on 2 stations?

Making Pictographs

Randy wants to make a pictograph of the number of students in the second, third, fourth, and fifth grades who want to be astronauts. First he writes a title for the graph. Then he draws a rocket for a picture key. Each rocket will stand for 2 students.

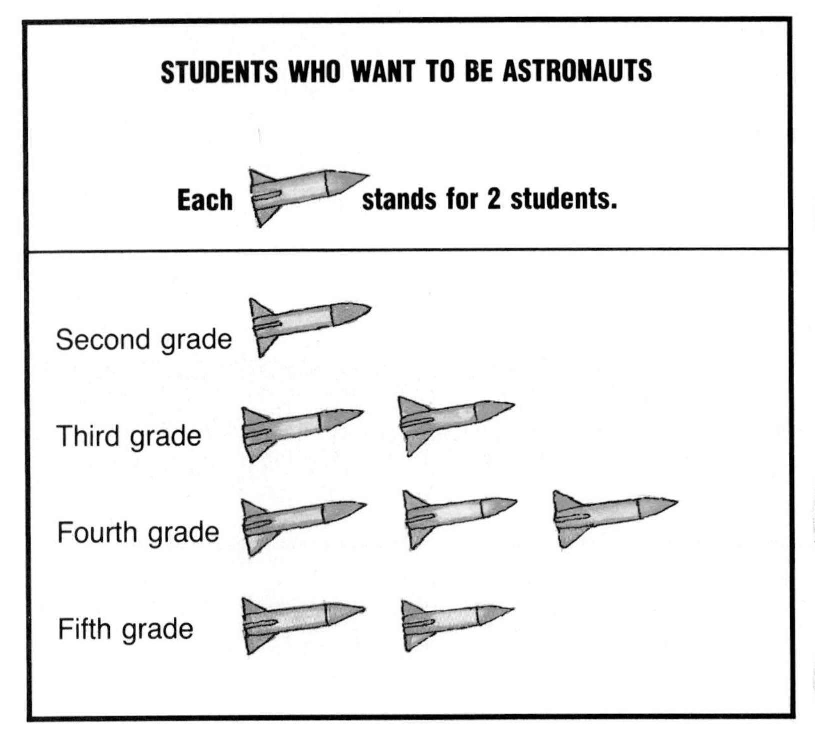

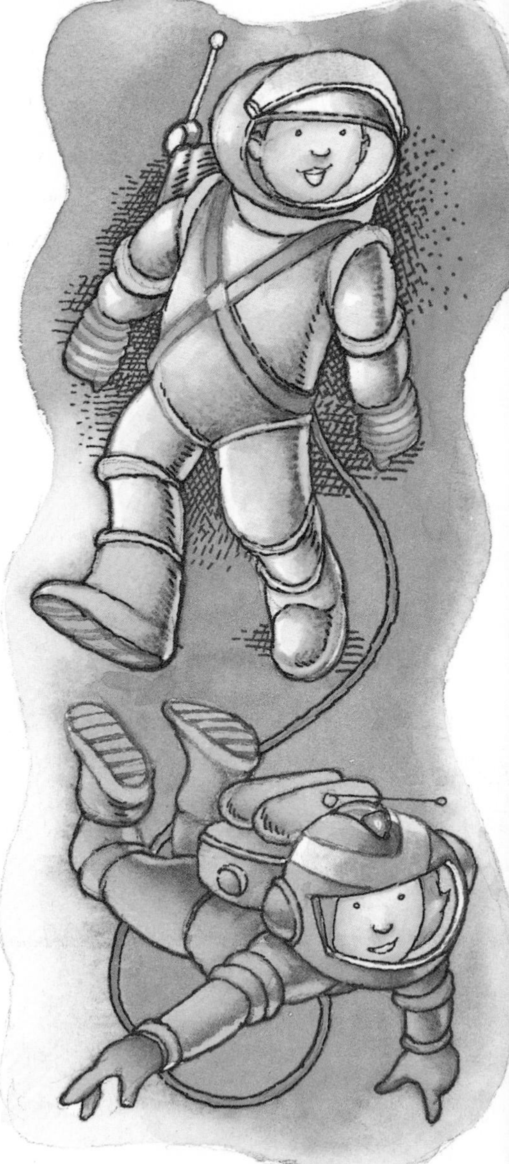

Use the pictograph to solve.

1. How many children in the fourth grade want to be astronauts?

2. How many children altogether want to be astronauts?

Solve. Use the Infobank on page 403. Include a title and a picture key.

3. Make a pictograph that shows the number of children in each grade who are thinking of becoming artists.

4. Make a pictograph that shows the number of children in each grade who are thinking of becoming athletes.

Making Bar Graphs

Jim wants to make a bar graph of what his friends want to be when they are older. He writes the number of friends on the left-hand side of the graph. Then he writes the jobs at the bottom of the graph. He draws bars to show how many friends want to do each job.

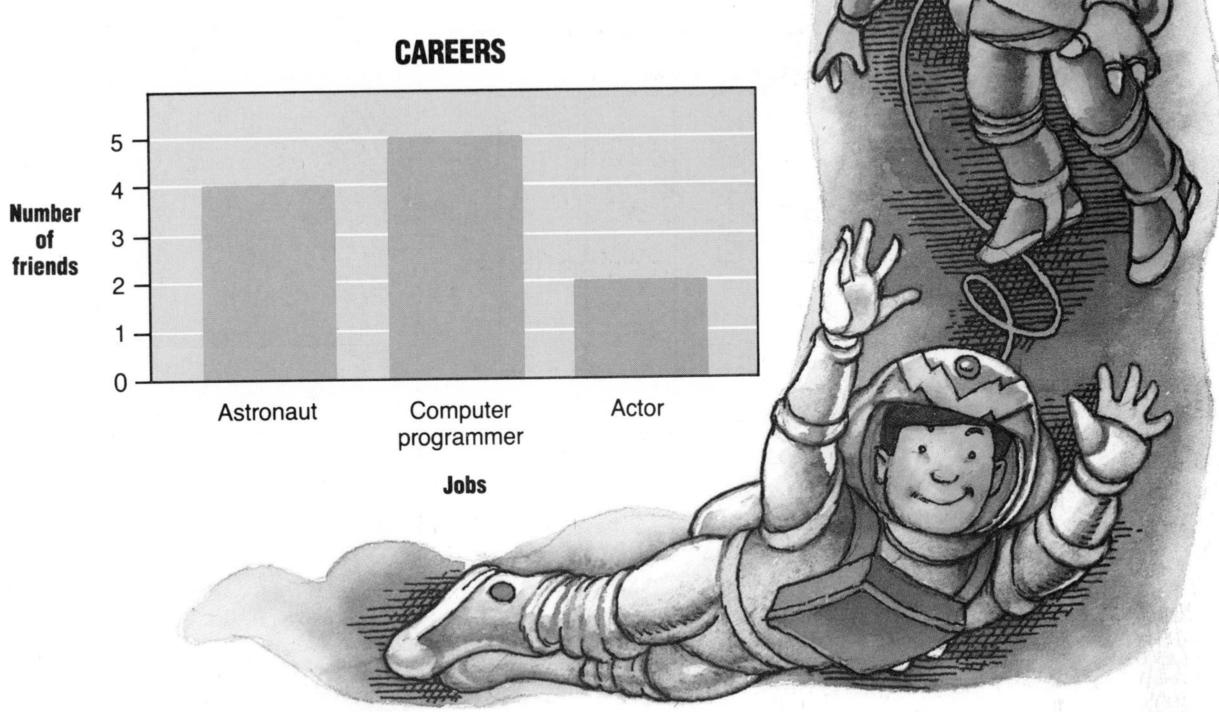

Use the bar graph to solve.

1. How many of Jim's friends want to be actors when they are older?

2. How many more friends want to be computer programmers than astronauts?

Use the information to make your own bar graphs. Remember to label the graphs.

3. There are 8 students who want to be pilots, 3 students who want to be doctors, 7 students who want to be athletes, and 6 students who want to be lawyers.

4. Take a survey in your class to find out what jobs your friends want to have when they grow up. Make a bar graph that shows the results.

Number Pairs

Number pairs tell you where a point can be found on a graph. Below is a graph that shows stars in a sky. Which star is at (1,4)?

You can find the star by counting.
Go to the right to 1 and up to 4.

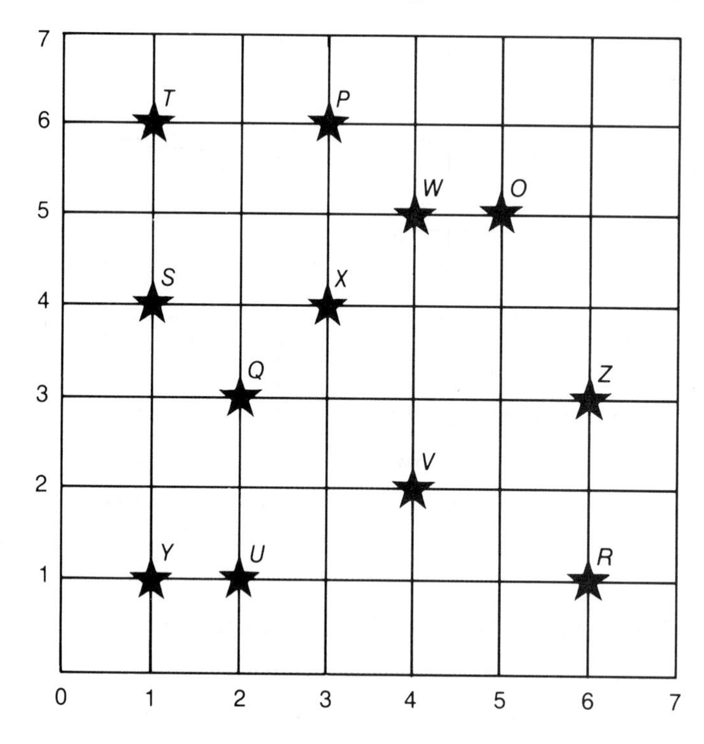

Star S is at (1,4).

Another example:
What is the number pair for Star U?
Star U is 2 to the right and up 1.
Star U is at (2,1).

Use the graph above to answer each question. What star is at

1. (3,4)? **2.** (4,5)? **3.** (1,6)? **4.** (2,3)?

5. (6,1)? **6.** (6,3)? **7.** (5,5)? **8.** (4,2)?

9. (3,6)? **10.** (1,1)?

Write the number pair for each.

11. the radio tower

12. the school

13. the library

14. the museum

15. the theater

16. the bank

17. the pond

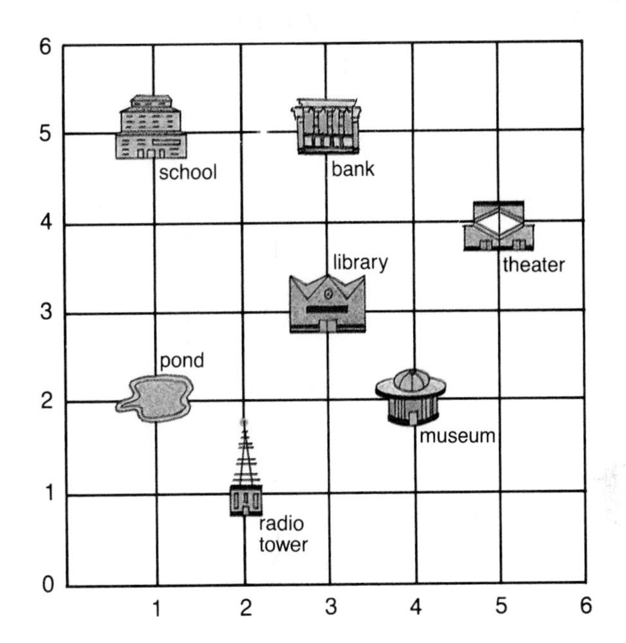

Copy the graph. Then draw points for each number pair.

18. (7,4)

19. (6,2)

20. (2,7)

21. (5,1)

22. (4,3)

23. (3,6)

24. (4,7)

25. (7,7)

26. (5,6)

27. (1,2)

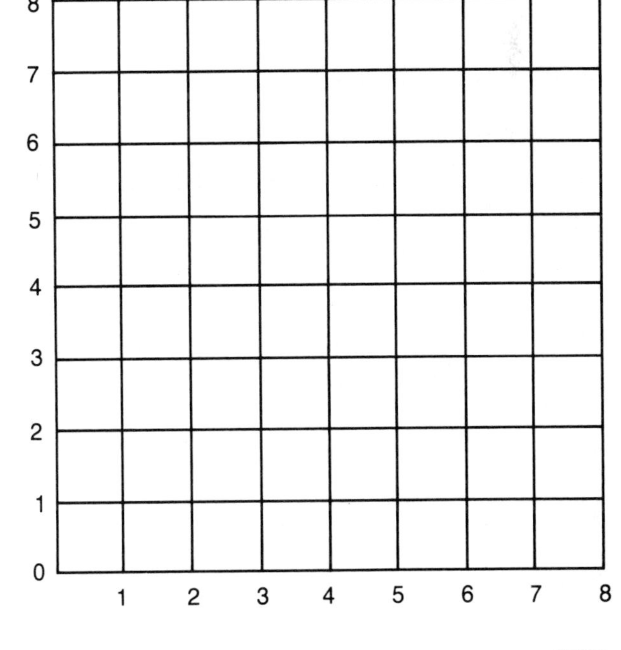

LOGICAL REASONING

You can use *all*, *some*, or *none* to describe how two sets of objects are related.

All of the squares are figures.
Some of the figures are squares.
None of the squares are triangles.

Notice that a triangle is always a figure, but a figure is not always a triangle.
A triangle is never a square.

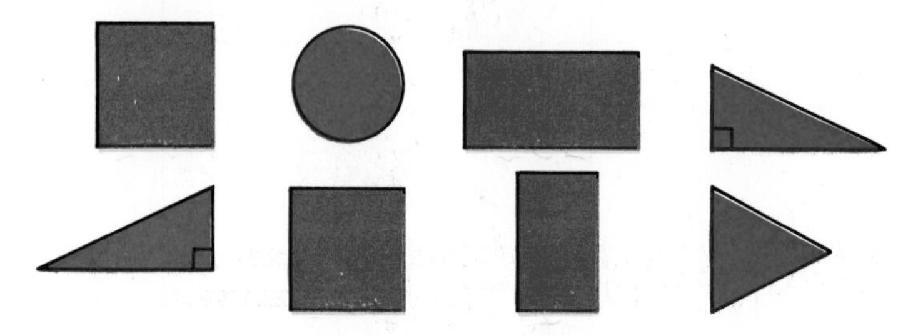

Complete. Write *All*, *Some*, or *None*.

1. ___ of the squares are red.
2. ___ of the red figures are squares.
3. ___ of the circles are red.
4. ___ of the triangles are congruent.
5. ___ of the squares are congruent.
6. ★ ___ of the squares are rectangles.

Write *always*, *sometimes*, or *never*.

7. The day after Monday is ___ Sunday.
8. A month ___ has 30 days.
9. The day before tomorrow is ___ yesterday.
10. The month before August is ___ July.
11. A 2-digit number is ___ greater than a 1-digit number.
12. The sum of three 1-digit numbers is ___ less than 20.

GROUP PROJECT

Space Palace

The problem: You have bought a new space house on the planet Zurg. You have 2,000 zurgons to spend on items for your house. Decide which items you can buy and still stay within your budget.

Key Facts

- Your house is far from any supply store.
- On Zurg, the weather is springlike all year long.
- The sun never sets on Zurg.
- You cannot bring furniture from Earth to Zurg. They don't allow it on the space shuttle.

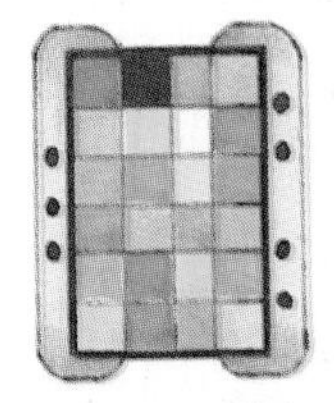

211 zurgons
zenatone clock

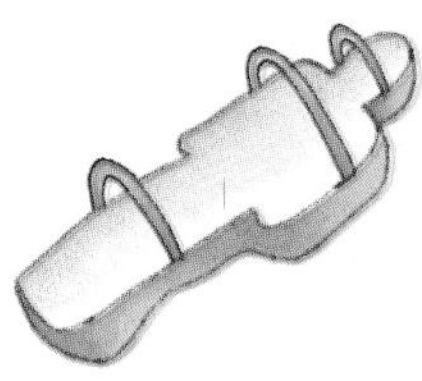

344 zurgons
sleep box

146 zurgons
galacto chair

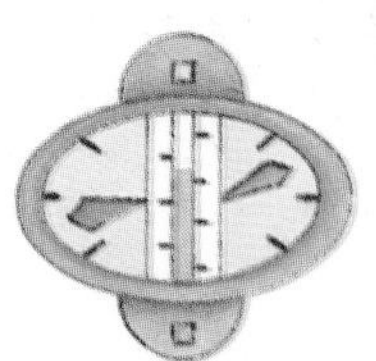

109 zurgons
tempatron control

561 zurgons
meal-o-rama

88 zurgons
floating planettable

578 zurgons
pluton jet pack

114 zurgons
lunar lamp

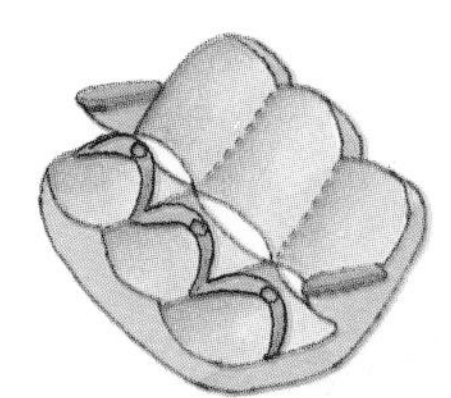

413 zurgons
cosmo couch

243 zurgons
astro dog robot

933 zurgons
gravity bike

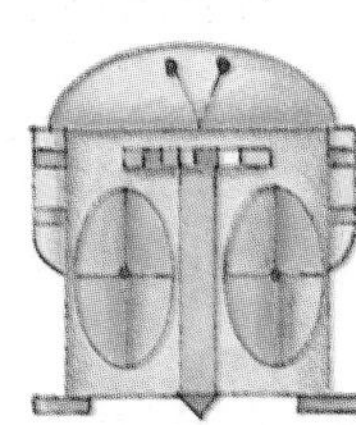

337 zurgons
sonic hifi system

CHAPTER TEST

Name each. (pages 304 and 308)

1.

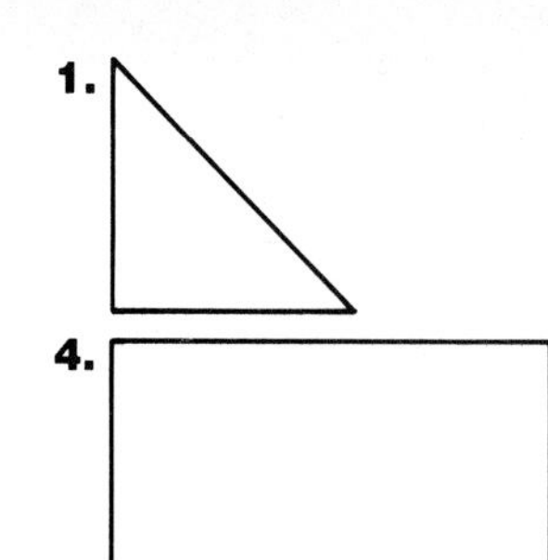

2.

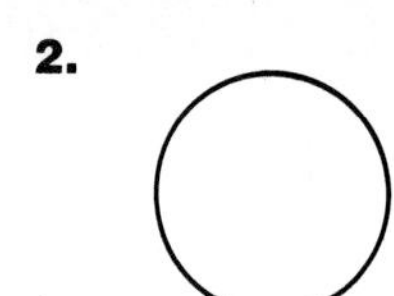

3.

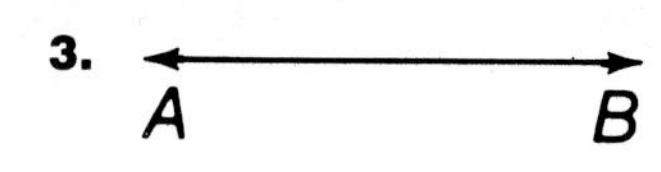

4.

5.

6.

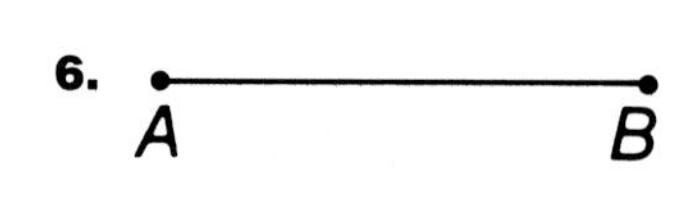

Write the number pair for each. (page 332)

7. point A
8. point B
9. point C
10. point D

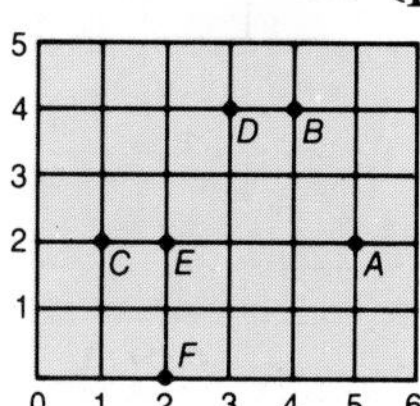

Are the figures congruent? Write *yes* or *no*. (page 312)

11.

12.

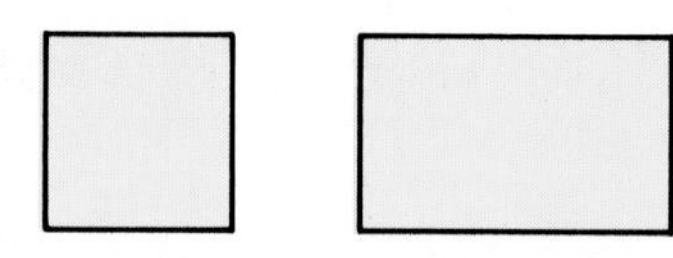

Name the space figure. Then write the number of faces, edges, and vertices. (pages 320, 322)

13.

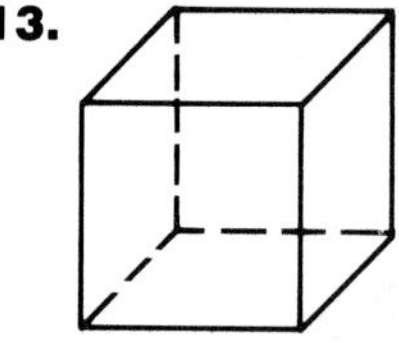

___ faces
___ edges
___ vertices

14.

___ faces
___ edges
___ vertices

15.

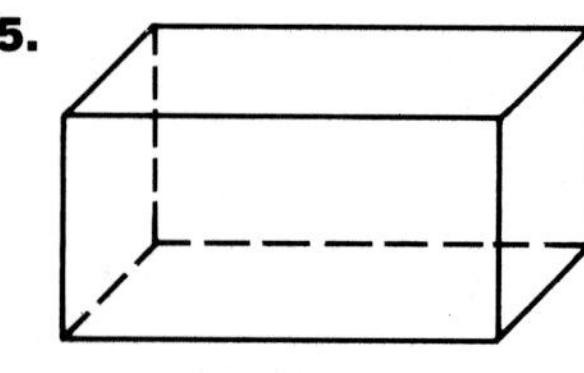

___ faces
___ edges
___ vertices

Is the line a line of symmetry? Write *yes* or *no*. (page 314)

16. 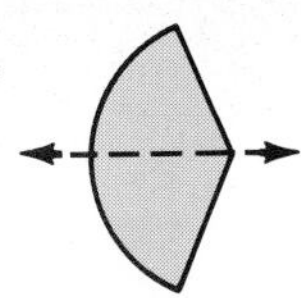

17. 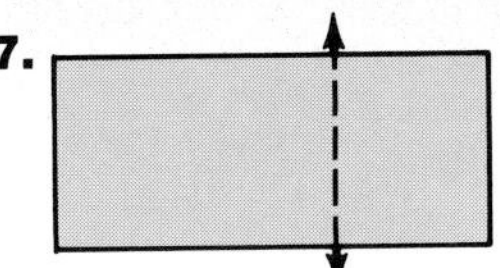

18. 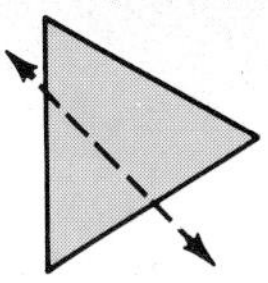

Find the area in square centimeters. (page 316)

19. 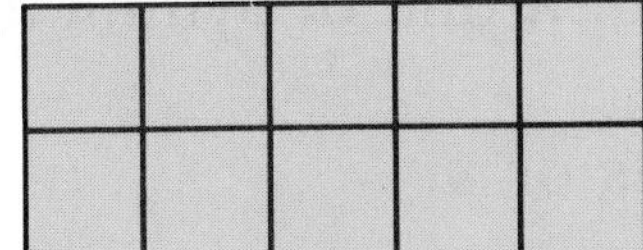

Find the volume in cubic centimeters. (page 324)

20. 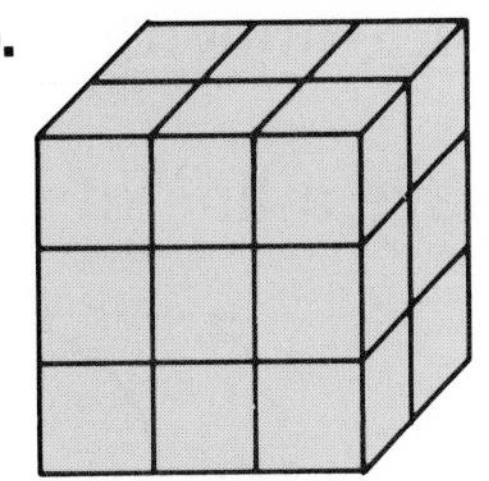

Use the graph to answer the questions. (page 328)

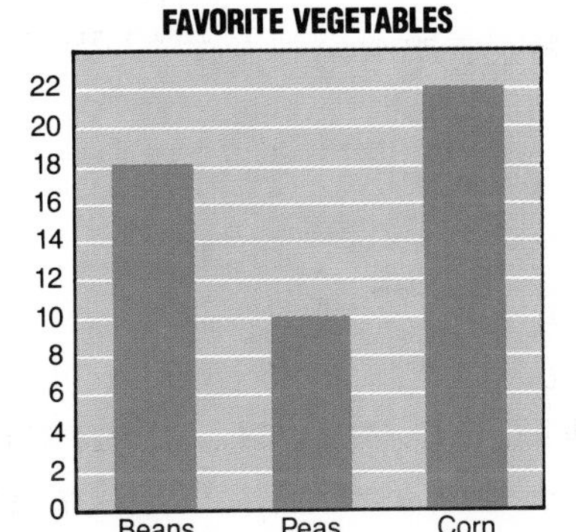

21. Students at Lee School voted for a favorite vegetable. How many children voted for beans?

22. How many children voted for peas?

23. How many children prefer corn to peas?

Use the picture to answer the questions. (page 310)

24. The building has the same number of windows on each side. How many windows does the building have in all?

25. Each apartment has 5 windows. How many apartments are in the building?

RETEACHING

A **line of symmetry** divides a figure into two equal parts. Each part is exactly the same size and shape.

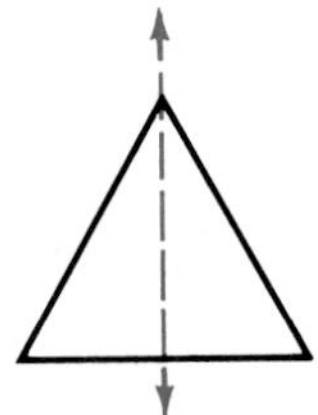

This line is a line of symmetry.

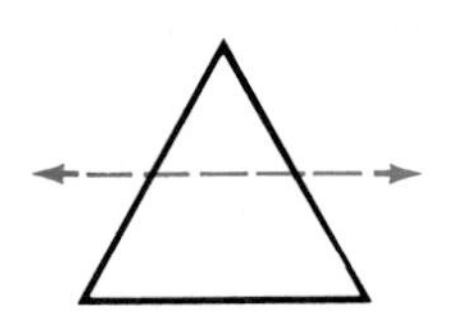

This line is not a line of symmetry.

Does each figure have a line of symmetry? Write *yes* or *no*.

1. 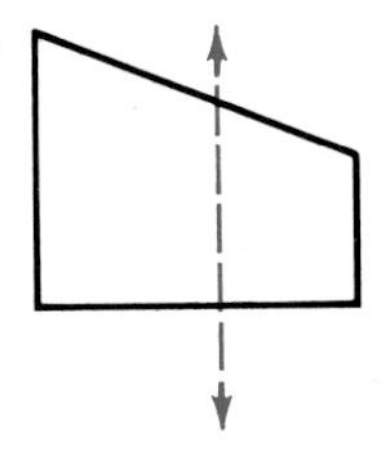

2. 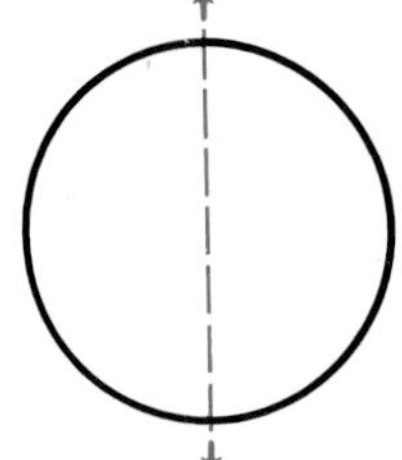

3. 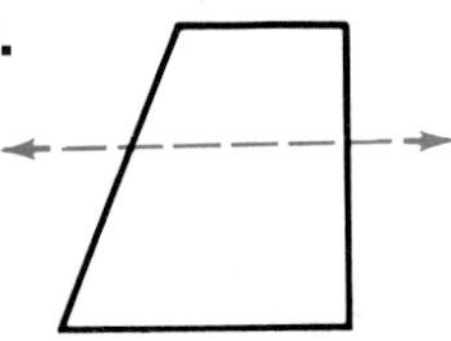

4. 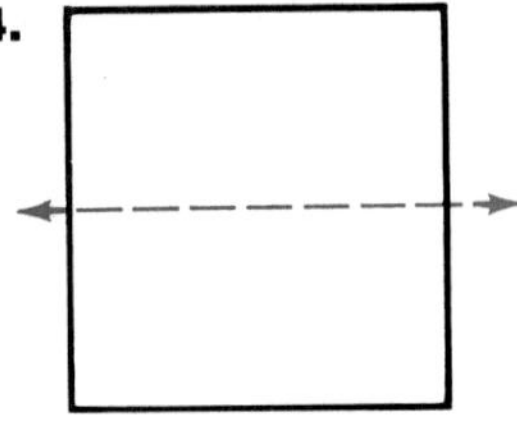

5. 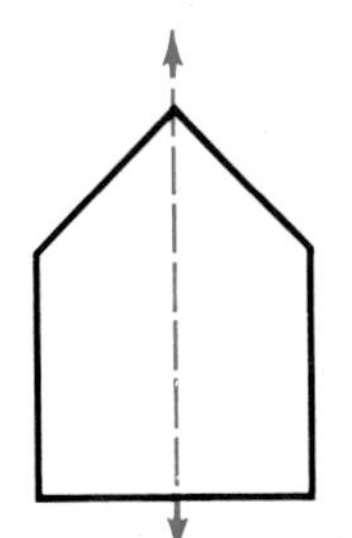

6. 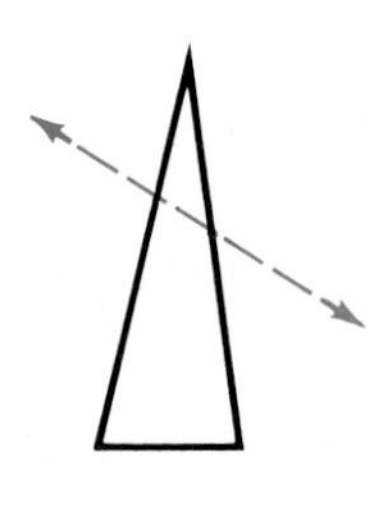

Is the line a line of symmetry? Write *yes* or *no*.

7. 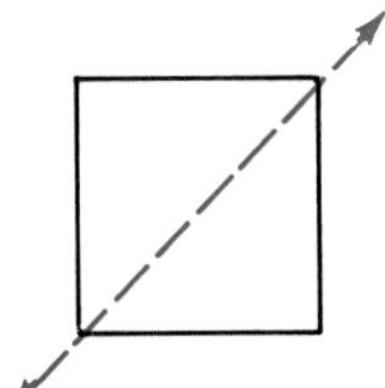

8. 

9. 

10. 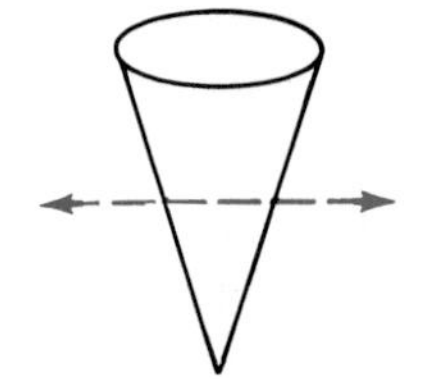

11. 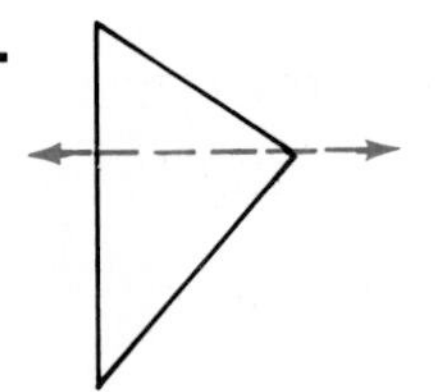

12. 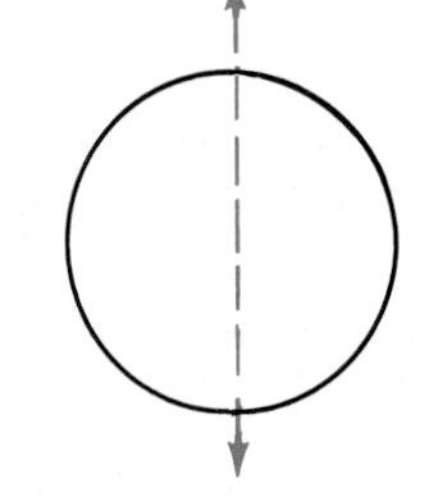

ENRICHMENT

Similarity

Similar figures have the same shape but can have different sizes.

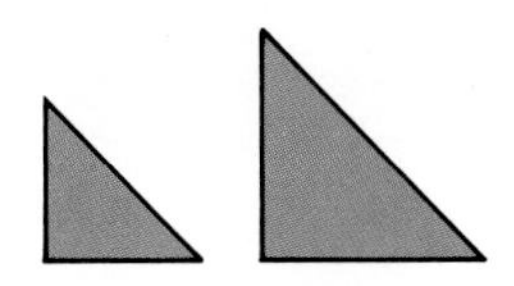

These figures are similar.

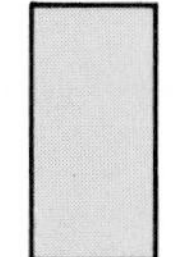

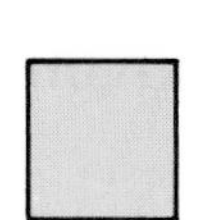

These figures are not similar.

Write the letter of the figure that is similar to the first figure.

1. 

a. 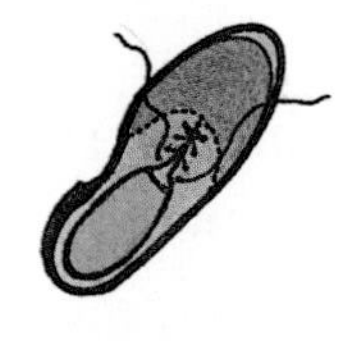 **b.** 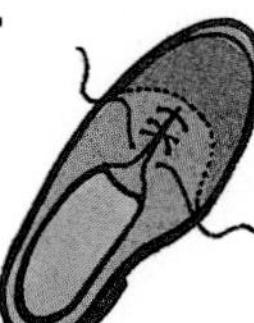

c. 

2. 

a.  **b.** 

c. 

3. 

a.  **b.** 

c. 

How many pairs of similar figures?

4. 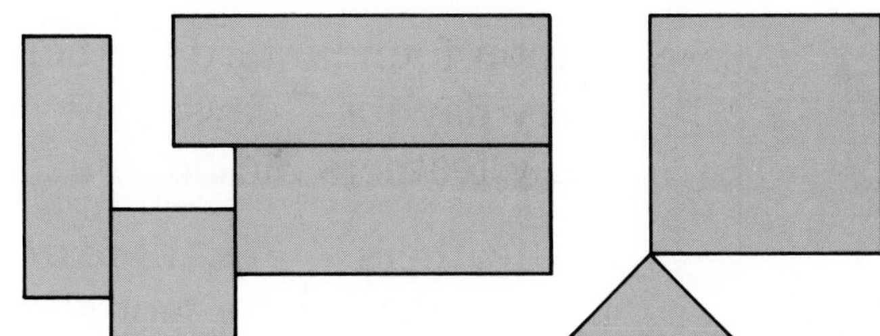

Copy the alphabet below. Choose each capital letter that is similar to the smaller letter. Ring them.

5. A B C D E F G H I J K L M N O P Q R S T U V W X Y Z

a b c d e f g h i j k l m n o p q r s t u v w x y z

CUMULATIVE REVIEW

Write the letter of the correct answer.

1. Complete the pattern.
   14, 16, 18, ___, ___, ___

   a. 20, 22, 24 b. 24, 26, 28
   c. 28, 38, 48 d. not given

2. $\frac{1}{3}$ of 18 = ___

   a. 6 b. 19
   c. 54 d. not given

3. Compare:
   $\frac{2}{8}$ ● $\frac{4}{8}$

   a. <
   b. >
   c. =
   d. not given

4. $6 \times 7 =$ ___

   a. 32 b. 42
   c. 48 d. not given

5. $4\overline{)32}$

   a. 6 b. 7
   c. 9 d. not given

6. Which unit would you use to measure the water?

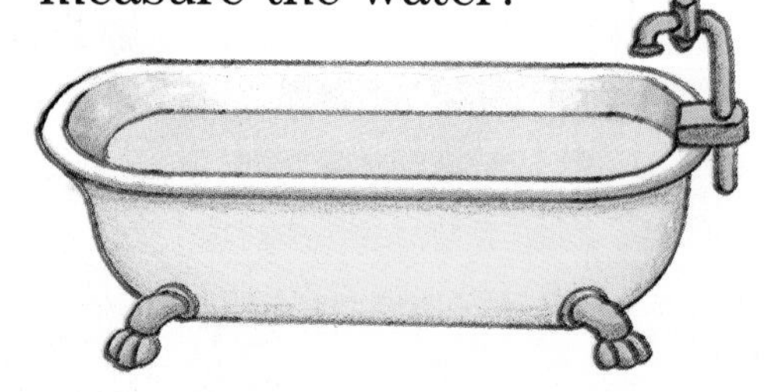

   a. in. b. pt c. gal d. not given

7. Which unit would you use to measure?

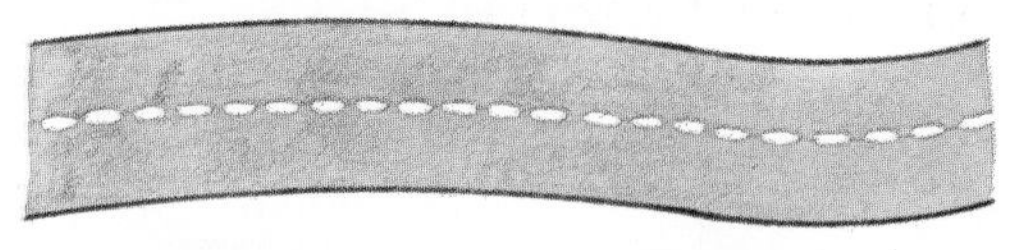

   a. mm b. cm
   c. km d. not given

8. Choose the temperature at which you could go sledding.

   a. $^{-}5$°C b. 25°C
   c. 100°C d. not given

9. Choose the best estimate for the height of the flag pole.

   a. 15 in.
   b. 15 ft
   c. 15 mi
   d. not given

10. Ramon found 5 bird feathers every day for 7 days. How many feathers did he find?

    a. 2 feathers b. 12 feathers
    c. 35 feathers d. not given

11. Tina builds a fence for her garden. The sides of the fence are 25 ft, 15 ft, 25 ft, and 15 ft. What is the fence's perimeter?

    a. 40 ft b. 50 ft
    c. 80 ft d. not given

11 MULTIPLICATION

The lights are on, but nobody is home. Can you help save electricity in your home? Keep a record of all the hours that lights are on in your home. How can you help to cut down on the use of electricity?

Multiplying Tens and Hundreds

John's dirt bike can travel 100 miles on 1 gallon of gasoline. How many miles can it travel on 6 gallons of gasoline?

You can mentally multiply the number of hundreds to find how many miles the dirt bike can travel.

Multiply 6×100.

$$\begin{array}{r} 100 \\ \times \quad 6 \\ \hline 600 \end{array}$$

1 hundred
× 6
6 hundreds

John's dirt bike can travel 600 miles on 6 gallons of gasoline.

Other examples:

$$\begin{array}{r} 10 \\ \times \ 4 \\ \hline 40 \end{array}$$

1 ten
× 4
4 tens

$$\begin{array}{r} 20 \\ \times \ 6 \\ \hline 120 \end{array}$$

2 tens
× 6
12 tens

$$\begin{array}{r} 500 \\ \times \quad 3 \\ \hline 1{,}500 \end{array}$$

5 hundreds
× 3
15 hundreds

Checkpoint Write the letter of the correct answer.

Multiply.

1. $\begin{array}{r} 10 \\ \times \ 3 \\ \hline \end{array}$	**2.** $2 \times 50 =$ ___	**3.** $\begin{array}{r} 100 \\ \times \quad 3 \\ \hline \end{array}$	**4.** $2 \times 600 =$ ___
a. 3	**a.** 10	**a.** 3	**a.** 12
b. 13	**b.** 52	**b.** 30	**b.** 120
c. 30	**c.** 100	**c.** 133	**c.** 1,200
d. 33	**d.** 102	**d.** 300	**d.** 1,222

Multiply.

1. $\begin{array}{r} 10 \\ \times\ 2 \\ \hline \end{array}$	**2.** $\begin{array}{r} 10 \\ \times\ 4 \\ \hline \end{array}$	**3.** $\begin{array}{r} 10 \\ \times\ 7 \\ \hline \end{array}$	**4.** $\begin{array}{r} 10 \\ \times\ 5 \\ \hline \end{array}$	**5.** $\begin{array}{r} 10 \\ \times\ 8 \\ \hline \end{array}$
6. $\begin{array}{r} 30 \\ \times\ 2 \\ \hline \end{array}$	**7.** $\begin{array}{r} 90 \\ \times\ 3 \\ \hline \end{array}$	**8.** $\begin{array}{r} 80 \\ \times\ 5 \\ \hline \end{array}$	**9.** $\begin{array}{r} 40 \\ \times\ 8 \\ \hline \end{array}$	**10.** $\begin{array}{r} 50 \\ \times\ 9 \\ \hline \end{array}$
11. $\begin{array}{r} 100 \\ \times\ 5 \\ \hline \end{array}$	**12.** $\begin{array}{r} 100 \\ \times\ 2 \\ \hline \end{array}$	**13.** $\begin{array}{r} 100 \\ \times\ 6 \\ \hline \end{array}$	**14.** $\begin{array}{r} 100 \\ \times\ 4 \\ \hline \end{array}$	**15.** $\begin{array}{r} 100 \\ \times\ 8 \\ \hline \end{array}$
16. $\begin{array}{r} 500 \\ \times\ 5 \\ \hline \end{array}$	**17.** $\begin{array}{r} 200 \\ \times\ 6 \\ \hline \end{array}$	**18.** $\begin{array}{r} 800 \\ \times\ 3 \\ \hline \end{array}$	**19.** $\begin{array}{r} 400 \\ \times\ 9 \\ \hline \end{array}$	**20.** $\begin{array}{r} 900 \\ \times\ 2 \\ \hline \end{array}$
21. $\begin{array}{r} 40 \\ \times\ 6 \\ \hline \end{array}$	**22.** $\begin{array}{r} 70 \\ \times\ 4 \\ \hline \end{array}$	**23.** $\begin{array}{r} 300 \\ \times\ 7 \\ \hline \end{array}$	**24.** $\begin{array}{r} 800 \\ \times\ 4 \\ \hline \end{array}$	**★25.** $\begin{array}{r} 1{,}200 \\ \times\ 4 \\ \hline \end{array}$

26. $4 \times 40 =$ ___ **27.** $2 \times 70 =$ ___ **28.** $8 \times 30 =$ ___

29. $2 \times 700 =$ ___ **30.** $8 \times 500 =$ ___ **31.** $9 \times 900 =$ ___

32. $8 \times 200 =$ ___ **33.** $9 \times 80 =$ ___ **34.** $5 \times 300 =$ ___

Solve. Use the information in the chart.

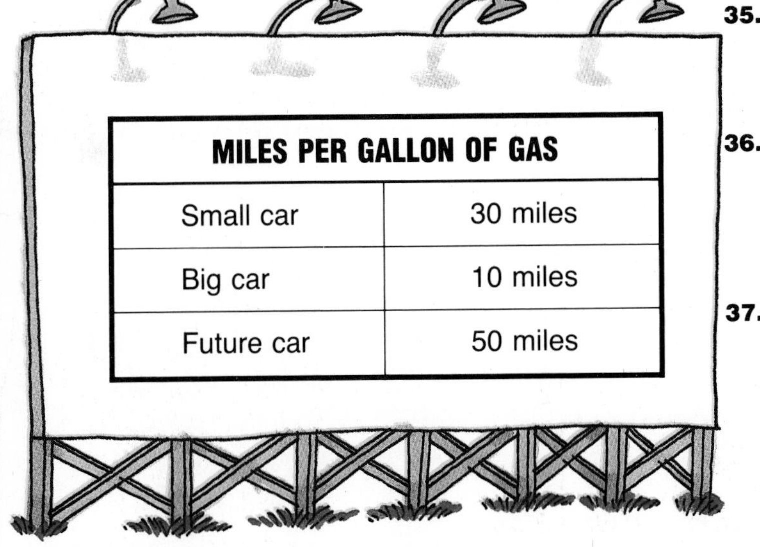

MILES PER GALLON OF GAS	
Small car	30 miles
Big car	10 miles
Future car	50 miles

35. How far can the small car travel on 5 gallons of gas?

36. How many more miles can the future car travel on a gallon of gas than the big car?

37. Suppose your family owns a small car. You buy 9 gallons of gas. Write a number sentence to show how many miles the car can travel on that amount of gas.

Estimating Products

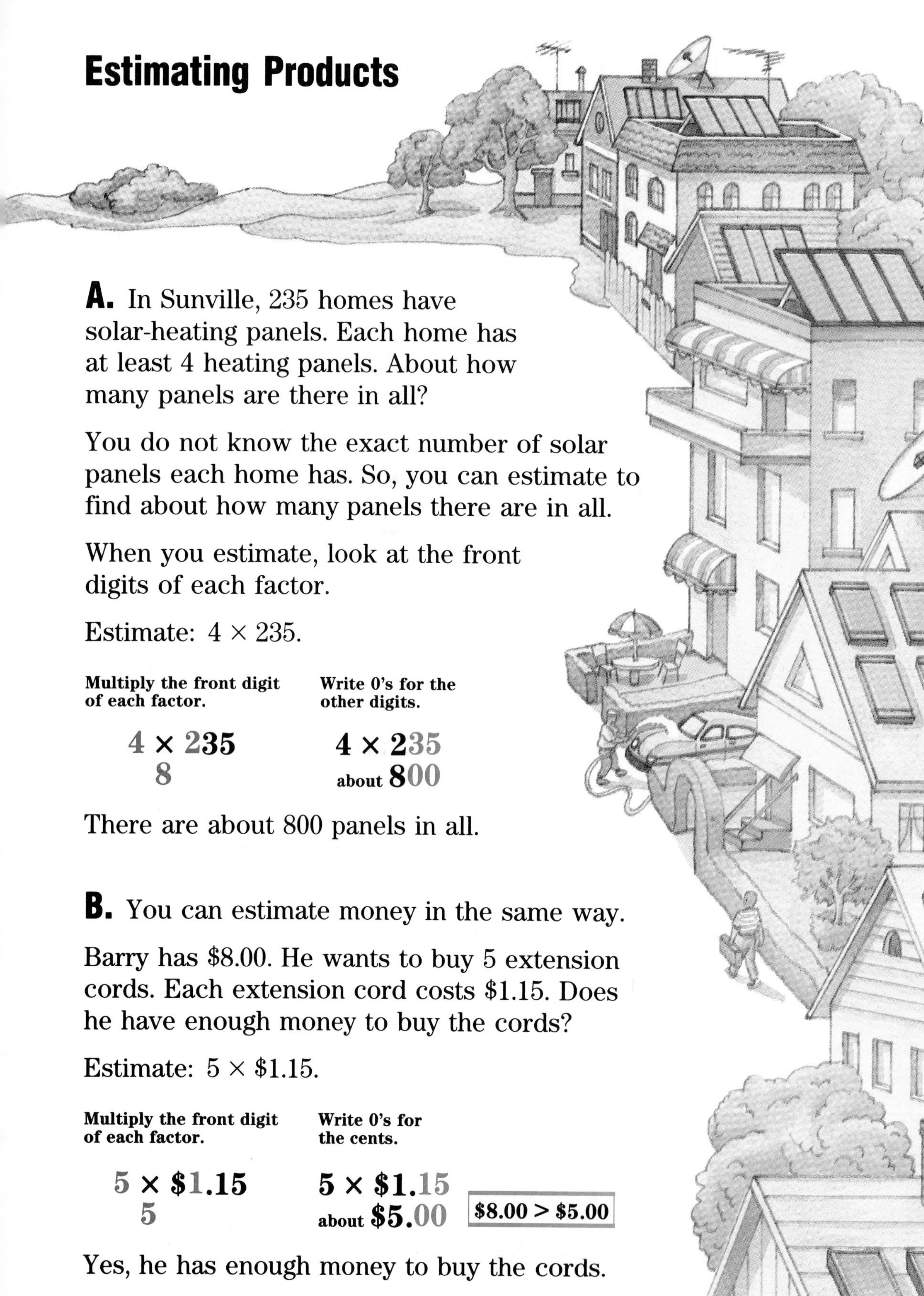

A. In Sunville, 235 homes have solar-heating panels. Each home has at least 4 heating panels. About how many panels are there in all?

You do not know the exact number of solar panels each home has. So, you can estimate to find about how many panels there are in all.

When you estimate, look at the front digits of each factor.

Estimate: 4×235.

Multiply the front digit of each factor.	**Write 0's for the other digits.**
4×235	4×235
8	about 800

There are about 800 panels in all.

B. You can estimate money in the same way.

Barry has $8.00. He wants to buy 5 extension cords. Each extension cord costs $1.15. Does he have enough money to buy the cords?

Estimate: 5 × \$1.15.

Multiply the front digit of each factor.	**Write 0's for the cents.**	
5 × \$1.15	5 × \$1.15	
5	about \$5.00	\$8.00 > \$5.00

Yes, he has enough money to buy the cords.

Estimate. Write the letter of the correct answer.

1. 34×5

a. 100 **b.** 150 **c.** 300

2. 82×6

a. 120 **b.** 480 **c.** 1,000

3. 36×3

a. 90 **b.** 900 **c.** 9,000

4. 413×6

a. 180 **b.** 240 **c.** 2,400

5. 825×4

a. 200 **b.** 2,000 **c.** 3,200

6. 76×3

a. 210 **b.** 560 **c.** 1,000

7. 52×7

a. 140 **b.** 200 **c.** 350

8. 94×9

a. 360 **b.** 810 **c.** 3,600

9. 319×7

a. 63 **b.** 1,000 **c.** 2,100

10. 737×7

a. 490 **b.** 4,000 **c.** 4,900

Estimate.

11. 63×4

12. 43×5

13. 56×7

14. 737×6

15. 978×4

16. Which two products are more than 500?

a. 7×82 **b.** 3×42

b. 5×62 **d.** 6×92

17. Which two products are more than 4,000?

a. 5×319 **b.** 8×523

c. 8×417 **d.** 7×821

Solve. Use the Infobank on page 404.

18. Can you buy 3 batteries for $10.00?

19. Can you buy 6 light bulbs for $10.00?

20. Can you buy 4 power cords for $10.00?

21. Can you buy 7 rolls of tape for $20.00?

22. Can you buy 5 batteries for $20.00?

More Practice, page 415

PROBLEM SOLVING

Estimation

Sometimes, estimating a product can be a quick way to solve a problem. At other times, you need to find the exact product.

A. Daniel's family wants to save $100.00 on their oil bill. They decide to use a wood stove part of the time. Their oil bill is $44.00 less each month when they use the stove. They use the stove for 3 months. Do they save $100?

Estimate 3 × $44.00. ⟶

> $44.00 can be written as $44.

Multiply the front digit of each factor.

3 × $44
12

Write 0's for the other digits

3 × $44
about $120

The family saves at least $100.

B. Fuel costs $36.00 per month. Daniel's family wants to buy enough fuel for 3 months. How much will they pay for the fuel?

They will not set aside the right amount of money for the fuel if they estimate the total. They need to find the exact answer.

Exact answer:

$$\begin{array}{r} \$36 \\ \times\ 3 \\ \hline \$108 \end{array}$$

They will pay $108.00 for the fuel.

Decide whether you should estimate to solve each problem. Write *estimate* or *exact answer.*

1. Jeremiah doesn't want to spend more than $250.00 per year on wood for the fireplace. He used $73.00 worth of wood each month for 3 months. How much more can he spend to stay within his budget?

2. Jeremiah spent $420.00 fixing his house so that it would stay cool. That summer, he saved $103.00. About how many summers will it take for his savings to equal the amount he spent?

Solve. Find an estimate or an exact answer as needed.

3. Emmy's family is planning a trip to her grandparents. The last time they drove there, they used 4 tanks of gas. A tank of gas costs about $20.00. Is $100.00 enough money to bring for gas?

4. The Slatniks save $95.00 each month to pay for fuel. At the end of 4 months, they receive the fuel bill at the right. Have they saved enough money to pay the bill?

Fuel Bill—Slatnik Family

Dec.	Jan.	Feb.	Mar.
$87.31	$113.22	$98.16	$99.11

5. Look at the Wilsons' bill. In which winter did the Wilsons spend the most money on electricity? About how much money should the Wilsons plan to spend on electricity this winter?

ELECTRIC BILL

Winter	Jan.	Feb.	Mar.
1	$85.00	$66.00	$72.00
2	$85.00	$72.00	$86.00
3	$91.00	$71.00	$72.00

Multiplying 2-Digit Numbers

A. A small radio needs 2 batteries. How many batteries are needed for 24 small radios?

You can multiply to find how many batteries are needed.

Multiply 2 × 24.	Multiply the ones.	Multiply the tens.

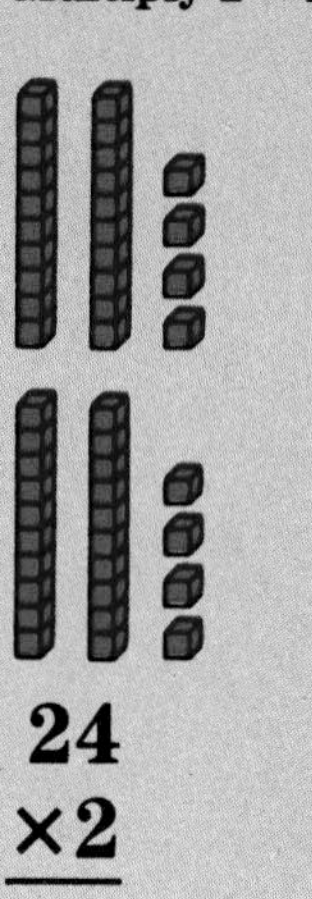

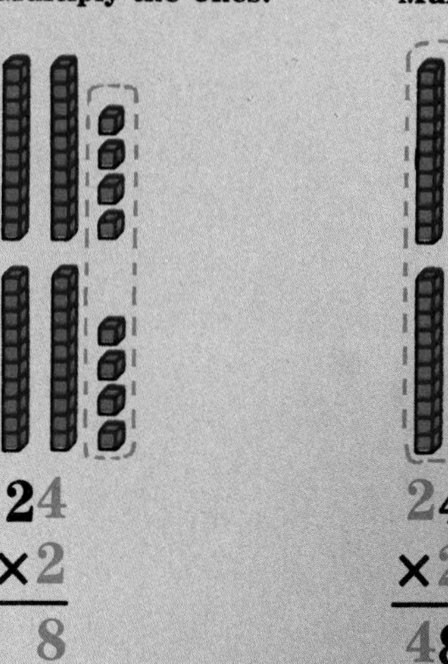

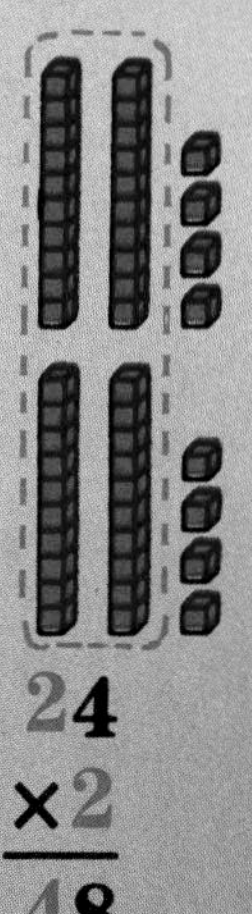

$$\begin{array}{r} 24 \\ \times 2 \\ \hline \end{array} \qquad \begin{array}{r} 24 \\ \times 2 \\ \hline 8 \end{array} \qquad \begin{array}{r} 24 \\ \times 2 \\ \hline 48 \end{array}$$

To play the radios, 48 batteries are needed.

B. You can multiply money in the same way.

Multiply 3 × \$0.23.

Multiply as you would whole numbers.

$$\begin{array}{r} \$0.23 \\ \times \quad 3 \\ \hline 69 \end{array}$$

Write the dollar sign and the cents point.

$$\begin{array}{r} \$0.23 \\ \times \quad 3 \\ \hline \$0.69 \end{array}$$

Checkpoint Write the letter of the correct answer.

Multiply.

1. $\begin{array}{r} 13 \\ \times \ 3 \\ \hline \end{array}$

a. 16 **b.** 19 **c.** 39 **d.** 93

2. 3 × 32 = ■

a. 35 **b.** 36 **c.** 69 **d.** 96

3. $\begin{array}{r} \$0.41 \\ \times \quad 2 \\ \hline \end{array}$

a. \$0.42 **b.** \$0.43 **c.** \$0.82 **d.** 82

Find the product.

1. 12×4
2. 43×2
3. 31×3
4. 21×4
5. 11×7
6. 87×1
7. 23×2
8. 11×3
9. 30×3
10. 42×2
11. 13×3
12. 20×2
13. $\$0.40 \times 2$
14. $\$0.11 \times 5$
15. $\$0.30 \times 3$
16. $\$0.12 \times 2$
17. $\$0.13 \times 2$
18. $\$0.14 \times 2$
19. 20×4
20. $\$0.03 \times 3$
21. $\$0.04 \times 2$
22. 10×2
23. $8 \times 11 =$ ■
24. $4 \times 11 =$ ■
25. $6 \times 10 =$ ■
26. $3 \times \$0.22 =$ ■
27. $2 \times \$0.44 =$ ■
28. $3 \times 33 =$ ■

Solve.

29. The wind is a source of energy. Using only the wind, a sailboat can travel 23 miles in 1 hour. How many miles can a sailboat travel in 3 hours?

30. People build dams across rivers to produce energy. The Hoover Dam is 726 feet high. The Cougar Dam is 519 feet high. How much higher is the Hoover Dam?

FOCUS: ESTIMATION

Choose the smallest bill you would use to pay for each. Write *$1*, *$5*, *$10*, or *$20*.

1. 3 candles for $1.25 each
2. 4 flashlights for $4.95 each
3. 9 reflectors for $1.10 each
4. 6 stickers for $0.15 each

Multiplying 2-Digit Numbers: Regrouping Once

A. It takes 25 gallons of water to fill a bathtub. How much water is used when a bathtub is filled 3 times?

You can multiply to find how much water is used.

Multiply 3 × 25.

Multiply the ones. Regroup the 15 ones.

$$\begin{array}{r} {}^{1} \\ 25 \\ \times\ \ 3 \\ \hline 5 \end{array} \qquad \begin{array}{r} 5 \\ \times 3 \\ \hline 15 \end{array}$$

Multiply the tens. Then add the 1 ten.

$$\begin{array}{r} {}^{1} \\ 25 \\ \times\ \ 3 \\ \hline 75 \end{array} \qquad \begin{array}{r} 2 \\ \times 3 \\ \hline 6 \\ +1 \\ \hline 7 \end{array}$$

When the bathtub is filled 3 times, 75 gallons are used.

B. Sometimes you regroup tens as hundreds.

Multiply 4 × 62.

Multiply the ones.

$$\begin{array}{r} 62 \\ \times\ \ 4 \\ \hline 8 \end{array}$$

Multiply the tens. Regroup. 24 tens = 2 hundreds + 4 tens

$$\begin{array}{r} 62 \\ \times\ \ 4 \\ \hline 248 \end{array}$$

Checkpoint

Write the letter of the correct answer.

Multiply.

1. $\begin{array}{r} 17 \\ \times\ 5 \\ \hline \end{array}$

   a. 55 b. 65 c. 85 d. 205

2. 6 × 13 = ___

   a. 68 b. 78 c. 128 d. 141

3. 3 × \$0.29 = ___

   a. \$0.67 b. \$0.77 c. \$0.87 d. \$1.27

Find the product.

1. 16×6
2. 19×5
3. 24×3
4. $\$0.15 \times 4$
5. $\$0.48 \times 2$
6. 54×2
7. 71×7
8. $\$0.80 \times 9$
9. $\$0.62 \times 3$
10. $\$0.51 \times 8$
11. 37×2
12. $\$0.73 \times 3$
13. 12×8
14. 29×3
15. 91×5

16. $5 \times 18 =$ ■
17. $3 \times 15 =$ ■
18. $2 \times \$0.36 =$ ■
19. $8 \times 61 =$ ■
20. $6 \times \$0.70 =$ ■
21. $2 \times 82 =$ ■
22. $7 \times \$0.51 =$ ■
23. $4 \times 23 =$ ■
24. $9 \times 71 =$ ■

Solve.

25. Suppose that a person uses 90 gallons of water per day. How much water would that person use in a week?

26. How much water does your family use per day? Multiply the number of people in your family by 90 gallons.

Copy and complete the sales receipt. Use a calculator to help you.

Amount	Item	Cost per item	Cost
2	Candles	$0.49	
3	Batteries	$0.53	
1	Light bulbs	$1.29	
		Total cost	

Multiplying 2-Digit Numbers: Regrouping Twice

A. You can save 15 gallons of water per day by taking a shower instead of a bath. How much water can you save in a week?

You can multiply to find out how much water you can save.

Multiply 7×15.

Multiply the ones. Regroup the 35 ones.

$$\begin{array}{r} {}^{3} \\ 15 \\ \times\ 7 \\ \hline 5 \end{array} \qquad \boxed{\begin{array}{r} 5 \\ \times 7 \\ \hline 35 \end{array}}$$

Multiply the tens. Then add the 3 tens. Regroup.

$$\begin{array}{r} {}^{3} \\ 15 \\ \times\ 7 \\ \hline 105 \end{array} \qquad \boxed{\begin{array}{r} 1 \\ \times 7 \\ \hline 7 \\ +3 \\ \hline 10 \end{array}}$$

You can save 105 gallons of water in a week.

B. You can multiply money in the same way.

Multiply $4 \times \$0.27$.

Multiply as you would whole numbers. Regroup.

$$\begin{array}{r} {}^{2} \\ \$0.27 \\ \times\ 4 \\ \hline 108 \end{array}$$

Write the dollar sign and the cents point.

$$\begin{array}{r} {}^{2} \\ \$0.27 \\ \times\ 4 \\ \hline \$1.08 \end{array}$$

Checkpoint Write the letter of the correct answer.

Multiply.

1. $\begin{array}{r} 47 \\ \times\ 6 \\ \hline \end{array}$

a. 242 **b.** 252 **c.** 282 **d.** 482

2. $3 \times 58 = \blacksquare$

a. 154 **b.** 164 **c.** 174 **d.** 214

3. $5 \times \$0.72 = \blacksquare$

a. \$3.50 **b.** \$3.60 **c.** \$4.00 **d.** \$360

Find the product.

1. $\begin{array}{r} 23 \\ \times\ 7 \\ \hline \end{array}$

2. $\begin{array}{r} 65 \\ \times\ 4 \\ \hline \end{array}$

3. $\begin{array}{r} 89 \\ \times\ 2 \\ \hline \end{array}$

4. $\begin{array}{r} 77 \\ \times\ 3 \\ \hline \end{array}$

5. $\begin{array}{r} 32 \\ \times\ 9 \\ \hline \end{array}$

6. $\begin{array}{r} \$0.29 \\ \times\ \quad 4 \\ \hline \end{array}$

7. $\begin{array}{r} \$0.94 \\ \times\ \quad 3 \\ \hline \end{array}$

8. $\begin{array}{r} \$0.75 \\ \times\ \quad 6 \\ \hline \end{array}$

9. $\begin{array}{r} \$0.98 \\ \times\ \quad 5 \\ \hline \end{array}$

10. $\begin{array}{r} \$0.44 \\ \times\ \quad 7 \\ \hline \end{array}$

11. $\begin{array}{r} 55 \\ \times\ 5 \\ \hline \end{array}$

12. $\begin{array}{r} \$0.48 \\ \times\ \quad 8 \\ \hline \end{array}$

13. $\begin{array}{r} 93 \\ \times\ 4 \\ \hline \end{array}$

14. $\begin{array}{r} \$0.37 \\ \times\ \quad 3 \\ \hline \end{array}$

15. $\begin{array}{r} 58 \\ \times\ 7 \\ \hline \end{array}$

16. $8 \times 55 =$ ■

17. $7 \times 69 =$ ■

18. $3 \times 66 =$ ■

19. $2 \times \$0.56 =$ ■

20. $4 \times \$0.76 =$ ■

21. $5 \times \$0.68 =$ ■

22. $9 \times \$0.45 =$ ■

23. $2 \times \$0.86 =$ ■

24. $3 \times 35 =$ ■

Solve.

25. Running the water while you brush your teeth uses about 43 gallons of water per week. How many gallons would you use in 4 weeks?

26. Running the water while you wash dishes uses about 45 gallons. If you wash dishes 5 times per week, how many gallons do you use?

★27. A dripping faucet can waste 1 gallon of water every 2 hours. How much does it waste in 1 hour?

★28. A hose carries 3 gallons of water in 3 minutes. Write a number sentence to show how much water it carries in 17 minutes.

MIDCHAPTER REVIEW

Multiply.

1. $8 \times 30 =$ ■

2. $9 \times 600 =$ ■

3. $5 \times 400 =$ ■

4. $3 \times 23 =$ ■

5. $5 \times 11 =$ ■

6. $2 \times \$0.34 =$ ■

7. $6 \times \$0.19 =$ ■

8. $3 \times 89 =$ ■

9. $7 \times \$0.51 =$ ■

PROBLEM SOLVING

Using a Recipe

A recipe tells you how to make a certain food. It lists the amount of each ingredient. It tells you how to mix ingredients and how long to cook or bake them. It also tells you the number of servings you can make.

> Minipizzas
>
> 4 English muffins, split in half
> $\frac{1}{2}$ cup tomato sauce
> 6 ounces grated mild cheese
>
> First, set the oven for 350°F. Place each muffin half face up on a cookie sheet. Put a spoonful of tomato sauce on each muffin half. Sprinkle the cheese on top. Put the cookie sheet into the oven. Bake for 15 to 20 minutes or until the cheese is melted. This recipe will serve 4.

How much cheese will you need for 8 servings?
How much cheese will you need for 2 servings?

Sometimes you may want to change the recipe by increasing or decreasing the number of servings. Find how many people the recipe serves.

If you want to double the number of servings, double the amount of each ingredient. To make half as many servings, divide the amount of each ingredient by 2.

$2 \times 6 = 12$

ounces of cheese

$6 \div 2 = 3$

ounces of cheese

You will need 12 ounces of cheese for 8 servings. For 2 servings, you will need 3 ounces of cheese.

Use this recipe for mushroom-barley soup to answer each question.

Ingredients	Directions
$\frac{1}{3}$ cup dried barley 3 cups clear beef broth $\frac{3}{4}$ cup fresh mushrooms 1 cup water 2 cups milk 1 carrot, thinly sliced	Cook the barley in the broth at a low boil (small bubbles) in a saucepan for 45 minutes. Add the mushrooms, water, milk, and carrot slices to the pan. Cook for another 15 minutes at a low boil. Stir while cooking. This recipe makes 6 servings.

1. How many kinds of liquid are listed? Which is used in the largest amount?

2. How many cups of liquid are needed in all for the soup?

3. Yolanda will use canned broth in her soup. The can she plans to buy holds 2 cups of broth. How many cans will she need? Will there be any broth left?

4. Water and broth both boil at a temperature of 212°F. If some broth has reached a temperature of 140°, how much hotter does it need to be in order to boil?

5. Molly wants to serve this soup to her friends at 1:00 P.M. By about what time should she start to prepare the soup?

6. Abel wants to make enough soup to feed 12 people, including himself. How many cups of milk will he need?

7. Instead of using water, Ivan wants to use 1 extra cup of milk. How much milk will he use in all?

8. Tracy is cooking soup for 18 people, including herself. How many cups of liquid will she need?

Multiplying 3-Digit Numbers: Regrouping Once

A. One gas truck can carry 601 gallons of oil. Gus's Gas Station orders 3 truckloads of oil. How many gallons of oil does Gus's station receive?

You can multiply to find how many gallons.

Multiply 3×601.

Multiply the ones.

$$\begin{array}{r} 601 \\ \times \quad 3 \\ \hline 3 \end{array}$$

Multiply the tens.

$$\begin{array}{r} 601 \\ \times \quad 3 \\ \hline 03 \end{array}$$

Multiply the hundreds. Regroup. 18 hundreds = 1 thousand + 8 hundreds

$$\begin{array}{r} 601 \\ \times \quad 3 \\ \hline 1,803 \end{array}$$

Gus's station receives 1,803 gallons of oil.

B. Sometimes you regroup in other places.

Regroup ones as tens.

$$\begin{array}{r} {}^{1} \\ 102 \\ \times \quad 5 \\ \hline 510 \end{array} \qquad \begin{array}{r} {}^{1} \\ 113 \\ \times \quad 5 \\ \hline 565 \end{array}$$

Regroup tens as hundreds.

$$\begin{array}{r} {}^{1} \\ 131 \\ \times \quad 5 \\ \hline 655 \end{array} \qquad \begin{array}{r} {}^{1} \\ 160 \\ \times \quad 2 \\ \hline 320 \end{array}$$

Checkpoint Write the letter of the correct answer.

Multiply.

1. $\begin{array}{r} 116 \\ \times \quad 6 \\ \hline \end{array}$

a. 122 **b.** 666
c. 696 **d.** 772

2. $\begin{array}{r} 273 \\ \times \quad 2 \\ \hline \end{array}$

a. 276 **b.** 446
c. 546 **d.** 556

3. $3 \times 713 = \blacksquare$

a. 716 **b.** 2,139
c. 2,149 **d.** 2,239

4. $6 \times 401 = \blacksquare$

a. 2,406 **b.** 2,407
c. 2,416 **d.** 2,516

Find the product.

1. $\begin{array}{r} 129 \\ \times \quad 3 \\ \hline \end{array}$

2. $\begin{array}{r} 228 \\ \times \quad 2 \\ \hline \end{array}$

3. $\begin{array}{r} 309 \\ \times \quad 3 \\ \hline \end{array}$

4. $\begin{array}{r} 270 \\ \times \quad 3 \\ \hline \end{array}$

5. $\begin{array}{r} 463 \\ \times \quad 2 \\ \hline \end{array}$

6. $\begin{array}{r} 721 \\ \times \quad 3 \\ \hline \end{array}$

7. $\begin{array}{r} 613 \\ \times \quad 2 \\ \hline \end{array}$

8. $\begin{array}{r} 911 \\ \times \quad 6 \\ \hline \end{array}$

9. $\begin{array}{r} 603 \\ \times \quad 3 \\ \hline \end{array}$

10. $\begin{array}{r} 820 \\ \times \quad 2 \\ \hline \end{array}$

11. $\begin{array}{r} 900 \\ \times \quad 8 \\ \hline \end{array}$

12. $\begin{array}{r} 123 \\ \times \quad 4 \\ \hline \end{array}$

13. $\begin{array}{r} 102 \\ \times \quad 6 \\ \hline \end{array}$

14. $\begin{array}{r} 217 \\ \times \quad 4 \\ \hline \end{array}$

15. $\begin{array}{r} 711 \\ \times \quad 8 \\ \hline \end{array}$

16. $\begin{array}{r} 802 \\ \times \quad 4 \\ \hline \end{array}$

17. $\begin{array}{r} 243 \\ \times \quad 3 \\ \hline \end{array}$

18. $\begin{array}{r} 107 \\ \times \quad 7 \\ \hline \end{array}$

19. $\begin{array}{r} 121 \\ \times \quad 8 \\ \hline \end{array}$

20. $\begin{array}{r} 804 \\ \times \quad 2 \\ \hline \end{array}$

21. $6 \times 141 =$ ▒

22. $4 \times 181 =$ ▒

23. $9 \times 105 =$ ▒

24. $2 \times 603 =$ ▒

25. $3 \times 933 =$ ▒

26. $8 \times 810 =$ ▒

27. $7 \times 801 =$ ▒

28. $4 \times 216 =$ ▒

29. $3 \times 283 =$ ▒

Solve.

30. A barrel of oil holds 161 liters. How many liters are there in 4 barrels of oil?

★31. Only 6 of every 100 oil wells that are drilled produce oil. Of every 100 wells, how many will not produce oil?

FOCUS: MENTAL MATH

To multiply $2 \times 7 \times 5$:

Look for a product that is a multiple of 10. $2 \times 5 = 10$
Multiply that product by the remaining factor. $7 \times 10 = 70$
So, $2 \times 7 \times 5 = 70$.

Compute mentally.

1. $4 \times 5 \times 3 =$ ▒

2. $5 \times 8 \times 2 =$ ▒

3. $6 \times 4 \times 5 =$ ▒

Multiplying 3-Digit Numbers: Regrouping Twice

The Thrifty Company uses 881 light bulbs in a year. How many light bulbs do they use in 3 years?

You can multiply to find how many light bulbs.

Multiply 3 × 881.

Multiply the ones.

$$\begin{array}{r} 881 \\ \times \quad 3 \\ \hline 3 \end{array}$$

Multiply the tens. Regroup the 24 tens.

$$\begin{array}{r} {\scriptstyle 2} \\ 881 \\ \times \quad 3 \\ \hline 43 \end{array} \qquad \boxed{\begin{array}{r} 8 \\ \times 3 \\ \hline 24 \end{array}}$$

Multiply the hundreds. Then add the 2 hundreds. Regroup.

$$\begin{array}{r} {\scriptstyle 2} \\ 881 \\ \times \quad 3 \\ \hline 2{,}643 \end{array} \qquad \boxed{\begin{array}{r} 8 \\ \times 3 \\ \hline 24 \\ +2 \\ \hline 26 \end{array}}$$

They use 2,643 light bulbs in 3 years.

Other examples:

$$\begin{array}{r} {\scriptstyle 1} \\ 340 \\ \times \quad 3 \\ \hline 1{,}020 \end{array} \qquad \begin{array}{r} {\scriptstyle 1\,1} \\ 234 \\ \times \quad 3 \\ \hline 702 \end{array} \qquad \begin{array}{r} {\scriptstyle 1} \\ 625 \\ \times \quad 2 \\ \hline 1{,}250 \end{array} \qquad \begin{array}{r} {\scriptstyle 2} \\ 305 \\ \times \quad 5 \\ \hline 1{,}525 \end{array}$$

Checkpoint Write the letter of the correct answer.

Multiply.

1. $\begin{array}{r} 157 \\ \times \quad 3 \\ \hline \end{array}$ **2.** $\begin{array}{r} 751 \\ \times \quad 6 \\ \hline \end{array}$ **3.** $\begin{array}{r} 724 \\ \times \quad 4 \\ \hline \end{array}$ **4.** 6 × 505 = ■

1.	2.	3.	4.
a. 371	**a.** 4,206	**a.** 2,886	**a.** 3,000
b. 451	**b.** 4,506	**b.** 2,896	**b.** 3,010
c. 471	**c.** 4,516	**c.** 2,996	**c.** 3,030
d. 911	**d.** 6,006	**d.** 3,626	**d.** 3,680

Multiply.

1. 125×5
2. 134×4
3. 268×2
4. 487×2
5. 159×3
6. 652×4
7. 743×3
8. 561×9
9. 381×5
10. 764×2
11. 707×7
12. 840×9
13. 319×6
14. 802×8
15. 517×3

16. $2 \times 258 =$ ■
17. $4 \times 197 =$ ■
18. $6 \times 146 =$ ■
19. $4 \times 452 =$ ■
20. $3 \times 763 =$ ■
21. $7 \times 581 =$ ■
22. $2 \times 652 =$ ■
23. $5 \times 908 =$ ■
24. $3 \times 166 =$ ■

Solve.

25. Joan buys 4 boxes of light bulbs. Each box costs $2.82. What is the total cost?

26. Ben's parents pay him $0.50 a week for turning off the lights. How much money can he earn in 6 weeks?

★27. Harry buys two 60-watt bulbs for $0.89 apiece. He buys three 100-watt bulbs for $1.05 apiece. How much money does he spend on light bulbs?

★28. Lights, Inc., sells 150-watt bulbs for $1.25 apiece. The same bulbs cost $1.15 at Bill's. How much money does Sue save if she buys 4 bulbs at Bill's?

CALCULATOR

Tell whether you would use mental math, pencil and paper, or a calculator to solve each. Tell why, then solve.

1. 7×8
2. 9×10
3. 8×986
4. 3×121
5. 6×700
6. $5 \times 1{,}111$
7. 4×576
8. $2 \times 10{,}000$

More Practice, page 415

PROBLEM SOLVING
Practice

Use the picture to solve.

1. If Mrs. Fry buys 4 gallons of regular gas and a can of motor oil, how much will she spend?

2. How much more expensive would 6 gallons of super gasoline be than 6 gallons of regular?

Estimate.

3. Mrs. Fry wants to travel 900 miles in 3 days. During the first 2 days, she drove 279 miles and 401 miles. If she traveled 352 miles on the third day, did she reach her 900-mile goal on time?

4. Mrs. Fry allowed $30 for meals on her trip. She spent $9.95 on the first day. On the next day, she spent $11.89. If she spent $15.97 on the last day, was she within the limit she set for herself?

Use the bar graph to solve.

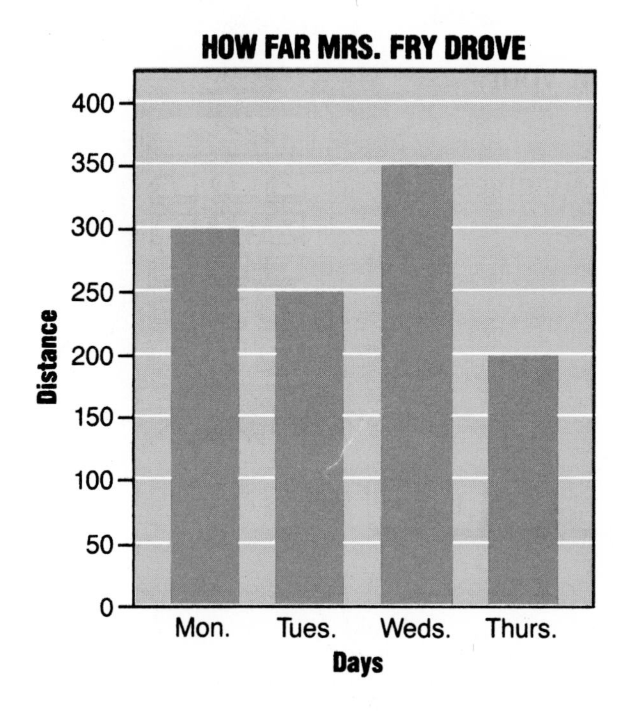

5. How many more miles did Mrs. Fry travel on Wednesday than she traveled on the fourth day?

6. On which 2 days did Mrs. Fry drive a total of 600 miles?

7. What was the total distance that Mrs. Fry drove?

8. On Friday, Mrs. Fry drove twice as far as she had on Tuesday. How far was that?

Use the map to solve.

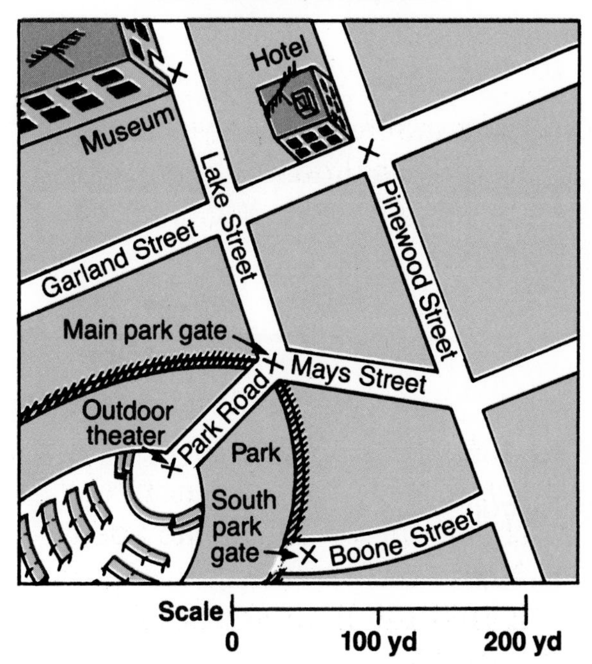

9. Which streets cross Garland Street?

10. How many yards must Mrs. Fry walk to go from her hotel to the theater? Measure from the X's. Mrs. Fry walks along only the streets or marked roads.

11. Which streets lead to the park gates?

12. How far is it from the museum to the main park gate? Measure from the X's.

CALCULATOR

You can use your calculator to sharpen your multiplication and estimation skills.

Which one-digit number multiplied by 61 gives a product between 430 and 520?

Estimate by using 60 instead of 61.

$5 \times 60 = 300$ (too small)
$9 \times 60 = 540$ (too great)
$8 \times 60 = 480$.

Use the calculator to compute.

So, 8×61 is between 430 and 520.

Estimate. Then use your calculator to complete.

	Product Range
1. ■ × 54	330 to 430
2. ■ × 69	210 to 310
3. ■ × 72	300 to 400
4. ■ × 47	250 to 300
5. ■ × 83	500 to 600
6. ■ × 71	450 to 550
7. 4 × 6 ■	249 to 253
8. 6 × 4 ■	283 to 293
9. 8 × ■ ■	445 to 455
10. 9 × ■ ■	775 to 790
11. ■ × 96	450 to 500
12. 7 × ■ ■	440 to 445

GROUP PROJECT

Buying a Bike

The problem: You want to buy a new bicycle. Your local bike shop is starting a month-long spring sale. You have to decide which bike you can buy.

Key Facts

- You can earn $0.50 per hour after school by running errands for your neighbors.
- You earn $3.00 on the weekends mowing lawns.
- You have $7.00 saved.
- You could sell your old bike, which is too little. It cost $57.00 when it was new.
- You need to buy a birthday present for a friend.

CHAPTER TEST

Multiply. (pages 342, 348, 350, and 352)

1. $\begin{array}{r} 10 \\ \times\ 3 \\ \hline \end{array}$ **2.** $\begin{array}{r} 80 \\ \times\ 2 \\ \hline \end{array}$ **3.** $\begin{array}{r} 12 \\ \times\ 5 \\ \hline \end{array}$

4. $\begin{array}{r} 61 \\ \times\ 1 \\ \hline \end{array}$ **5.** $\begin{array}{r} 45 \\ \times\ 2 \\ \hline \end{array}$ **6.** $\begin{array}{r} 59 \\ \times\ 3 \\ \hline \end{array}$

7. $3 \times \$0.43 =$ ___ **8.** $5 \times 51 =$ ___

9. $7 \times \$0.86 =$ ___ **10.** $5 \times 62 =$ ___

Estimate. Use the front digits to choose the best answer. (page 344)

11. $\begin{array}{r} 38 \\ \times\ 6 \\ \hline \end{array}$ **12.** $\begin{array}{r} 94 \\ \times\ 5 \\ \hline \end{array}$ **13.** $\begin{array}{r} 414 \\ \times\ 7 \\ \hline \end{array}$ **14.** $\begin{array}{r} 868 \\ \times\ 7 \\ \hline \end{array}$

11.	12.	13.	14.
a. 100	**a.** 45	**a.** 280	**a.** 56
b. 180	**b.** 200	**b.** 2,800	**b.** 560
c. 360	**c.** 450	**c.** 28,000	**c.** 5,600

Estimate. (page 344)

15. $\begin{array}{r} 86 \\ \times\ 4 \\ \hline \end{array}$ **16.** $\begin{array}{r} 48 \\ \times\ 9 \\ \hline \end{array}$ **17.** $\begin{array}{r} 360 \\ \times\ 7 \\ \hline \end{array}$ **18.** $\begin{array}{r} 939 \\ \times\ 9 \\ \hline \end{array}$

Multiply. (pages 356 and 358)

19. $\begin{array}{r} 300 \\ \times\ 8 \\ \hline \end{array}$ **20.** $\begin{array}{r} 421 \\ \times\ 2 \\ \hline \end{array}$ **21.** $\begin{array}{r} 512 \\ \times\ 3 \\ \hline \end{array}$

22. $\begin{array}{r} 709 \\ \times\ 4 \\ \hline \end{array}$ **23.** $\begin{array}{r} 248 \\ \times\ 5 \\ \hline \end{array}$ **24.** $\begin{array}{r} 626 \\ \times\ 3 \\ \hline \end{array}$

Multiply. (pages 356 and 358)

25. $4 \times 192 = $ ____

26. $3 \times 613 = $ ____

27. $8 \times 194 = $ ____

28. $6 \times 289 = $ ____

Solve. Find an estimate or an exact answer as needed. (page 346)

29. Neil saves $2.50 per week. He wants to buy a record for $6.50. Will he have enough money in 3 weeks?

30. Kate saved $6.75. Can she buy 5 hair clips that cost $2.15 each? If not, how many can she buy?

Use the recipe to answer each question. (page 354)

31. How many cups of fruit juice are used?

32. How many cups of fruit juice are needed to make enough punch to serve 8?

33. How many cups of juice, soda, and ginger ale are needed to make enough fruit punch to serve 8?

FRUIT PUNCH

2 c grape juice	1 c club soda
3 c orange juice	2 c ginger ale
15 strawberries	mint leaves

Combine the liquids in a large punch bowl. Add the strawberries. Crush some mint leaves and add. Serve with ice. Serves 4.

BONUS

Multiply.

1. $1{,}234 \times 2$

2. $4{,}562 \times 3$

3. $3{,}147 \times 2$

4. $7{,}592 \times 8$

5. $2{,}687 \times 6$

6. $5{,}948 \times 5$

RETEACHING

Multiply 4×352.

Multiply the ones.

$$\begin{array}{r} 352 \\ \times\ \ 4 \\ \hline 8 \end{array}$$

Multiply the tens.
Regroup the 2 hundreds.

$$\begin{array}{r} {\scriptstyle 2}\ \ \\ 352 \\ \times\ \ 4 \\ \hline 08 \end{array}$$

Multiply the hundreds.
Then add the 2 hundreds.
Regroup.

$$\begin{array}{r} {\scriptstyle 2}\ \ \ \ \\ 352 \\ \times\ \ 4 \\ \hline 1{,}408 \end{array}$$

Complete.

1. 381×4
2. 617×2
3. 807×5
4. 275×3
5. 521×6
6. 261×5
7. 392×4
8. 608×2
9. 417×3
10. 572×3
11. 729×3
12. 615×4
13. 852×3
14. 805×4
15. 653×2
16. 227×4
17. 541×6
18. 709×3
19. 403×7
20. 387×2

21. $8 \times 116 =$ ■
22. $6 \times 209 =$ ■
23. $5 \times 195 =$ ■
24. $7 \times 761 =$ ■
25. $5 \times 516 =$ ■
26. $6 \times 148 =$ ■
27. $2 \times 861 =$ ■
28. $9 \times 209 =$ ■
29. $4 \times 167 =$ ■
30. $8 \times 450 =$ ■
31. $7 \times 409 =$ ■
32. $6 \times 137 =$ ■

ENRICHMENT

Multiplication

You can use a grid to help you multiply. To multiply 8×64, follow these steps.

Step 1: Draw a grid like the one shown. Write the first factor to the right of the grid and the second factor at the top of the grid.

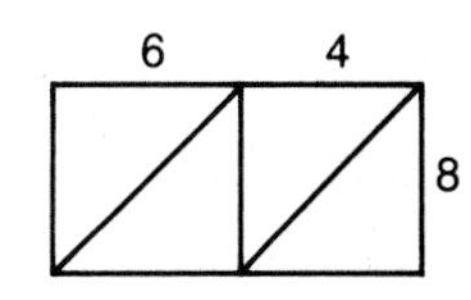

Step 2: Multiply $8 \times 4 = (32)$. Write the 3 in the top half of the square. Write the 2 in the bottom half of the square.

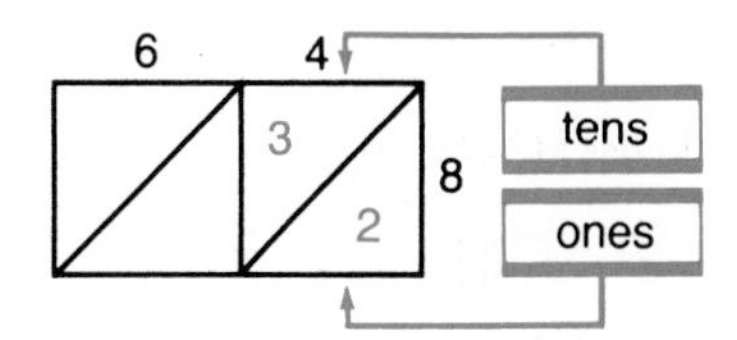

Step 3: Do the same for 8×6. Add along the diagonals. Regroup when necessary.
$8 \times 64 = 512$

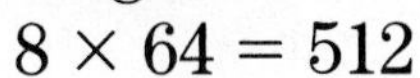

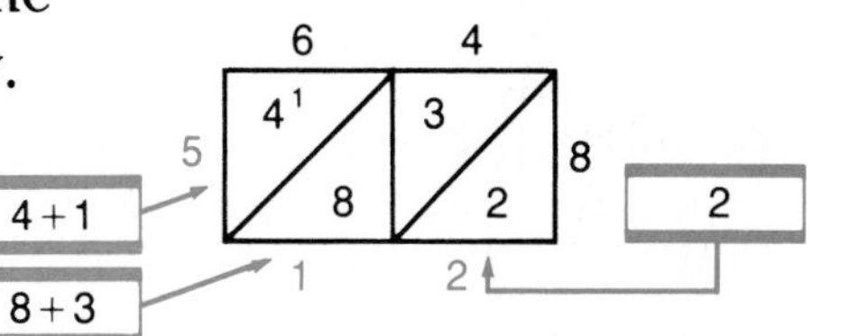

Another example:
Multiply 7×429.
$7 \times 429 = 3{,}003$

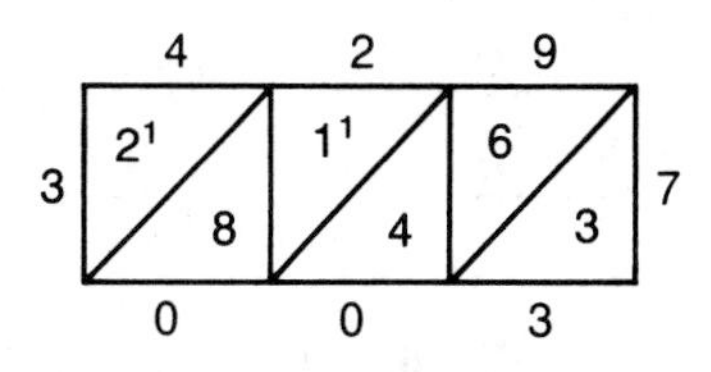

Use the grid to find the product.

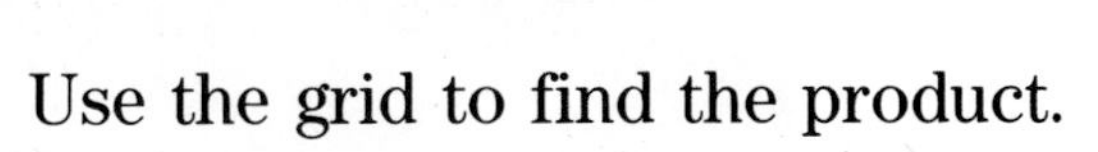

1. $8 \times 33 =$ ■

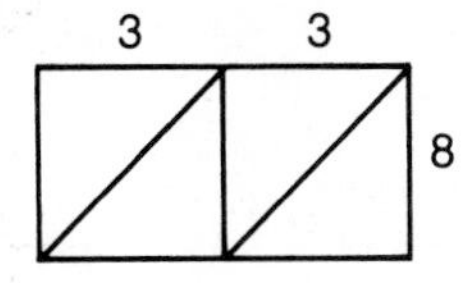

2. $9 \times 25 =$ ■

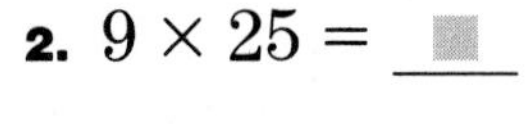

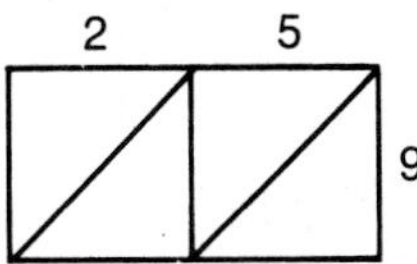

3. $7 \times 65 =$ ■

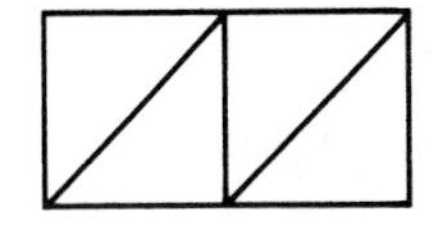

4. $7 \times 415 =$ ■

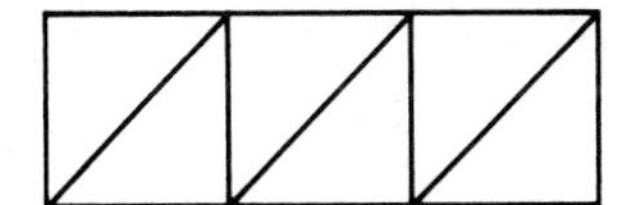

5. $6 \times 325 =$ ■

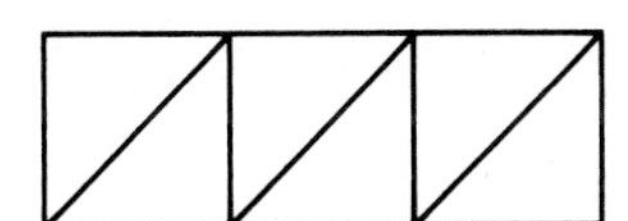

6. $3 \times 965 =$ ■

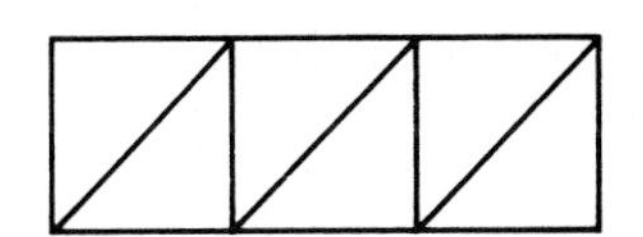

TECHNOLOGY

If the LOGO turtle draws two squares, each of which has sides of 10 steps, then the squares are congruent. They will have exactly the same size and shape.

If one square the turtle draws has sides of 10 steps, and another square has sides of 15 steps, the squares are not congruent. One square is larger than the other.

A line of symmetry divides a shape into two congruent parts.

Here is a procedure for drawing a square with sides of 40 steps.

TO SQUARE

FD 40 RT 90 FD 40 RT 90 FD 40 RT 90 FD 40
END

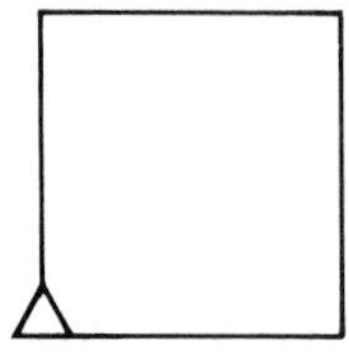

1. What would happen if you told the turtle TO SQUARE twice in a row? Copy the picture. Then repeat the procedure. Draw a ring around the line of symmetry. How many congruent squares are there?

2. Finish writing commands that would draw this figure. Ring the commands that draw the line of symmetry. The first few commands are written for you.

LT 90 FD 30
RT 120 FD 60
RT 120

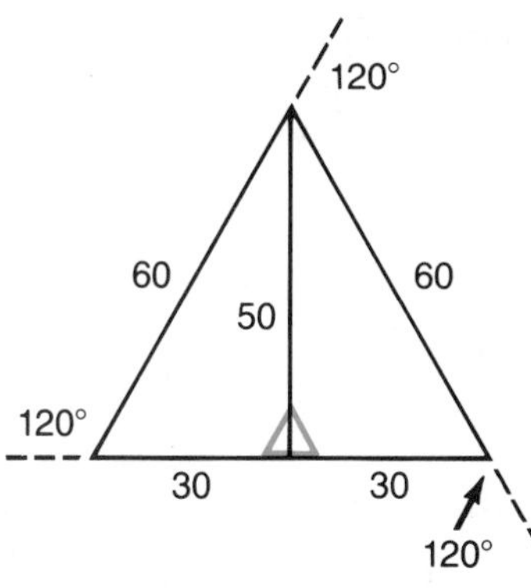

3. Copy the figure. Follow the commands to finish drawing the figure. Draw a line of symmetry.

RT 60 FD 35
RT 60 FD 35
RT 85 FD 70
RT 130 FD 70

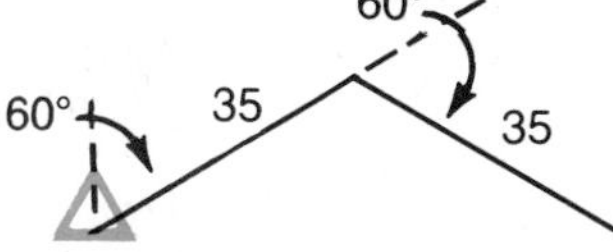

4. Write commands to draw this figure. Then ring the commands that draw the line of symmetry.

CUMULATIVE REVIEW

Write the letter of the correct answer.

1. Which sentence completes the fact family?

   $7 \times 6 = 42$
   $6 \times 7 = 42$
   $42 \div 6 = 7$

   a. $7 - 6 = 1$
   b. $42 \div 7 = 6$
   c. $42 \times 7 = 6$
   d. not given

2. Compare. 3.0 ● 1.9

   a. $>$
   b. $<$
   c. $=$
   d. not given

3. $\begin{array}{r} 4.97 \\ -\,2.48 \\ \hline \end{array}$

   a. 2.48
   b. 2.49
   c. 2.45
   d. not given

4. This is ____.

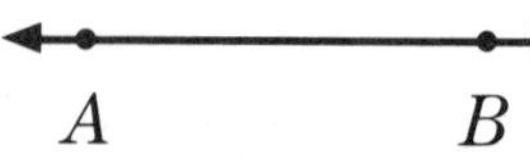

   a. a line
   b. a line segment
   c. an angle
   d. not given

5. A ____ has four equal sides and four equal angles.

   a. right angle
   b. square
   c. triangle
   d. not given

6. What is the volume?

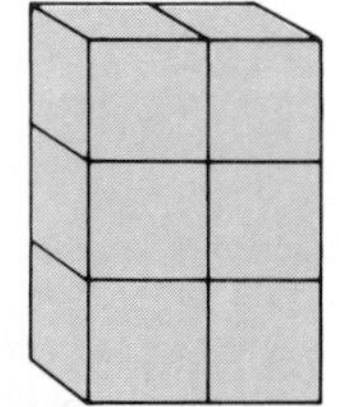

   a. 6 cubic centimeters
   b. 7 cubic centimeters
   c. 11 cubic centimeters
   d. not given

7. Two congruent figures must have the same shape and ____.

   a. perimeter
   b. a different perimeter
   c. size
   d. not given

8. What is the area?

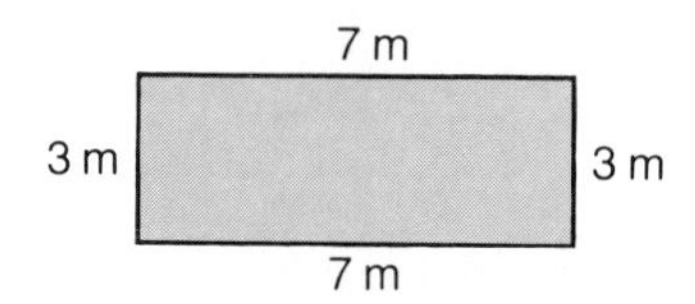

   a. 10 square meters
   b. 21 square meters
   c. 26 square meters
   d. not given

9. 3,508 fans see Tuesday's game. 4,116 fans see Wednesday's game. How many more fans see Wednesday's game?

   a. 1,292
   b. 1,392
   c. 7,624
   d. not given

10. 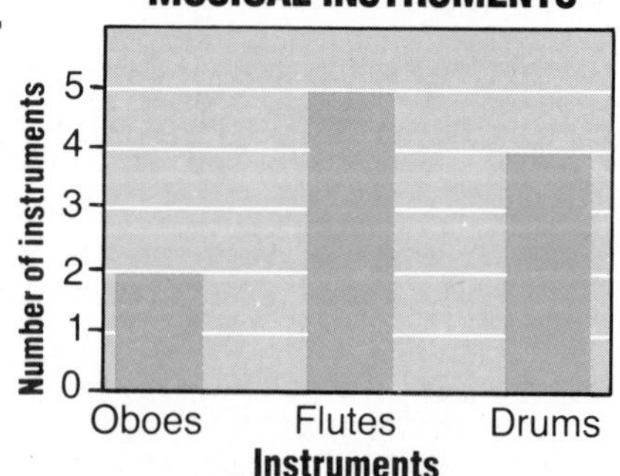

   How many flutes are there?

   a. 1
   b. 10
   c. 5
   d. not given

12 DIVISION

What if your class discovered a treasure castle? Inside there are many rooms. Each room has many gems, including diamonds, rubies, and emeralds. The gems all have different values. How would you plan to search for them? What would be a fair way to divide the treasure?

Showing Remainders

Hal collects rocks. He finds 22 rocks that he wants to put in cases. He puts 5 rocks in each case. How many cases does he fill? How many rocks are left?

You can divide to find how many cases Hal fills with rocks.

Divide $22 \div 5$.

Think: ___ $\times 5$ is close to 22.

$1 \times 5 = 5$
$2 \times 5 = 10$
$3 \times 5 = 15$
$4 \times 5 = 20$
$5 \times 5 = 25$ Too great. So, use 4.

Divide.	**Multiply.**	**Subtract.**	**Write the remainder.**
$5\overline{)22}$ quotient 4	$5\overline{)22}$ quotient 4; 20 (4 × 5)	$5\overline{)22}$ quotient 4; -20; 2	$5\overline{)22}$ quotient 4 R2; -20; 2

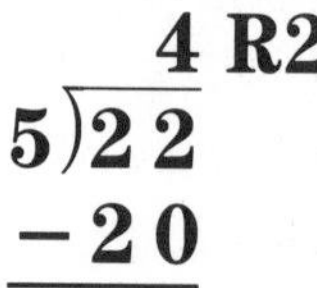

He fills 4 cases. There are 2 rocks left.

Divide. Show the remainder.

1. $2\overline{)19}$
2. $3\overline{)10}$
3. $5\overline{)18}$
4. $9\overline{)42}$
5. $7\overline{)31}$
6. $8\overline{)23}$
7. $6\overline{)17}$
8. $9\overline{)80}$
9. $4\overline{)11}$
10. $3\overline{)20}$
11. $25 \div 3 =$ ___
12. $11 \div 2 =$ ___
13. $19 \div 5 =$ ___
14. $55 \div 9 =$ ___
15. $43 \div 8 =$ ___
16. $35 \div 4 =$ ___

Divide. Show the remainder.

17. $3\overline{)29}$ **18.** $4\overline{)31}$ **19.** $6\overline{)47}$ **20.** $2\overline{)17}$ **21.** $7\overline{)52}$

22. $5\overline{)34}$ **23.** $6\overline{)27}$ **24.** $3\overline{)26}$ **25.** $8\overline{)68}$ **26.** $6\overline{)19}$

27. $2\overline{)9}$ **28.** $4\overline{)17}$ **29.** $9\overline{)65}$ **30.** $6\overline{)20}$ **31.** $8\overline{)62}$

32. $4\overline{)24}$ **33.** $6\overline{)56}$ **34.** $7\overline{)49}$ **35.** $8\overline{)75}$ **36.** $5\overline{)45}$

37. $27 \div 4 =$ ___ **38.** $46 \div 5 =$ ___ **39.** $20 \div 3 =$ ___

40. $33 \div 6 =$ ___ **41.** $67 \div 7 =$ ___ **42.** $59 \div 8 =$ ___

43. $13 \div 2 =$ ___ **44.** $19 \div 5 =$ ___ **45.** $38 \div 9 =$ ___

46. $39 \div 4 =$ ___ **47.** $87 \div 9 =$ ___ **48.** $56 \div 7 =$ ___

Solve.

49. John has 26 butterflies. He has 3 glass cases. He puts the same number of butterflies into each glass case. How many are there in each case? How many are left?

50. There are 4 maps used on a treasure hunt. Each map is used by 15 campers. How many campers are there?

51. Plan a treasure hunt. Decide how you will divide the people into groups. How many groups of hunters will there be? How many will there be in each group?

★52. Edna finds 9 unusual rocks. Li finds 6. They divide the rocks so that each has the same number. How many rocks does each have now? How many are left?

ANOTHER LOOK

Multiply.

1. $9 \times 6 =$ ___ **2.** $5 \times 7 =$ ___ **3.** $8 \times 3 =$ ___ **4.** $2 \times 4 =$ ___

Dividing Tens and Hundreds

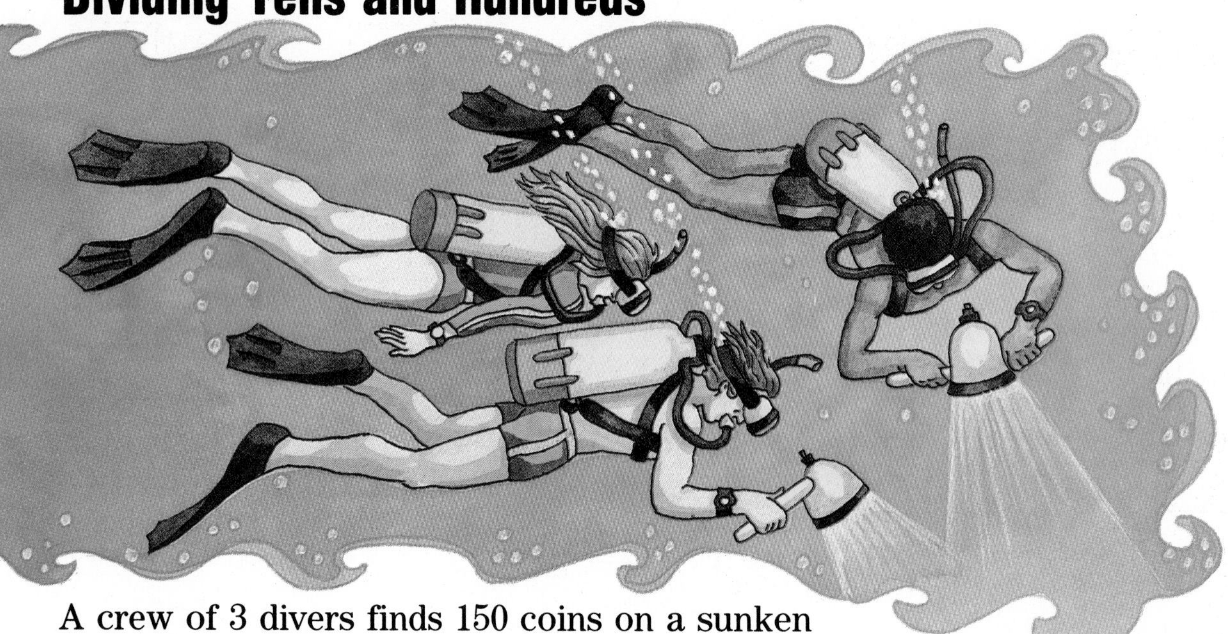

A crew of 3 divers finds 150 coins on a sunken ship. Each diver takes the same number of coins. How many coins does each diver take?

You can divide mentally to find how many coins each diver takes.

$150 \div 3 = ■$
$150 \div 3 = 50$ | $15 \div 3 = 5$

Each diver takes 50 coins.

Another example:

$4{,}500 \div 5 = ■$
$4{,}500 \div 5 = 900$ | $45 \div 5 = 9$

Checkpoint

Write the letter of the correct answer.

Divide.

1. $6\overline{)420}$	**2.** $5\overline{)150}$	**3.** $9\overline{)2{,}700}$	**4.** $2{,}800 \div 4 =$ ■
a. 7	**a.** 3	**a.** 3	**a.** 70
b. 70	**b.** 30	**b.** 30	**b.** 200
c. 71	**c.** 31	**c.** 300	**c.** 700
d. 700	**d.** 300	**d.** 3,000	**d.** 7,000

Divide.

1. $4\overline{)160}$ 2. $3\overline{)240}$ 3. $6\overline{)420}$ 4. $5\overline{)300}$ 5. $2\overline{)180}$

6. $8\overline{)6{,}400}$ 7. $6\overline{)1{,}800}$ 8. $7\overline{)2{,}800}$ 9. $9\overline{)5{,}400}$ 10. $4\overline{)3{,}600}$

11. $5\overline{)200}$ 12. $7\overline{)5{,}600}$ 13. $6\overline{)3{,}000}$ 14. $8\overline{)720}$ 15. $2\overline{)1{,}400}$

16. $8\overline{)320}$ 17. $5\overline{)350}$ 18. $3\overline{)2{,}700}$ 19. $9\overline{)6{,}300}$ 20. $3\overline{)180}$

21. $630 \div 9 =$ ▇ 22. $640 \div 8 =$ ▇ 23. $400 \div 5 =$ ▇

24. $1{,}600 \div 2 =$ ▇ 25. $1{,}200 \div 3 =$ ▇ 26. $1{,}600 \div 4 =$ ▇

27. $240 \div 6 =$ ▇ 28. $350 \div 7 =$ ▇ 29. $3{,}600 \div 9 =$ ▇

Solve. Use the Infobank on page 404.

30. Suppose the same number of silver bars were stored in 9 safes. How many bars would there be in each safe?

31. If the divers gave the same number of gold items to each of 4 museums, how many items did each museum receive?

32. Suppose the divers collect the lead balls in 5 trips. If they carried up the same number of pounds on each trip, how many pounds did they collect on each trip?

33. If the same number of chests of indigo were shown in each of 5 museums, how many chests were shown in each museum?

FOCUS: MENTAL MATH

Write the missing dividend.

1. $7\overline{)}$ = 8 $7\overline{)}$ = 80 $7\overline{)}$ = 800

2. $4\overline{)}$ = 9 $4\overline{)}$ = 90 $4\overline{)}$ = 900

PROBLEM SOLVING
Writing a Number Sentence

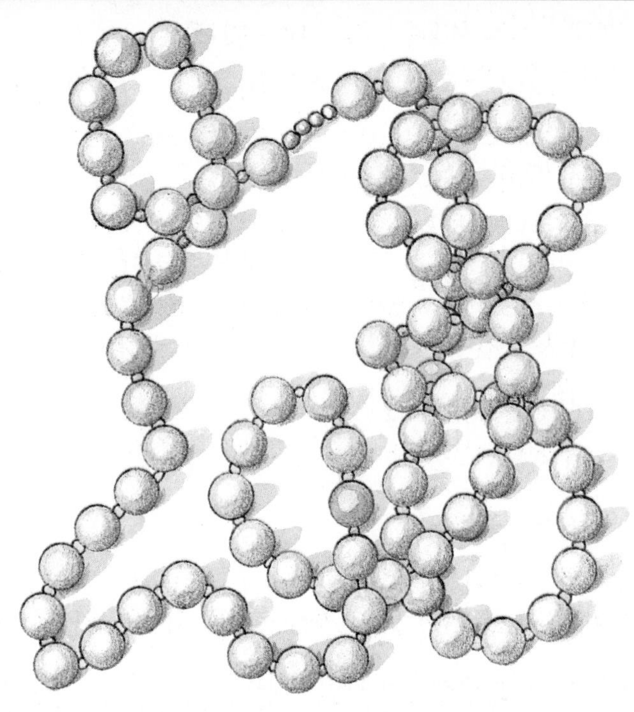

Try to write a number sentence for the problem below. Review the three steps if you need to.

> One jewelry company had 240 pearls. From these pearls, 4 necklaces were made. Each necklace was the same length. How many pearls were there on each necklace?

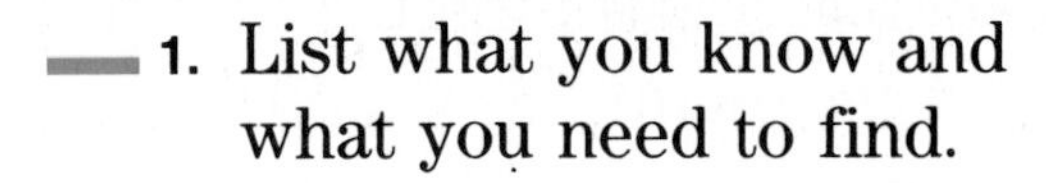

1. List what you know and what you need to find.

know There were 240 pearls. There were 4 necklaces made from them. Each necklace was the same length.

find How many pearls were there on each necklace?

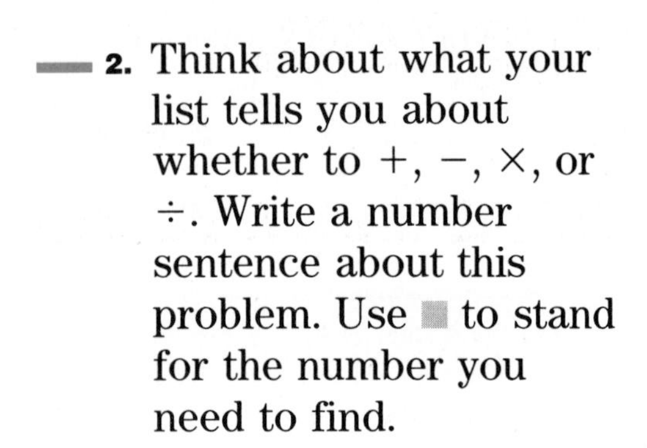

2. Think about what your list tells you about whether to $+$, $-$, $\times$, or $\div$. Write a number sentence about this problem. Use ■ to stand for the number you need to find.

240	$\div$	4	$=$	■
pearls in all		necklaces		pearls on each necklace

3. Solve. Write the answer.

$240 \div 4 = ■$
$240 \div 4 = 60$

There were 60 pearls on each necklace.

Write the letter of the correct number sentence.

1. A jeweler made rings in the shape of a curled snake. Each snake had 2 tiny amber eyes. The jeweler has 15 orders for rings. How many amber eyes will she need for the rings?

   a. $15 \times 2 = \blacksquare$
   b. $15 + 2 = \blacksquare$

2. Jewels are weighed in units called carats. One famous diamond weighed 100 carats. Suppose that gem were cut into 5 smaller stones of equal weight. How many carats would each weigh?

   a. $100 \div 5 = \blacksquare$
   b. $100 \times 5 = \blacksquare$

Write a number sentence, and solve. Use the chart for Problems 3–6.

WEIGHTS OF SOME FAMOUS GEMS

Name	Weight
De Long Ruby	100 carats
Star of India (sapphire)	563 carats
Star of Africa (diamond)	530 carats
Midnight Star (sapphire)	116 carats
Hope Diamond	45 carats
Cullinan Diamond	3,106 carats

3. How much do all the diamonds weigh together?

4. How much would a stone weigh if it were twice as heavy as the Midnight Star?

5. Suppose a stone the weight of the Hope Diamond were cut into stones that weighed 9 carats each. How many stones would there be?

6. How much more does the ruby weigh than the Hope Diamond?

7. In 1725, diamonds were found in Brazil. They were found in South Africa 141 years after that. In which year were diamonds found in South Africa?

8. The city of Mandalay, Burma, is 75 miles from Mogok's ruby mines. A Mandalay trader made 6 trips to Mogok and back. How many miles did he travel in all?

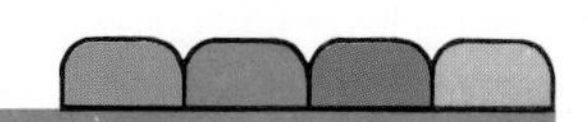

Dividing 2-Digit Numbers

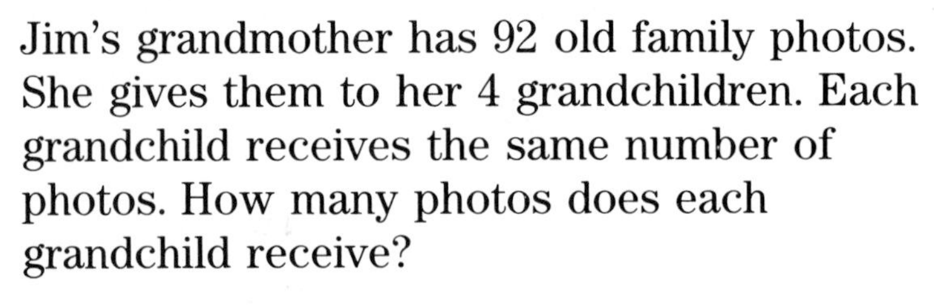

Jim's grandmother has 92 old family photos. She gives them to her 4 grandchildren. Each grandchild receives the same number of photos. How many photos does each grandchild receive?

You can divide to find how many photos each grandchild receives.

Find the quotient: $92 \div 4$.

Divide the tens.
Think: $4\overline{)9}$.
Write 2.

$$\begin{array}{r} 2 \\ 4\overline{)92} \end{array}$$

Multiply.
$2 \times 4 = 8$

$$\begin{array}{r} 2 \\ 4\overline{)92} \\ 8 \end{array}$$

Subtract and compare.
$9 - 8 = 1$; $1 < 4$

$$\begin{array}{r} 2 \\ 4\overline{)92} \\ -8 \\ \hline 1 \end{array}$$

The remainder must be less than the divisor.

Divide the ones.
Bring down the 2.
Think: $4\overline{)12}$.
Write 3.

$$\begin{array}{r} 23 \\ 4\overline{)92} \\ -8\downarrow \\ \hline 12 \end{array}$$

Multiply.
$3 \times 4 = 12$

$$\begin{array}{r} 23 \\ 4\overline{)92} \\ -8\downarrow \\ \hline 12 \\ 12 \end{array}$$

Subtract and compare.
$12 - 12 = 0$; $0 < 4$

$$\begin{array}{r} 23 \\ 4\overline{)92} \\ -8\downarrow \\ \hline 12 \\ -12 \\ \hline 0 \end{array}$$

Each grandchild receives 23 photos.

Divide.

1. $4\overline{)84}$ 2. $2\overline{)64}$ 3. $3\overline{)96}$ 4. $8\overline{)88}$ 5. $2\overline{)68}$

6. $5\overline{)65}$ 7. $4\overline{)76}$ 8. $7\overline{)91}$ 9. $3\overline{)48}$ 10. $6\overline{)72}$

11. $7\overline{)84}$ 12. $2\overline{)46}$ 13. $4\overline{)68}$ 14. $5\overline{)80}$ 15. $3\overline{)63}$

16. $48 \div 4 =$ ___ 17. $69 \div 3 =$ ___ 18. $26 \div 2 =$ ___

19. $51 \div 3 =$ ___ 20. $75 \div 5 =$ ___ 21. $56 \div 4 =$ ___

22. $39 \div 3 =$ ___ 23. $66 \div 6 =$ ___ 24. $57 \div 3 =$ ___

Solve.

25. At a party, Jay, Duane, and Bob had 36 photos taken. They each left the party with the same number of photos. How many photos did each one leave with?

26. Maureen found an old photo album. There are 8 photos on each of the album's 20 pages. How many photos are there in the album?

27. Pretend you are a photographer for your class party. Decide how many photos you will take of each person. You can buy rolls of film to take 12, 20, or 36 photos. How many rolls will you buy?

★28. José has 4 photo albums. There are 12 photos in each album. He gives all the photos to 3 friends. Each friend receives the same number of photos. How many photos does each friend receive?

FOCUS: REASONING

agate, beryl, coral, ___

1. Would the next stone in the pattern be an emerald or a diamond?

2. Explain your answer.

More Practice, page 416

Dividing 2-Digit Numbers with Remainders

A. Tandy makes necklaces from shells. She has 95 shells. She uses 7 shells for each necklace. How many necklaces can she make? How many shells will be left?

You can divide to find how many necklaces Tandy can make.

Divide $95 \div 7$.

Divide the tens.
Think $7\overline{)9}$.
Write 1.

$$\begin{array}{r} 1 \\ 7\overline{)95} \\ -7 \\ \hline 2 \end{array}$$

Multiply. $1 \times 7 = 7$
Subtract. $9 - 7 = 2$
Compare. $2 < 7$

Divide the ones.
Bring down the 5.
Think $7\overline{)25}$.
Write 3.

$$\begin{array}{r} 13 \text{ R4} \\ 7\overline{)95}\phantom{\text{ R4}} \\ -7\downarrow\phantom{\text{ R4}} \\ \hline 25\phantom{\text{ R4}} \\ -21\phantom{\text{ R4}} \\ \hline 4\phantom{\text{ R4}} \end{array}$$

Multiply. $3 \times 7 = 21$
Subtract. $25 - 21 = 4$
Compare. $4 < 7$
Write the remainder.

Tandy can make 13 necklaces.
There will be 4 shells left.

B. You can check division by multiplying.

$$\begin{array}{r} 13 \\ 7\overline{)95} \\ -7\downarrow \\ \hline 25 \\ -21 \\ \hline 4 \end{array}$$

$$\begin{array}{rl} 13 & \longleftarrow \textbf{quotient} \\ \times\ 7 & \longleftarrow \textbf{divisor} \\ \hline 91 & \\ +\ 4 & \longleftarrow \textbf{remainder} \\ \hline 95 & \longleftarrow \textbf{dividend} \end{array}$$

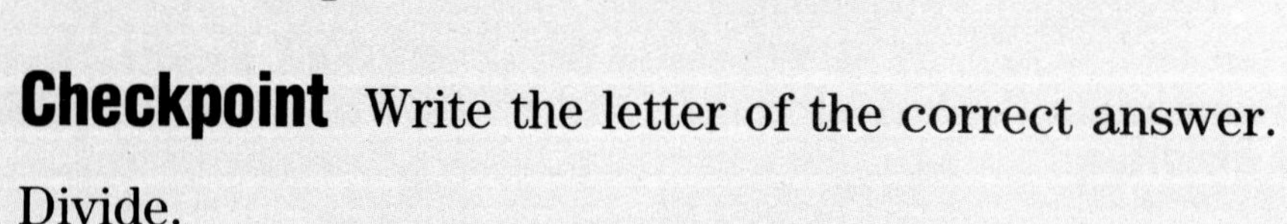

Checkpoint Write the letter of the correct answer.

Divide.

1. $2\overline{)81}$

a. 4 **b.** 4 R1
c. 40 **d.** 40 R1

2. $5\overline{)77}$

a. 11 **b.** 15
c. 15 R2 **d.** 10 R7

3. $95 \div 4 = \blacksquare$

a. 23 **b.** 23 R3
c. 32 R3 **d.** 32 R5

Divide.

1. $4\overline{)69}$ 2. $7\overline{)90}$ 3. $5\overline{)68}$ 4. $2\overline{)57}$ 5. $8\overline{)87}$

6. $3\overline{)79}$ 7. $6\overline{)89}$ 8. $7\overline{)86}$ 9. $4\overline{)95}$ 10. $9\overline{)98}$

11. $2\overline{)67}$ 12. $8\overline{)92}$ 13. $6\overline{)77}$ 14. $3\overline{)35}$ 15. $4\overline{)97}$

16. $65 \div 2 =$ ■ 17. $91 \div 8 =$ ■ 18. $86 \div 4 =$ ■

19. $74 \div 3 =$ ■ 20. $50 \div 4 =$ ■ 21. $71 \div 7 =$ ■

22. $93 \div 6 =$ ■ 23. $40 \div 3 =$ ■ 24. $76 \div 6 =$ ■

Solve.

25. Jay has 50 scallop shells. He has 4 sacks. He puts the same number of shells into each sack. How many shells does he put into each sack? How many are left?

26. Ned uses 13 shells to make a picture frame. He has enough shells to make 8 frames. How many shells does Ned have?

27. Chip and Leo went clamming. They dug 23 clams. They each ate the same number of clams. How many clams did they each eat? How many were left?

★28. Grace finds 71 shells. She gives 9 to her brother. She divides the other shells evenly among her 3 friends. How many does each friend receive? How many are left?

MIDCHAPTER REVIEW

Divide.

1. $6\overline{)47}$ 2. $5\overline{)48}$ 3. $3\overline{)23}$ 4. $4\overline{)18}$ 5. $2\overline{)17}$

6. $8\overline{)160}$ 7. $5\overline{)4{,}500}$ 8. $9\overline{)180}$ 9. $4\overline{)2{,}800}$ 10. $6\overline{)3{,}600}$

11. $4\overline{)52}$ 12. $7\overline{)98}$ 13. $5\overline{)60}$ 14. $3\overline{)93}$ 15. $9\overline{)99}$

More Practice, page 416

PROBLEM SOLVING

Practice

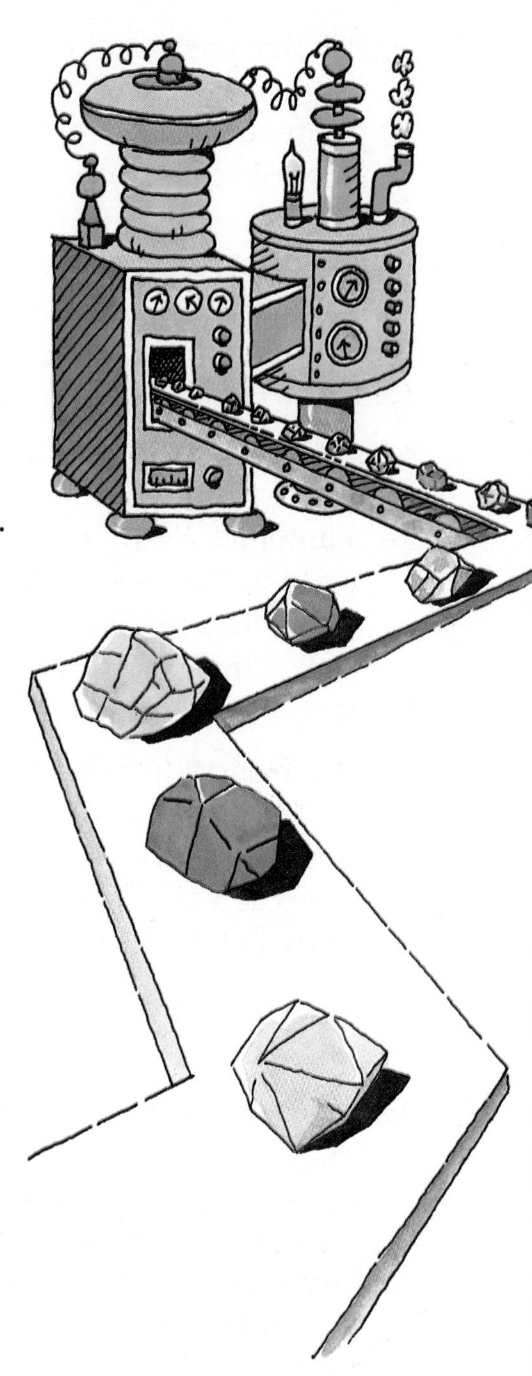

Make a table, and complete the pattern. Then solve.

1. Ms. Weems is treasure hunting underwater. Her sub travels 8 miles every 15 minutes. How far will it travel in an hour?

2. Mr. Bry's new gem maker makes diamonds and rubies. Every time it makes 5 diamonds, it makes 6 rubies. Mr. Bry now has 36 rubies. How many diamonds does he have?

Make a list to solve.

3. Mr. Bry put 8 diamonds and 8 rubies into a bag. Without looking, he took out 8 gems. How many possible combinations of gems could he have picked?

4. Mr. Bry sells Ms. Weems a gem. She pays him with the first 3 coins she pulls out of her pocket. There were 2 dimes, 2 quarters, and 1 nickel in her pocket. What are all the possible prices she could pay?

Draw a picture to solve.

5. Ms. Weems is standing on the middle rung of a ladder. She goes down 2 rungs. Then she goes up 5 rungs to the top rung. How many rungs does the ladder have?

To solve, make a guess and check your answer.

6. Mr. Bry has 2 sons. The sum of their ages is 19. The product of their ages is 84. How old is each boy?

7. A book is open to 2 facing pages. The sum of the page numbers is 55. Neither number is above 30. What are the page numbers?

Find the pattern to solve.

8. Mr. Bry lines up some gems in a row. He puts them in this order: 1 pearl, 2 rubies, 2 diamonds, 1 pearl, 2 rubies, 2 diamonds. What are the next gems in the row? What is the eighteenth gem?

9. Mr. Bry's machine makes gems in shapes. It makes a square, a triangle, a square, a circle, a square, a triangle, a square, and a circle. What is the fifteenth shape it makes?

Solve.

10. Ms. Weems traded some pearls for Mr. Bry's gems. He gave 5 gems in return for 4 pearls. Now, he has 32 pearls. How many gems did he trade?

11. Mr. Bry has a pearl, a ruby, a diamond, and a sapphire. He has 2 boxes. Each box holds 2 jewels. How many possible combinations of jewels can he choose from in packing the boxes?

12. There are 7 gems in a row. They are numbered left to right, from 1 to 7. A ruby is number 5. A diamond is 2 gems to the right of the ruby. A sapphire is 4 gems away from the diamond. A pearl is 4 gems away from the ruby. What number is the pearl?

★13. A ruby's weight is measured in carats. The combined weight of 2 rubies is 100 carats. Each is a different weight. One of the numbers divides evenly by 7. The second is an odd number under 55. The sum of the second number's digits is 6. What does each ruby weigh?

Dividing 3-Digit Numbers

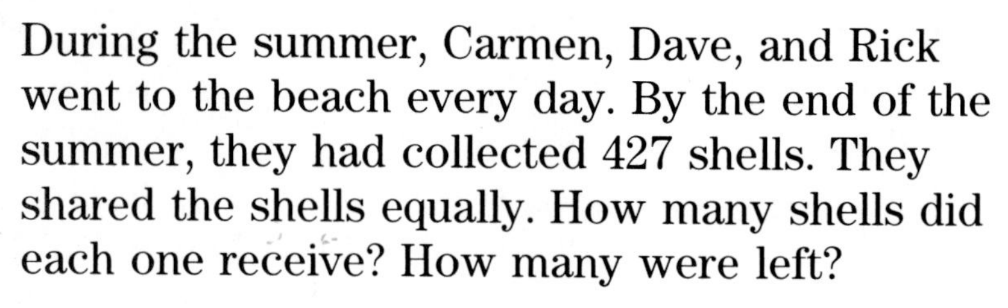

During the summer, Carmen, Dave, and Rick went to the beach every day. By the end of the summer, they had collected 427 shells. They shared the shells equally. How many shells did each one receive? How many were left?

You can divide to find how many shells.

Divide $427 \div 3$.

Divide the hundreds.
Think: $3\overline{)4}$.
Write 1.

```
   1
3)427
 -3
  1
```

Multiply. $1 \times 3 = 3$
Subtract. $4 - 3 = 1$
Compare. $1 < 3$

Divide the tens.
Bring down the 2.
Think: $3\overline{)12}$.
Write 4.

```
   14
3)427
 -3↓
  12
 -12
   0
```

Multiply. $4 \times 3 = 12$
Subtract. $12 - 12 = 0$

Divide the ones.
Bring down the 7.
Think: $3\overline{)7}$.
Write 2.

```
   142 R1
3)427
 -3
  12
 -12↓
   07
 -  6
    1
```

Multiply. $2 \times 3 = 6$
Subtract. $7 - 6 = 1$
Compare. $1 < 3$
Write the remainder.

They each have 142 shells. There is 1 shell left.

Divide.

1. $7\overline{)875}$
2. $8\overline{)984}$
3. $4\overline{)692}$
4. $6\overline{)972}$
5. $5\overline{)815}$
6. $2\overline{)493}$
7. $4\overline{)853}$
8. $7\overline{)893}$
9. $3\overline{)796}$
10. $5\overline{)676}$
11. $5\overline{)627}$
12. $3\overline{)732}$
13. $8\overline{)976}$
14. $6\overline{)805}$
15. $2\overline{)596}$
16. $6\overline{)804}$

★17. $7\overline{)714}$

★18. $9\overline{)963}$

★19. $9\overline{)937}$

★20. $4\overline{)837}$

Solve.

21. Hilary and Phil find 407 shells. They put the same number of shells into 3 knapsacks. How many shells are there in each knapsack? How many are left?

22. Stella collects 88 sand dollars. The Beach Boutique pays her 2¢ for each sand dollar. How much money does she earn from selling the sand dollars?

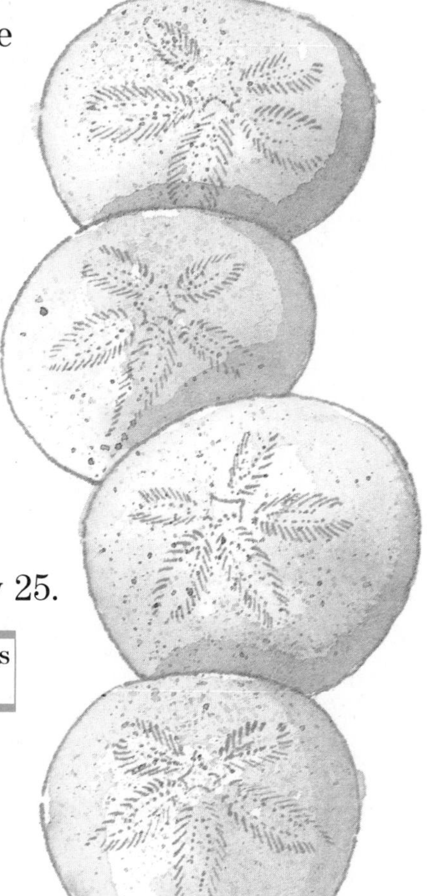

FOCUS: MENTAL MATH

You can think of quarters in order to divide by 25.

Here is how: $200 \div 25$. How many quarters are there in \$2.00?

\$1.00 = 4 quarters
\$2.00 = 8 quarters
OR
$2 \times 4 = 8$ quarters
So, $200 \div 25 = 8$.

Try these.

1. $100 \div 25 =$ ____
2. $300 \div 25 =$ ____
3. $400 \div 25 =$ ____
4. $500 \div 25 =$ ____
5. $600 \div 25 =$ ____
6. $800 \div 25 =$ ____
7. $150 \div 25 =$ ____
8. $350 \div 25 =$ ____
9. $550 \div 25 =$ ____

More Practice, page 416

Mixed Practice

Complete.

1. $288 + 705$
2. $774 - 524$
3. 102×5
4. $3.4 + 5.8$
5. $8{,}233 + 1{,}845$
6. $9.53 - 5.45$
7. $1{,}545 + 7{,}566$
8. $8.9 + 7.9$
9. $922 - 284$
10. 82×2
11. $\$58.53 + \$39.69$
12. $4.7 + 1.7$
13. $853 - 414$
14. $\$80.14 + \$24.35$
15. 242×4
16. 37×7
17. $8.4 - 3.5$
18. $736 - 311$
19. $3.28 + 1.41$
20. $\$4.56 + \$4.92$
21. $2{,}736 + 1{,}749$
22. $9{,}992 - 4{,}631$
23. 850×9
24. $4.11 + 2.35$
25. $869 - 294$
26. $7.73 - 3.47$
27. $\$2.23 \times 3$
28. $41 + 25$
29. $4{,}674 - 2{,}547$
30. $484 + 954$
31. $\$49.66 + \$43.67 = \_\_\_$
32. $80 \div 6 = \_\_\_$
33. $881 \div 3 = \_\_\_$
34. $736 - 311 = \_\_\_$
35. $2 \times 469 = \_\_\_$
36. $3.21 + 5.89 = \_\_\_$
37. $38 \div 5 = \_\_\_$
38. $6\overline{)763}$
39. $6{,}393 - 2{,}281 = \_\_\_$
40. $\$9.25 - \$6.21 = \_\_\_$
41. $9 \times 570 = \_\_\_$
42. $3\overline{)88}$
43. $869 - 294 = \_\_\_$
44. $\$1.99 + \$7.13 = \_\_\_$
45. $617 \div 3 = \_\_\_$
46. $9\overline{)972}$
47. $5.32 + 4.32 = \_\_\_$
48. $\$7.81 - \$5.59 = \_\_\_$
49. $17 \div 3 = \_\_\_$
50. $4.9 + 6.9 = \_\_\_$
51. $3.88 - 1.66 = \_\_\_$

Complete.

52. $3.58 + 4.06$

53. $27 + 62$

54. 82×8

55. $97 - 16$

56. $6.75 - 5.04$

57. $\$6.11 - 4.05$

58. 312×2

59. $18 + 98$

60. $8.14 + 1.27$

61. 253×4

62. $873 - 792$

63. $2{,}748 + 1{,}717$

64. $1.43 + 5.36$

65. $\$9.52 - 8.75$

66. $169 + 284$

67. $5.09 + 1.45$

68. $857 - 741$

69. 438×2

70. $8.7 - 7.8$

71. 189×5

72. $3.35 + 6.09 =$ ■

73. $577 + 399 =$ ■

74. $\$7.13 - \$2.34 =$ ■

75. $3{,}841 + 1{,}903 =$ ■

76. $\$7.84 + \$2.98 =$ ■

77. $565 \div 5 =$ ■

78. $2.51 - 1.95 =$ ■

79. $3\overline{)47}$

80. $96 \div 4 =$ ■

81. $2 \times 292 =$ ■

82. $\$6.19 - \$5.45 =$ ■

83. $4.86 + 2.76 =$ ■

84. $4.14 - 1.56 =$ ■

85. $805 \div 7 =$ ■

86. $6 \times 650 =$ ■

Solve.

87. Jake's father has 75 arrowheads. Jake gives his father 9 more. How many arrowheads does his father have now?

88. Sio has 147 old pennies. He trades 28 of them. How many does he have left?

89. Tia has 65 old bottles. She gives each of her 3 sisters an equal number of them. How many does each sister receive? How many are left?

PROBLEM SOLVING
Interpreting the Quotient and the Remainder

Sometimes when you divide to solve a problem, the answer is not a whole number. If the answer is a quotient with a remainder, read the question again. Be sure that the answer you write really answers the question.

You may need to:

1. drop the remainder, or
2. round the quotient to the next-greater whole number.

On the Space Hunters' trip, each spaceship cabin holds 4 people. There are 30 people signed up for the trip.

$$\text{Divide: } \begin{array}{r} 7\ \text{R}2 \\ 4\overline{)30} \\ \underline{28} \\ 2 \end{array}$$

Read each question below. Think about how the answers differ for each question.

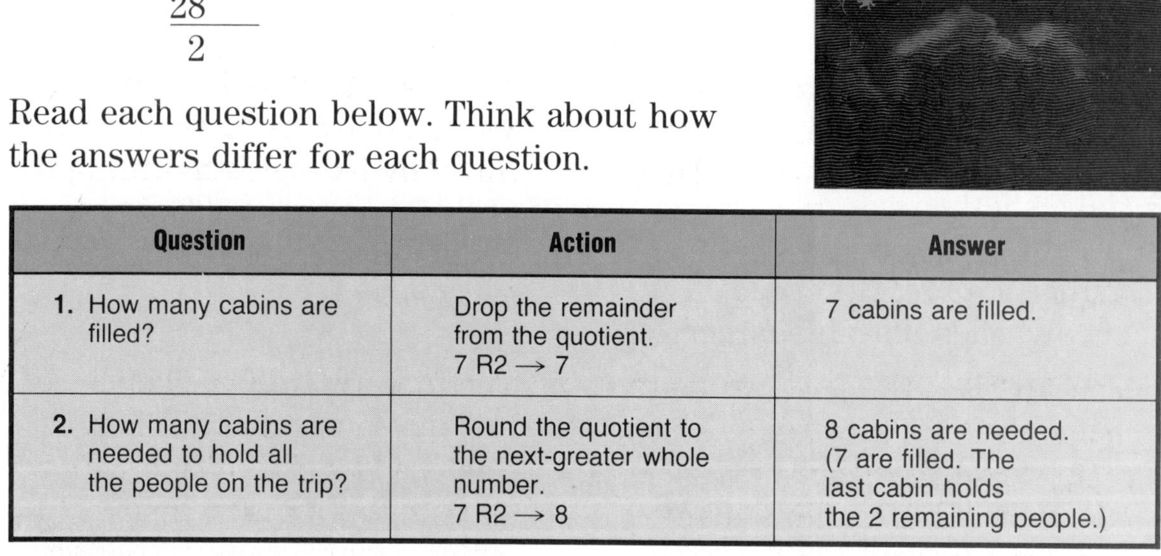

Question	Action	Answer
1. How many cabins are filled?	Drop the remainder from the quotient. 7 R2 → 7	7 cabins are filled.
2. How many cabins are needed to hold all the people on the trip?	Round the quotient to the next-greater whole number. 7 R2 → 8	8 cabins are needed. (7 are filled. The last cabin holds the 2 remaining people.)

Write the letter of the correct answer.

1. A group travels in space pods to the planet Corsi. Each pod holds 6 people. There are 40 people in the group. How many pods are needed to transport everyone?

$$\begin{array}{r} 6 \text{ R}4 \\ 6\overline{)40} \end{array}$$

a. 4 pods b. 6 pods
c. 7 pods

2. A group digs in a diamond field. There are 2 people needed to work each digging tool. There are 37 people digging for diamonds. How many tools are needed?

$$\begin{array}{r} 18 \text{ R}1 \\ 2\overline{)37} \end{array}$$

a. 1 tool b. 18 tools
c. 19 tools

Solve.

3. There are 86 treasure hunters mining for gold. An elevator takes people down to the mine. The elevator holds only 9 people. How many trips must the elevator make?

4. One group hunts for rubies that can power rocket ships. It takes 3 rubies to power a ship. The group finds 79 rubies. How many ships can these rubies power?

5. On Corsi, 5 hunters pick 49 jewel fruits. Each hunter takes an equal number of fruits. They give the rest away. How many fruits does each hunter take?

6. There are 62 people who hunt for cloud jewels. They need to travel by jet platform during the hunt. Each platform holds 4 people. How many platforms are needed?

7. Mr. Blix uses the jewel fruits to make a salad. If he follows his recipe, he will make six 10-ounce servings. How many 4-ounce servings will the same amount of fruit make?

★8. The food robot has 49 slices of bread to make sandwiches. Each sandwich is made with 2 slices of bread. How many whole sandwiches will be left if 19 people take 1 whole sandwich each?

CALCULATOR

Connie buys 27 oranges. The clerk puts them in cartons. Each carton holds 6 oranges. How many oranges are left for the last carton?

Find $6\overline{)27}$. Use your calculator to find the quotient and the remainder

Press: 

The display should show 4.5.

The 4 in the quotient means that there are 4 full cartons. These cartons hold $4 \times 6 = 24$ oranges. The .5 means that there is a remainder. To find the remainder subtract $27 - 24 = 3$. Connie has 3 oranges left for the last carton.

Copy and complete. Use your calculator.

	Number of oranges	Number in each box	Calculator division	Number of oranges in full boxes	Number of oranges left for the last carton
Example	32	6	$32 \div 6 = 5.333$	$5 \times 6 = 30$	$32 - 30 = 2$
1.	37	4			
2.	57	8			
3.	76	10			
4.	87	6			
5.	107	8			
6.	173	4			
7.	243	12			
8.	527	16			
9.	649	25			
10.	837	36			

GROUP PROJECT

Amazon Adventure

The problem: You and three friends have a map. It shows a buried treasure in the rain forest. You have $700 to order the items you need to search for the treasure. Look at the Key Facts. Order what the four of you will need for your trip. Stay within the budget.

Key Facts

- You will be traveling deep into the rain forest.
- The rain forest is very hot.
- You will be traveling for many days with the items you take.
- There are many insects and animals in the rain forest.
- You must carry what you take with you.

ADVENTURE EQUIPMENT COMPANY

Item	Price	Item	Price
Backpack	$15	Hat	$13
Canteen	$6	Knife	$15
Climbing Boots	$40 a pair	Lantern	$43
Compass	$8	Mosquito Net	$45
Cot	$30	Portable Stove	$59
Diving Mask	$12	Rope	$21
First-Aid Kit	$28	Rubber Raft	$56
Flashlight	$10	Shovel	$25
Flippers	$15	Sleeping Bag	$19
Hiking Jungle Boots	$37 a pair	Tent	$82

CHAPTER TEST

Divide. Show the remainder. (page 372)

1. $4\overline{)23}$
2. $6\overline{)19}$
3. $3\overline{)25}$
4. $8\overline{)19}$
5. $75 \div 9 =$ ■
6. $36 \div 8 =$ ■
7. $11 \div 7 =$ ■

Write a number sentence and solve. (page 376)

8. There are 560 children eating lunch in the Grant School cafeteria. They sit at 8 tables. How many children sit at each table?
9. The 34 students in Mr. Kay's class sell tickets to the Grant School Carnival. Each child sells 8 tickets. How many tickets are sold in all?

Find the quotient. (pages 378 and 380)

10. $7\overline{)92}$
11. $4\overline{)56}$
12. $3\overline{)57}$
13. $2\overline{)35}$
14. $9\overline{)92}$
15. $8\overline{)99}$
16. $3\overline{)80}$
17. $4\overline{)96}$
18. $93 \div 9 =$ ■
19. $72 \div 4 =$ ■
20. $80 \div 5 =$ ■

Solve. (page 388)

21. Steve buys a packet of 100 cactus seeds. Each pot of sand can hold no more than 6 seeds. How many pots does he need to plant all the seeds?
22. The boating club has 145 members. Each boat can hold 7 people. How many boats do they need to take all the members sailing?

Divide. (page 384)

23. $6\overline{)618}$ **24.** $4\overline{)840}$ **25.** $8\overline{)927}$ **26.** $5\overline{)783}$ **27.** $6\overline{)788}$

28. $746 \div 6 =$ ■ **29.** $644 \div 4 =$ ■ **30.** $919 \div 9 =$ ■

31. $742 \div 7 =$ ■ **32.** $989 \div 7 =$ ■ **33.** $942 \div 2 =$ ■

BONUS

Copy and complete.

1.

```
   1,9■4 R4
5)9,624
  5
  46
  45
   12
   10
    24
    20
     4
```

2.

```
   2,485 R2
3)7,457
  6
  14
  12
   2■
   24
    17
    1■
     2
```

3.

```
   1,2■3 R■
7)8,984
  7
  19
  14
   58
   56
    2■
    21
     3
```

Divide.

4. $5\overline{)6,148}$ **5.** $2\overline{)6,547}$ **6.** $8\overline{)9,385}$

7. $5\overline{)6,240}$ **8.** $7\overline{)8,176}$ **9.** $4\overline{)5,894}$

RETEACHING

Divide $382 \div 3$.

Divide the hundreds.
Think: $3\overline{)3}$.
Write 1.

$$\begin{array}{r} 1 \\ 3\overline{)382} \\ -3 \\ \hline 0 \end{array}$$

Multiply. $1 \times 3 = 3$
Subtract. $3 - 3 = 0$

Divide the tens.
Bring down the 8.
Think: $3\overline{)8}$.
Write 2.

$$\begin{array}{r} 12 \\ 3\overline{)382} \\ -3\downarrow \\ \hline 08 \\ -6 \\ \hline 2 \end{array}$$

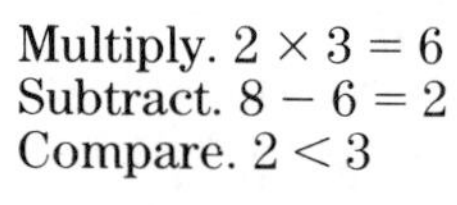

Multiply. $2 \times 3 = 6$
Subtract. $8 - 6 = 2$
Compare. $2 < 3$

Divide the ones.
Bring down the 2.
Think: $3\overline{)22}$.
Write 7.

$$\begin{array}{r} 127 \text{ R1} \\ 3\overline{)382} \\ -3 \\ \hline 08 \\ -6\downarrow \\ \hline 22 \\ -21 \\ \hline 1 \end{array}$$

Multiply. $7 \times 3 = 21$
Subtract. $22 - 21 = 1$
Compare. $1 < 3$
Write the remainder.

Complete.

1. $2\overline{)391}$
2. $4\overline{)502}$
3. $3\overline{)423}$
4. $5\overline{)566}$
5. $7\overline{)733}$
6. $2\overline{)254}$
7. $3\overline{)935}$
8. $4\overline{)656}$
9. $6\overline{)714}$
10. $4\overline{)425}$
11. $9\overline{)976}$
12. $5\overline{)973}$
13. $6\overline{)899}$
14. $3\overline{)731}$
15. $5\overline{)879}$

ENRICHMENT

Divisibility

A. When a number is divided by another number and there is no remainder, we say that the first number is evenly divisible by the second number.

$24 \div 3 = 8$ — There is no remainder. 24 is evenly divisible by 3.

$25 \div 3 = 8$ R1 — There is a remainder. 25 is not evenly divisible by 3.

B. Here's an easy way to find if a number is evenly divisible by 3.
Is 642 evenly divisible by 3?
Add the digits of the first number.

$$6 + 4 + 2 = 12$$

642 is evenly divisible by 3.

If the sum is evenly divisible by 3, then the number itself is evenly divisible by 3.

$12 \div 3 = 4$ — There is no remainder. 12 is evenly divisible by 3.

C. Here are two more rules of divisibility.

Any number that ends in 0, 2, 4, 6, or 8 is evenly divisible by 2.

Any number that ends in 0 or 5 is evenly divisible by 5.

$2\overline{)3{,}876}$ — The number ends in 6. It is evenly divisible by 2.

$5\overline{)7{,}005}$ — The number ends in 5. It is evenly divisible by 5.

Copy and complete the chart. Is each number evenly divisible by 2, 3, or 5? Write *yes* or *no.*

Evenly divisible by:	622	573	982	1,115	6,095	12,498
2						
3						
5						

CUMULATIVE REVIEW

Write the letter of the correct answer.

1. $\begin{array}{r} 824 \\ \times \quad 2 \\ \hline \end{array}$

   a. 148
   b. 208
   c. 1,648
   d. not given

2. $91 \times 6 = \_\_\_$

   a. 85　b. 97
   c. 546　d. not given

3. $7 \times \$0.75 = \_\_\_$

   a. $5.25　b. $52.50
   c. $525.00　d. not given

4. $108 \times 3 = \_\_\_$

   a. 54　b. 324
   c. 3,024　d. not given

5. Estimate: $5 \times 337 = \_\_\_$

   a. 1,500　b. 2,500
   c. 3,500　d. 15,000

6. $8\overline{)66}$

   a. 7
   b. 8
   c. 8 R2
   d. not given

7. $35 \div 4 = \_\_\_$

   a. 6　b. 8
   c. 8 R3　d. not given

8. $5\overline{)68}$

   a. 12
   b. 13
   c. 13 R3
   d. not given

9. $3\overline{)675}$

   a. 125
   b. 235
   c. 325
   d. not given

10. $915 \div 4 = \_\_\_$

    a. 228　b. 228 R3
    c. 229　d. not given

11. Gas costs $1.19 per gallon. About how much does 6 gallons cost?

    a. $5.00　b. $6.00
    c. $8.00　d. $12.00

12. Choose the correct number sentence:
    There are 96 seats in the auditorium. There are 8 seats in each row. How many rows are there?

    a. $96 + 8 = 104$
    b. $96 - 8 = 88$
    c. $96 \div 8 = 12$
    d. not given

Help File

If you have trouble understanding the question, use these hints.

Rewriting the question
If you're not sure of what you are looking for, rewrite the question in your own words.

Organizing information
It is easier to understand the question if you organize the information.

Finding needed information
Sometimes you need information that is not given in the problem. Find this information in resource books, newspapers, or magazines.

Crossing out information
Some problems contain more information than you need. List the information in the problem. Cross out the information you do not need.

Making a list
Making a list of the information in a problem can help you understand the question.

Drawing a picture
Drawing a picture of the information given in a problem can help you understand the question.

Choosing the operation
Look at the numbers in the problem. Think about how to use the numbers to answer the question. Choose the correct operation.

Finding information in the problem
Ask yourself what the question is asking you to find. Write this down. Then find information in the problem that will help you answer the question.

Help File

You understand the question. To find out what you need to do, use these hints.

Writing a number sentence
A number sentence can help you use the numbers in the problem to find the answer.

Making a plan
You need to do more than one step to solve some problems. Making a plan helps you find all the steps you need to do to solve the problem.

Estimating
Sometimes you can solve a problem by estimating. Sometimes you need to find an exact answer.

Finding the pattern
Sometimes you are asked to find what comes next in a group of numbers, shapes, or letters. Looking at the pattern can help you tell what comes next.

Making a table/list
Making a table or a list can sometimes help you find a pattern that will help you solve a problem.

Guessing an answer
Sometimes you can guess at an answer. Your guess can help you find the correct answer.

Finding needed information
You may need information that is not given in a problem. Use resource materials to find this information.

Using graphs, schedules, recipes
Sometimes you have to study a graph, a recipe, or a schedule to find the information you need.

Help File

If you have trouble while you're solving the problem, use these hints. If you need more help, use the Table of Contents or the Index.

Comparing
To compare fractions that have the same denominator, compare the numerators. See pages 234–237.

$\frac{1}{4} < \frac{3}{4}$

Regrouping
When two digits in a column add up to more than 10, remember to regroup. See pages 92–97.

$$\begin{array}{r} {}^{1} \\ 18 \\ +12 \\ \hline 30 \end{array}$$

In subtraction, you have to regroup when the top number in a column is smaller than the bottom number. See pages 126–129.

$$\begin{array}{r} {}^{2}\,{}^{1} \\ \not{3}1 \\ -29 \\ \hline 2 \end{array}$$

In multiplication, when the product in a column is greater than 10, you must remember to regroup. See pages 350–353.

$$\begin{array}{r} {}^{3} \\ 16 \\ \times\ \ 6 \\ \hline 96 \end{array}$$

Adding and subtracting decimals
Be sure to keep the decimal points in a column. See pages 256–259.

$$2.5 + 1.4 = \begin{array}{r} 2.5 \\ +1.4 \\ \hline 3.9 \end{array}$$

Multiplying
You can multiply when you want to find out how many there are in all and you know how many groups there are and that each group has the same number in it. Knowing your multiplication tables makes multiplying easier. See pages 156–177.

4 groups of 3 children

$4 \times 3 = 12$

Dividing
When dividing, be sure each number in the answer is in the correct column. See pages 378–385.

$$3\overline{)396}\ \ \text{answer: } 132$$

Help File

When you want to be sure your answer is correct, use these hints.

Checking for a reasonable answer
Does your answer make sense? If you aren't sure, try estimating. If your estimated answer is very different from your solution, compute the problem again.

Checking addition
Check your answer by subtracting one addend from the answer. The difference should be the other addend.

$$\begin{array}{r} 18 \\ +\ 3 \\ \hline 21 \end{array} \qquad \begin{array}{r} 21 \\ -3 \\ \hline 18 \end{array}$$

Checking subtraction
Add your answer to the smaller number in the problem. The sum should be the larger number in the problem.

$$\begin{array}{r} 17 \\ -\ 9 \\ \hline 8 \end{array} \qquad \begin{array}{r} 8 \\ +9 \\ \hline 17 \end{array}$$

Checking multiplication
Divide your answer by one of the numbers in the problem. The quotient should be the other number in the problem.

$$\begin{array}{r} 12 \\ \times\ 3 \\ \hline 36 \end{array} \qquad 3\overline{)36}\ \text{= } 12$$

Checking division
Multiply your answer by the divisor. The product should be the dividend.

$$3\overline{)18}\ \text{= } 6 \qquad 6 \times 3 = 18$$

When your quotient has a remainder, multiply as above. Add the remainder to your answer.

$$3\overline{)19}\ \text{= } 6\text{ R1}$$

There are several ways to use the remainder. Be sure you answer the question that is asked.

$$3 \times 6 = 18 + 1 = 19$$

Using outside sources
Sometimes you can use outside sources, such as an almanac or an encyclopedia, to be sure your answer is correct.

Infobank

FABULOUS FIRSTS

First	Year
First book printed (using blocks)	868
First printed map	1472
First adding machine	1642
First children's magazine	1751
First crossword puzzle	1875

DISTANCES TRAVELED IN ONE HOUR

Greyhound	36 miles
Jackrabbit	45 miles
Turkey	55 miles
Pelican	29 miles
Chicken	9 miles
Pig	11 miles

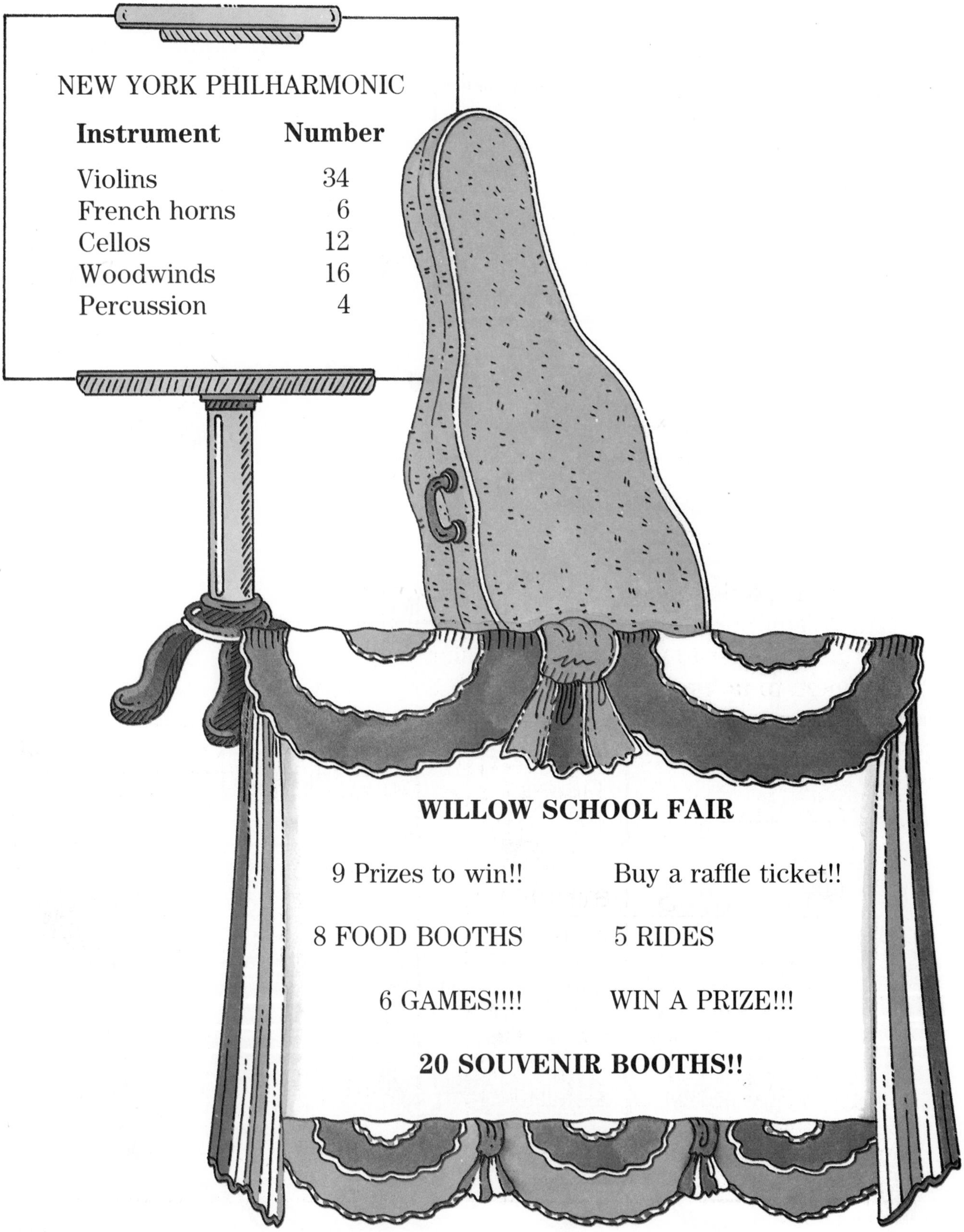

NEW YORK PHILHARMONIC

Instrument	Number
Violins	34
French horns	6
Cellos	12
Woodwinds	16
Percussion	4

WILLOW SCHOOL FAIR

9 Prizes to win!!

Buy a raffle ticket!!

8 FOOD BOOTHS

5 RIDES

6 GAMES!!!!

WIN A PRIZE!!!

20 SOUVENIR BOOTHS!!

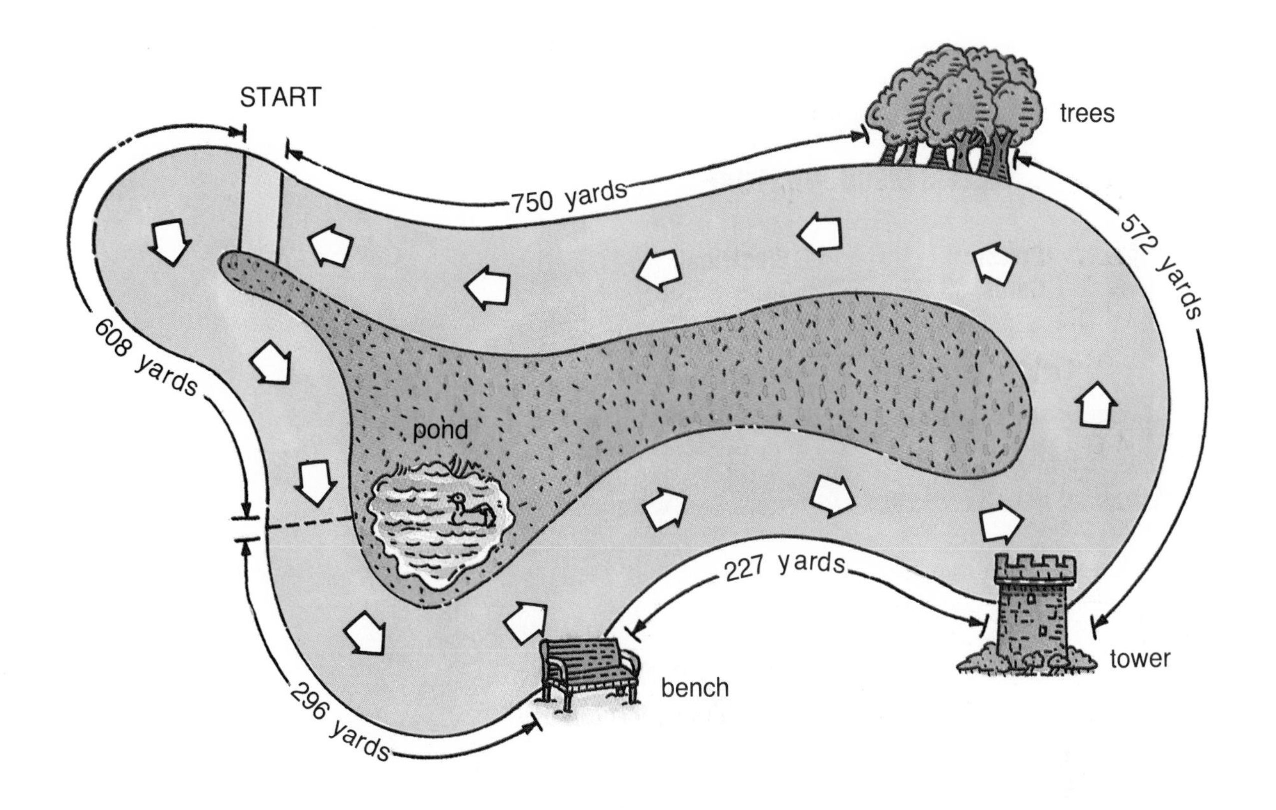

An artist, a doctor, and a baseball player came to Wilson School Career Day. The students signed a sheet to show which speaker they wanted to talk to.

Name	Grade	Speaker
Sue	3	artist
Amy	5	athlete
Ricky	3	athlete
Greg	4	artist
Darryl	4	artist
Mark	4	athlete
Reggie	2	artist
Don	4	doctor
Arthur	4	artist

Name	Grade	Speaker
Peg	4	doctor
John	4	artist
Samson	5	athlete
Jeremiah	3	athlete
Libby	4	athlete
Ollie	3	artist
Billy	5	doctor
June	2	artist

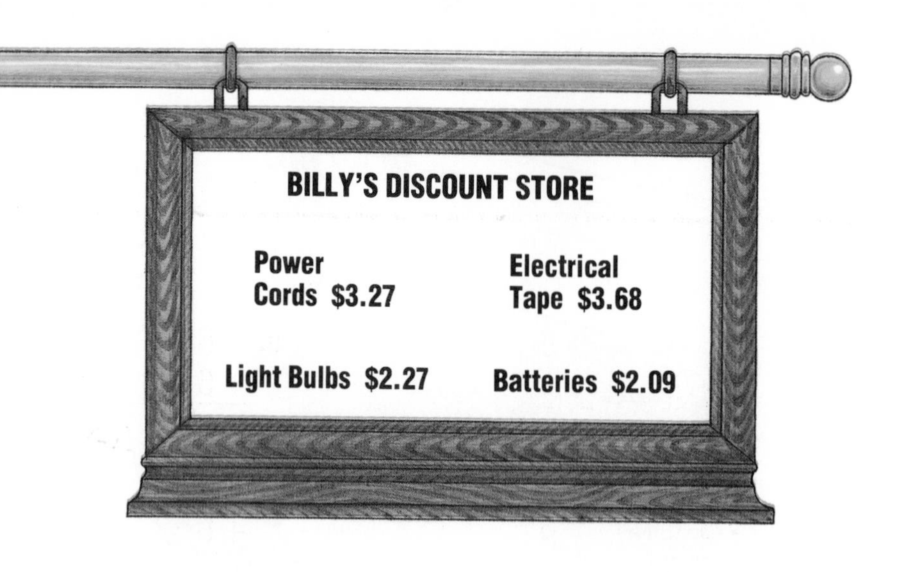

In 1622, the Spanish sailing ship *Atocha* sank. On it was a great treasure. For more than 300 years, divers tried to recover the treasure. In 1979, one diver and his crew were successful.

More Practice

Chapter 1, Page 5

Add.

1. $\begin{array}{r} 7 \\ +6 \\ \hline \end{array}$ $\begin{array}{r} 6 \\ +7 \\ \hline \end{array}$

2. $\begin{array}{r} 3 \\ +8 \\ \hline \end{array}$ $\begin{array}{r} 8 \\ +3 \\ \hline \end{array}$

3. $\begin{array}{r} 8 \\ +6 \\ \hline \end{array}$ $\begin{array}{r} 6 \\ +8 \\ \hline \end{array}$

4. $\begin{array}{r} 8 \\ +4 \\ \hline \end{array}$ $\begin{array}{r} 4 \\ +8 \\ \hline \end{array}$

5. $\begin{array}{r} 9 \\ +8 \\ \hline \end{array}$ $\begin{array}{r} 8 \\ +9 \\ \hline \end{array}$

6. $\begin{array}{r} 9 \\ +6 \\ \hline \end{array}$ $\begin{array}{r} 6 \\ +9 \\ \hline \end{array}$

7. $5 + 8 =$ ___
8. $8 + 8 =$ ___
9. $5 + 7 =$ ___
10. $7 + 4 =$ ___
11. $6 + 6 =$ ___
12. $3 + 9 =$ ___
13. $9 + 9 =$ ___
14. $7 + 7 =$ ___

Chapter 1, Page 11

Subtract.

1. $\begin{array}{r} 11 \\ -\ 3 \\ \hline \end{array}$
2. $\begin{array}{r} 15 \\ -\ 8 \\ \hline \end{array}$
3. $\begin{array}{r} 11 \\ -\ 9 \\ \hline \end{array}$
4. $\begin{array}{r} 14 \\ -\ 9 \\ \hline \end{array}$
5. $\begin{array}{r} 11 \\ -\ 4 \\ \hline \end{array}$
6. $\begin{array}{r} 15 \\ -\ 8 \\ \hline \end{array}$
7. $\begin{array}{r} 12 \\ -\ 6 \\ \hline \end{array}$
8. $\begin{array}{r} 16 \\ -\ 7 \\ \hline \end{array}$
9. $\begin{array}{r} 6 \\ -0 \\ \hline \end{array}$
10. $\begin{array}{r} 13 \\ -\ 5 \\ \hline \end{array}$
11. $\begin{array}{r} 14 \\ -\ 8 \\ \hline \end{array}$
12. $\begin{array}{r} 17 \\ -\ 9 \\ \hline \end{array}$

13. $15 - 7 =$ ___
14. $12 - 6 =$ ___
15. $11 - 2 =$ ___
16. $16 - 8 =$ ___
17. $12 - 5 =$ ___
18. $11 - 6 =$ ___
19. $12 - 3 =$ ___
20. $13 - 4 =$ ___

Chapter 1, Page 17

Find the missing addend.

1. $2 +$ ___ $= 4$
2. ___ $+ 3 = 5$
3. $7 +$ ___ $= 10$
4. $3 +$ ___ $= 7$
5. ___ $+ 8 = 9$
6. $8 +$ ___ $= 11$
7. $9 +$ ___ $= 14$
8. ___ $+ 5 = 5$
9. $3 +$ ___ $= 12$
10. ___ $+ 8 = 16$
11. $3 +$ ___ $= 10$
12. ___ $+ 1 = 9$
13. ___ $+ 7 = 14$
14. $0 +$ ___ $= 8$
15. ___ $+ 6 = 14$
16. $9 +$ ___ $= 18$

Chapter 2, Page 43

Write > or < for ●.

1. 17 ___ 71
2. 143 ___ 98
3. 211 ___ 111
4. 67 ___ 71

Write the numbers in order from the least to the greatest.

5. 6, 11, 3, 9
6. 76, 62, 91, 37
7. 101, 98, 146, 183

Write the numbers in order from the greatest to the least.

8. 749; 274; 1,002; 630
9. 557; 3,201; 5,000; 3,199

Chapter 2, Page 47

Round to the nearest ten.

1. 7
2. 49
3. 65
4. 52
5. 86
6. 74

Round to the nearest hundred.

7. 380
8. 721
9. 562
10. 97
11. 853
12. 471

Chapter 2, Page 49

Write the number.

1. three hundred fifty-two thousand, sixty one
2. seven hundred sixty-nine thousand, five hundred thirty-four
3. nine hundred thirty-two thousand, one hundred twenty-three
4. four hundred twenty thousand, seven hundred eight

Write the value of the blue digit.

5. 721
6. 459
7. 1,038
8. 972
9. 6,526
10. 10,476
11. 31,104
12. 95,288
13. 207,682
14. 783,939

Chapter 3, Page 65

Write the time.

1. 

2. 

3. 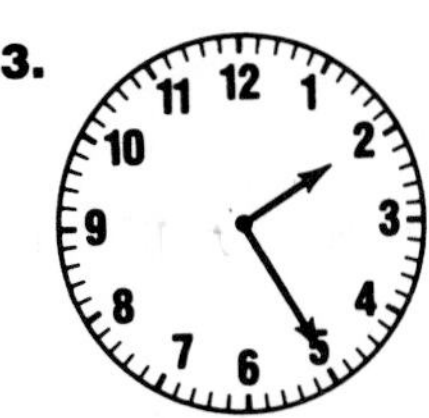

4. 

Write the letter of the clock that shows the time.

5. half past seven

6. six minutes after two

7. one minute to one

a. 

b. 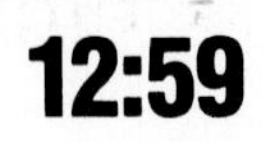

c. 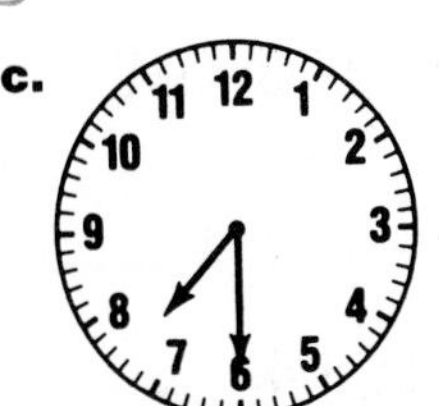

Chapter 3, Page 67

How many minutes are between

1. 

12:30 and 12:45?

2. 

1:07 and 1:52?

3. 

5:29 and 6:01?

4. **3:11**

3:11 and 3:39?

Chapter 3, Page 77

Write the amount. Write the dollar sign ($) and the cents point (.).

1. 2 dollars and 61 cents
2. 8 dollars and 2 cents
3. ninety-eight cents
4. 5 dollars and 40 cents

Chapter 4, Page 95

Add.

1. $\begin{array}{r} 15 \\ +37 \\ \hline \end{array}$
2. $\begin{array}{r} 63 \\ +17 \\ \hline \end{array}$
3. $\begin{array}{r} 49 \\ +23 \\ \hline \end{array}$
4. $\begin{array}{r} 56 \\ +35 \\ \hline \end{array}$
5. $\begin{array}{r} 78 \\ +18 \\ \hline \end{array}$
6. $\begin{array}{r} 67 \\ +27 \\ \hline \end{array}$
7. $\begin{array}{r} 18 \\ +28 \\ \hline \end{array}$
8. $\begin{array}{r} 44 \\ +46 \\ \hline \end{array}$
9. $\begin{array}{r} 35 \\ +19 \\ \hline \end{array}$
10. $\begin{array}{r} 49 \\ +28 \\ \hline \end{array}$
11. $\begin{array}{r} 22 \\ +59 \\ \hline \end{array}$
12. $\begin{array}{r} 76 \\ +18 \\ \hline \end{array}$
13. 62 + 18 = ■
14. 58 + 26 = ■
15. 45 + 29 = ■

Chapter 4, Page 111

Add.

1. $\begin{array}{r} 387 \\ +153 \\ \hline \end{array}$
2. $\begin{array}{r} 342 \\ +589 \\ \hline \end{array}$
3. $\begin{array}{r} 237 \\ +575 \\ \hline \end{array}$
4. $\begin{array}{r} 679 \\ +165 \\ \hline \end{array}$
5. $\begin{array}{r} 582 \\ +278 \\ \hline \end{array}$
6. $\begin{array}{r} \$4.64 \\ +\ 1.57 \\ \hline \end{array}$
7. $\begin{array}{r} 609 \\ +608 \\ \hline \end{array}$
8. $\begin{array}{r} 729 \\ +683 \\ \hline \end{array}$
9. $\begin{array}{r} 852 \\ +629 \\ \hline \end{array}$
10. $\begin{array}{r} \$8.06 \\ +\ 0.95 \\ \hline \end{array}$
11. 765 + 415 + 120 = ■
12. 873 + 519 + 386 = ■

Chapter 4, Page 113

Add.

1. $\begin{array}{r} 1{,}220 \\ +3{,}475 \\ \hline \end{array}$
2. $\begin{array}{r} 3{,}532 \\ +5{,}121 \\ \hline \end{array}$
3. $\begin{array}{r} 2{,}063 \\ +6{,}529 \\ \hline \end{array}$
4. $\begin{array}{r} 7{,}157 \\ +2{,}363 \\ \hline \end{array}$
5. $\begin{array}{r} 6{,}477 \\ +1{,}359 \\ \hline \end{array}$
6. $\begin{array}{r} \$50.68 \\ +\ 23.48 \\ \hline \end{array}$
7. $\begin{array}{r} 4{,}879 \\ +5{,}654 \\ \hline \end{array}$
8. $\begin{array}{r} \$29.77 \\ +\ 85.48 \\ \hline \end{array}$
9. 4,506 + 7,893 = ■
10. $71.83 + 99.39 = ■

Chapter 5, Page 129

Subtract.

1. 27 − 19
2. 36 − 19
3. 54 − 35
4. 72 − 16
5. 38 − 29
6. 83 − 64
7. 22 − 17
8. 94 − 46
9. 77 − 39
10. 96 − 59
11. 65 − 36
12. 96 − 48
13. 82 − 56 = ____
14. 71 − 66 = ____
15. 47 − 29 = ____

Chapter 5, Page 143

Subtract.

1. 702 − 356
2. 309 − 163
3. 540 − 229
4. 804 − 346
5. $4.50 − 1.28
6. 907 − 538
7. $5.00 − 2.46
8. 708 − 445
9. $6.20 − 3.53
10. 300 − 183
11. $9.04 − $7.42 = ____
12. 604 − 375 = ____
13. $8.00 − $2.56 = ____

Chapter 5, page 147

Subtract.

1. 4,238 − 1,124
2. 7,339 − 5,018
3. 5,708 − 2,356
4. $65.93 − 23.65
5. 9,230 − 6,169
6. $84.23 − 35.64
7. 3,176 − 1,488
8. 5,268 − 3,979
9. 4,064 − 1,578 = ____
10. $10.96 − $5.99 = ____
11. 7,453 − 3,685 = ____

Chapter 6, Page 171

Multiply.

1. 6×2
2. 1×6
3. 3×6
4. 6×4
5. 6×6
6. 5×6
7. 6×8
8. 6×5
9. 6×0
10. 2×6
11. 6×3
12. 4×6
13. 6×9
14. 6×7
15. $6 \times 5 =$ ■
16. $6 \times 1 =$ ■
17. $9 \times 6 =$ ■
18. $4 \times 6 =$ ■

Chapter 6, Page 175

Multiply.

1. 7×2
2. 1×7
3. 7×3
4. 4×7
5. 7×7
6. 5×7
7. 7×8
8. 6×7
9. 7×0
10. 2×7
11. 3×7
12. 7×4
13. 7×9
14. 7×6
15. $7 \times 5 =$ ■
16. $7 \times 1 =$ ■
17. $9 \times 7 =$ ■
18. $7 \times 4 =$ ■

Chapter 6, Page 177

Multiply.

1. 8×2
2. 1×8
3. 8×3
4. 9×1
5. 2×9
6. 9×3
7. 4×8
8. 8×5
9. 4×9
10. 9×5
11. 8×6
12. 8×7
13. 8×8
14. 8×9
15. $9 \times 6 =$ ■
16. $9 \times 7 =$ ■
17. $9 \times 8 =$ ■
18. $9 \times 9 =$ ■

Chapter 7, Page 203

Find the quotient.

1. $6\overline{)18}$ 2. $6\overline{)12}$ 3. $6\overline{)24}$ 4. $6\overline{)30}$ 5. $6\overline{)42}$

6. $6\overline{)36}$ 7. $4\overline{)24}$ 8. $6\overline{)48}$ 9. $6\overline{)54}$ 10. $2\overline{)18}$

11. $30 \div 5 = \underline{\quad}$ 12. $54 \div 6 = \underline{\quad}$ 13. $42 \div 6 = \underline{\quad}$ 14. $48 \div 6 = \underline{\quad}$

15. $36 \div 6 = \underline{\quad}$ 16. $18 \div 6 = \underline{\quad}$ 17. $24 \div 6 = \underline{\quad}$ 18. $27 \div 3 = \underline{\quad}$

Chapter 7, Page 209

Find the quotient.

1. $7\overline{)21}$ 2. $7\overline{)14}$ 3. $7\overline{)28}$ 4. $7\overline{)35}$ 5. $7\overline{)42}$

6. $7\overline{)49}$ 7. $4\overline{)28}$ 8. $7\overline{)56}$ 9. $7\overline{)63}$ 10. $3\overline{)21}$

11. $35 \div 7 = \underline{\quad}$ 12. $63 \div 7 = \underline{\quad}$ 13. $0 \div 7 = \underline{\quad}$ 14. $7 \div 1 = \underline{\quad}$

15. $56 \div 7 = \underline{\quad}$ 16. $7 \div 7 = \underline{\quad}$ 17. $14 \div 7 = \underline{\quad}$ 18. $42 \div 7 = \underline{\quad}$

Chapter 7, Page 211

Find the quotient.

1. $8\overline{)24}$ 2. $8\overline{)16}$ 3. $8\overline{)32}$ 4. $9\overline{)27}$ 5. $2\overline{)18}$

6. $9\overline{)36}$ 7. $1\overline{)8}$ 8. $8\overline{)0}$ 9. $5\overline{)40}$ 10. $9\overline{)9}$

11. $45 \div 9 = \underline{\quad}$ 12. $48 \div 8 = \underline{\quad}$ 13. $56 \div 8 = \underline{\quad}$ 14. $64 \div 8 = \underline{\quad}$

15. $54 \div 9 = \underline{\quad}$ 16. $72 \div 8 = \underline{\quad}$ 17. $63 \div 9 = \underline{\quad}$ 18. $81 \div 9 = \underline{\quad}$

Chapter 8, Page 237

Write >, <, or = for ●.

1. 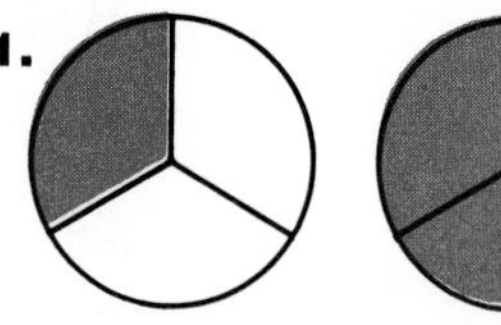

$\frac{1}{3}$ ● $\frac{2}{3}$

2. 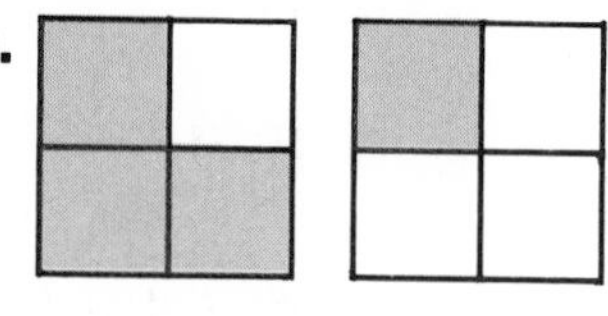

$\frac{3}{4}$ ● $\frac{1}{4}$

3. 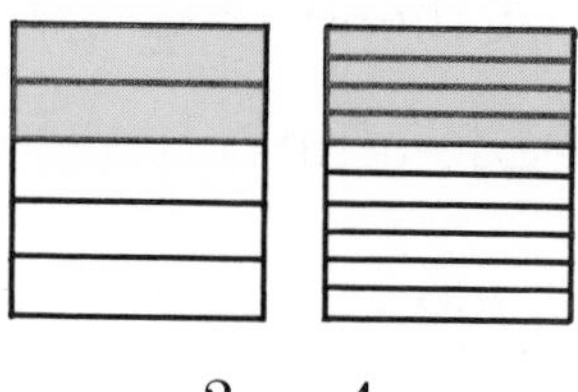

$\frac{2}{5}$ ● $\frac{4}{10}$

4. $\frac{2}{3}$ ● $\frac{1}{3}$
5. $\frac{1}{4}$ ● $\frac{2}{4}$
6. $\frac{5}{6}$ ● $\frac{1}{6}$
7. $\frac{1}{8}$ ● $\frac{3}{8}$
8. $\frac{7}{10}$ ● $\frac{2}{10}$

Chapter 8, Page 257

Add.

1. $2.1 + 0.3$
2. $0.5 + 1.2$
3. $5.6 + 1.3$
4. $0.9 + 2.0$
5. $1.8 + 2.3$
6. $\$0.40 + 0.39$
7. $1.47 + 3.28$
8. $2.39 + 0.75$
9. $\$3.58 + 9.78$
10. $0.69 + 2.73$
11. $3.6 + 4.8 = $ ■
12. $\$0.79 + \$0.54 = $ ■
13. $6.05 + 3.95 = $ ■

Chapter 8, Page 259

Subtract.

1. $0.6 - 0.3$
2. $2.8 - 1.5$
3. $\$0.94 - 0.54$
4. $5.6 - 2.8$
5. $0.52 - 0.37$
6. $\$3.64 - 1.49$
7. $7.37 - 6.56$
8. $0.83 - 0.56$
9. $4.59 - 3.79$
10. $\$6.47 - 3.68$
11. $8.2 - 6.9 = $ ■
12. $\$5.07 - \$2.38 = $ ■
13. $\$9.24 - \$6.76 = $ ■

Chapter 9, Page 278

Choose the unit you would use to measure each of these. Write *g* or *kg*.

1. a table-tennis ball
2. a car
3. an apple
4. a horse
5. an anchor
6. a grape

Chapter 9, Page 285

Choose the unit you would use to measure each of these. Write *in.*, *ft*, *yd*, or *mi*.

1. the length of a pencil
2. the length of a baseball bat
3. the length of a road
4. the height of a doorway
5. the length of a hair ribbon
6. the length of a football field

Chapter 9, Page 290

Choose the unit you would use to measure each. Write *oz* or *lb*.

1. a handful of raisins
2. a person
3. a paper notebook
4. a zebra
5. a typewriter
6. a pair of eyeglasses

Chapter 10, Page 309

Count the sides and vertices of each shape.
Then write the name of each shape.

1.

___ sides
___ vertices
This is a ___.

2. 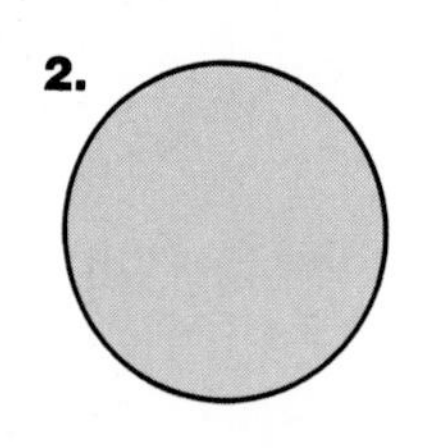

___ sides
___ vertices
This is a ___.

3. 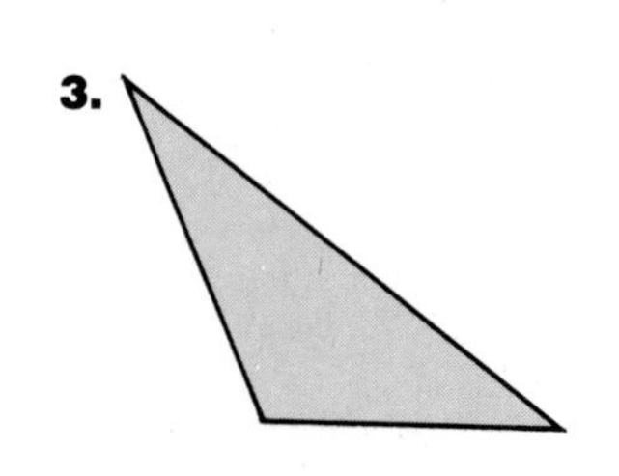

___ sides
___ vertices
This is a ___.

4. 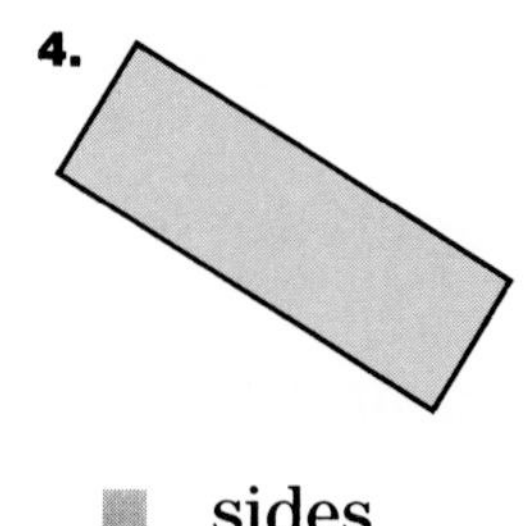

___ sides
___ vertices
This is a ___.

Chapter 10, Page 313

Are the figures congruent? Write *yes* or *no*.

1. 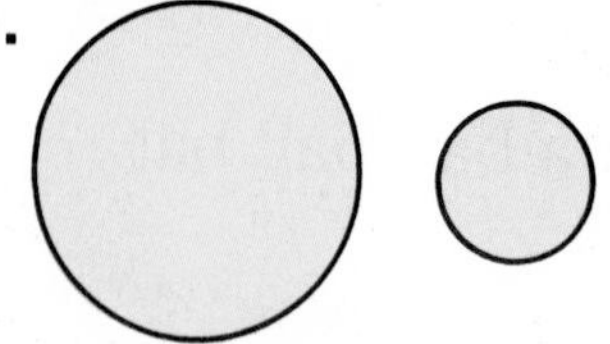

2. 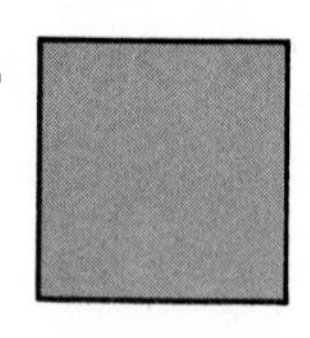 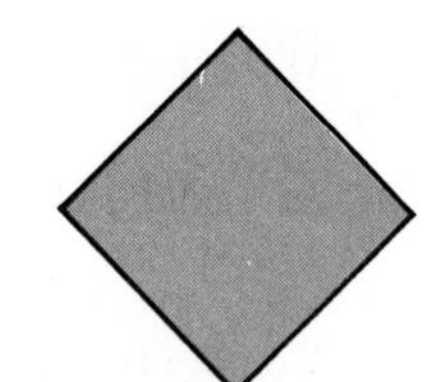

3. 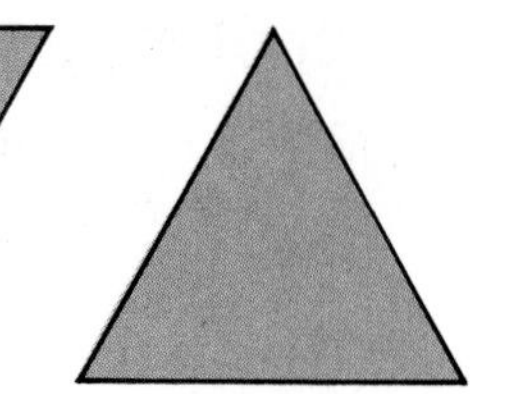

4. 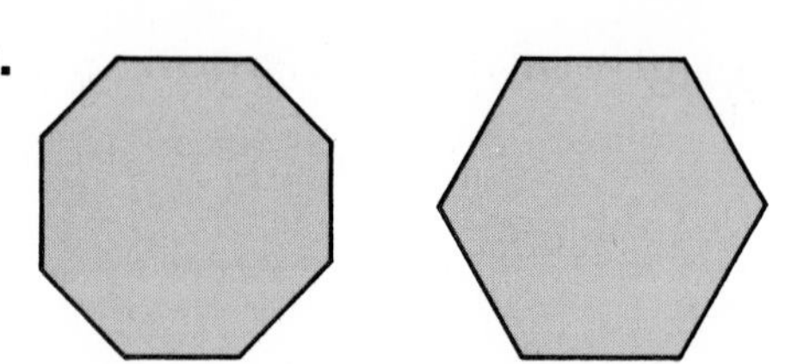

Chapter 10, Page 321

Name the space figure. What shape is

1. a tree trunk?

2. a basketball?

3. a square box?

4. a book?

5. an orange?

6. a cereal box?

7. a drinking straw?

8. a soup can?

Chapter 11, Page 345

Estimate.

1. 14×3
2. 17×6
3. 24×9
4. 58×2
5. 35×4
6. 271×5
7. 462×2
8. 620×7
9. 349×6
10. 528×9
11. $6 \times 85 =$ ■
12. $743 \times 4 =$ ■
13. $9 \times 667 =$ ■

Chapter 11, Page 353

Find the product.

1. 22×6
2. 43×4
3. 56×6
4. 78×7
5. 93×8
6. $\$0.64 \times 9$
7. 86×5
8. $\$0.73 \times 4$
9. $\$0.54 \times 8$
10. $\$0.99 \times 2$
11. $36 \times 7 =$ ■
12. $8 \times \$0.82 =$ ■
13. $6 \times 72 =$ ■

Chapter 11, Page 359

Multiply.

1. 156×2
2. 184×4
3. 129×7
4. 366×2
5. 438×3
6. 624×5
7. 578×7
8. 289×6
9. 736×8
10. 305×9
11. $4 \times 979 =$ ■
12. $780 \times 7 =$ ■
13. $8 \times 678 =$ ■

Chapter 12, Page 379

Divide.

1. $3\overline{)72}$
2. $2\overline{)86}$
3. $4\overline{)88}$
4. $5\overline{)55}$
5. $5\overline{)70}$
6. $2\overline{)94}$
7. $3\overline{)99}$
8. $5\overline{)65}$
9. $7\overline{)98}$
10. $3\overline{)81}$
11. $44 \div 4 = \blacksquare$
12. $72 \div 2 = \blacksquare$
13. $80 \div 5 = \blacksquare$
14. $72 \div 4 = \blacksquare$
15. $38 \div 2 = \blacksquare$
16. $96 \div 8 = \blacksquare$

Chapter 12, Page 381

Divide.

1. $2\overline{)37}$
2. $4\overline{)59}$
3. $3\overline{)46}$
4. $2\overline{)73}$
5. $4\overline{)57}$
6. $6\overline{)86}$
7. $8\overline{)92}$
8. $4\overline{)63}$
9. $2\overline{)77}$
10. $5\overline{)79}$
11. $86 \div 3 = \blacksquare$
12. $93 \div 4 = \blacksquare$
13. $25 \div 2 = \blacksquare$
14. $67 \div 4 = \blacksquare$
15. $79 \div 3 = \blacksquare$
16. $88 \div 7 = \blacksquare$

Chapter 12, Page 385

Divide.

1. $5\overline{)685}$
2. $2\overline{)746}$
3. $4\overline{)848}$
4. $3\overline{)794}$
5. $6\overline{)837}$
6. $2\overline{)948}$
7. $4\overline{)627}$
8. $5\overline{)729}$
9. $8\overline{)976}$
10. $7\overline{)804}$
11. $278 \div 2 = \blacksquare$
12. $793 \div 6 = \blacksquare$
13. $864 \div 7 = \blacksquare$
14. $620 \div 6 = \blacksquare$
15. $858 \div 8 = \blacksquare$
16. $927 \div 9 = \blacksquare$

Table of Measures

TIME	1 minute (min) = 60 seconds (s)
	1 hour (h) = 60 minutes
	1 day (d) = 24 hours

METRIC UNITS

Length	1 meter (m) = 100 centimeters (cm)
	1 kilometer (km) = 1,000 meters
Mass	1 kilogram (kg) = 1,000 grams (g)
Capacity	1 liter (L) = 1,000 milliliters (mL)
Temperature	0° Celsius (C) .... Water freezes
	100° Celsius (C) .... Water boils

CUSTOMARY UNITS

Length	1 foot (ft) = 12 inches (in.)
	1 yard (yd) = 36 inches
	1 yard = 3 feet
	1 mile (mi) = 5,280 feet
	1 mile = 1,760 yards
Weight	1 pound (lb) = 16 ounces (oz)
	1 ton (T) = 2,000 pounds
Capacity	1 pint (pt) = 2 cups (c)
	1 quart (qt) = 2 pints
	1 quart = 4 cups
	1 gallon (gal) = 4 quarts
Temperature	32° Fahrenheit (F) .... Water freezes
	212° Fahrenheit (F) .... Water boils

Symbols

$<$	is less than	$^\circ$	degree	$\overline{AB}$	line segment AB
$>$	is greater than	$\bullet A$	point A	$4 \div 2$	4 divided by 2
$=$	is equal to	$\overleftrightarrow{AB}$	line AB	(5,3)	the ordered pair 5,3

Glossary

Addends Numbers that are added.
Example: $3 + 6 = 9$
↑ ↑
addends

Angle A figure formed when two lines meet at the same endpoint. This is angle *DEF*, angle *FED*, or $\angle E$.

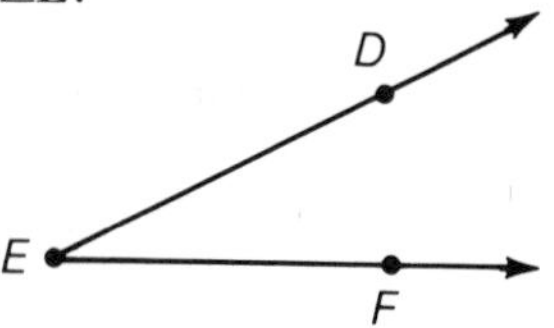

Area The area of a figure is the number of square units that cover its surface. The area of this figure is 8 square units.

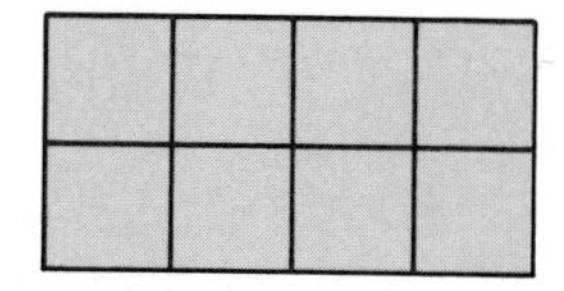

Bar graph A graph that shows number information by using bars of different lengths.

Circle A path that begins and ends at the same point. All the points of the circle are the same distance from a point inside, called the *center.*

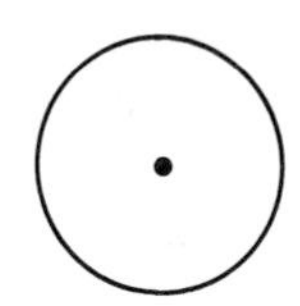

Cone A solid figure that has one face that is a circle.

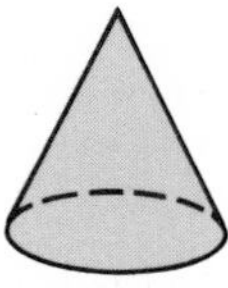

Congruent Line segments or figures that are the same size and shape.

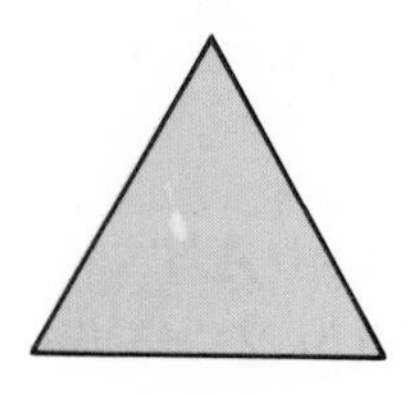

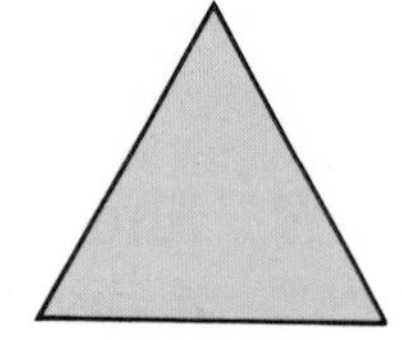

Cube A solid figure that has six square faces.

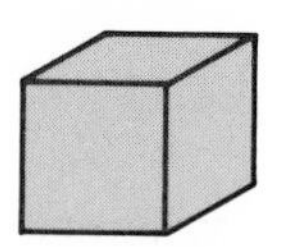

Cylinder A solid figure that has two faces that are circles.

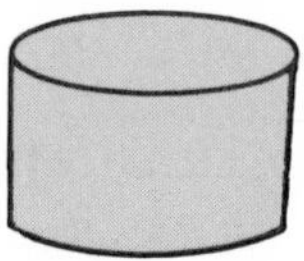

Decimal A number that uses place value and a decimal point to show tenths and hundredths.
Examples: 0.8, 1.76, 23.04

Denominator The bottom number in a fraction.
Example: $\frac{1}{4}$ ← denominator

Diameter A line segment that goes through the center of a circle and has its two endpoints on the circle.

Difference The answer in subtraction.
Example: $7 - 5 = 2$ 2 is the difference

Digit Any of the ten symbols 0, 1, 2, 3, 4, 5, 6, 7, 8, 9 used to write numerals.

Dividend A number to be divided.
Example: $8\overline{)32}^{\,4}$ ← dividend → $32 \div 8 = 4$

Divisible A number is divisible by another number if the quotient is a whole number and the remainder is zero.

Divisor The number that divides the dividend.
Example: $3\overline{)18}^{\,6}$ $18 \div 3 = 6$
← divisor →

Edge An edge is the place where two faces of a solid figure meet.

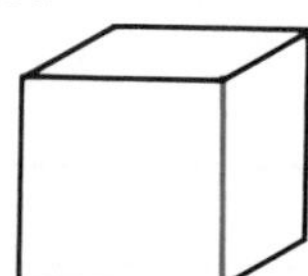

Equivalent fractions Fractions that name the same amount.
Example: $\frac{1}{3} = \frac{2}{6} = \frac{4}{12}$

Even number A number that ends in 0, 2, 4, 6, or 8.

Expanded form A way to show a number as the sum of the value of its digits.
Example: $79 = 70 + 9$

Face A flat surface of a solid figure.

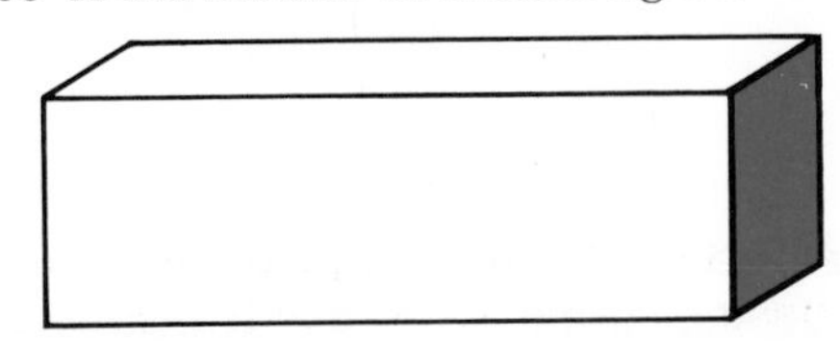

Factors The numbers that you multiply to find a product.
Examples: $3 \times 2 = 6$. 3 and 2 are factors.

Fraction Numbers that name parts of a whole or parts of a set, such as $\frac{2}{3}, \frac{4}{5}, \frac{8}{6}$.

Graph Information shown by using pictures, lines, or bars.

Grouping Property of Addition Changing the grouping of the addends does not change the sum.
Example: $(4 + 3) + 5 = 4 + (3 + 5)$
$7 \quad + 5 = 4 + \quad 8$
$12 \quad = \quad 12$

Grouping Property of Multiplication Changing the grouping of factors does not change the product.
Example: $(3 \times 2) \times 4 = 3 \times (2 \times 4)$
$6 \quad \times 4 = 3 \times \quad 8$
$24 = 24$

Line A straight path that goes in two directions without end.

A B

line AB, or $\overleftrightarrow{AB}$

Line of symmetry A line of folding so that the two halves of a figure match.

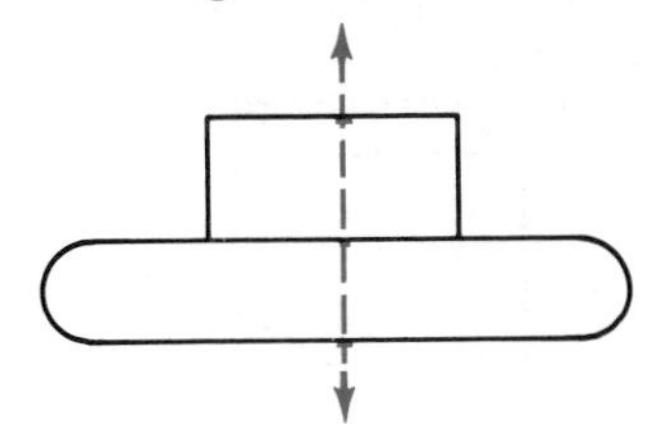

Line segment A straight path that has two endpoints.

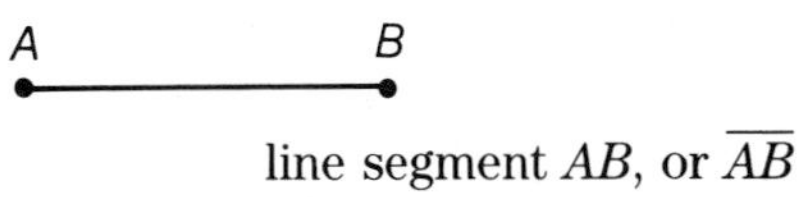

line segment AB, or $\overline{AB}$

Mixed number A number that is made up of a whole number and a fraction.
Example: $3\frac{3}{4}$

Multiple The product of a number and a given number.
Example: 12 is a multiple of 3 because $3 \times 4 = 12$.

Numeral A symbol used to name a number. A numeral for the number seven is 7.

Numerator In $\frac{5}{6}$, 5 is the numerator.

Odd number A number that ends in 1, 3, 5, 7, or 9.

Order Property of Addition Changing the order of the addends does not change the sum.
Example: $6 + 7 = 7 + 6$

Order Property of Multiplication Changing the order of factors does not change the product.
Example: $6 \times 8 = 8 \times 6$

Perimeter The distance around a figure. The perimeter of this figure is 10 cm.

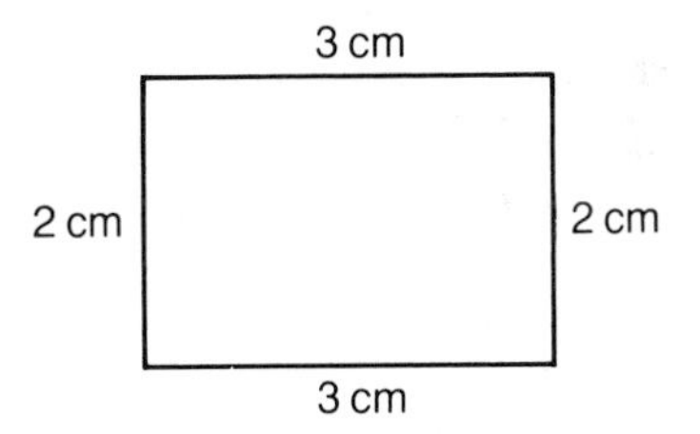

Period A group of three digits in a numeral set off by commas: In 178,630, the digits 178 are in the thousands period. The digits 630 are in the ones period.

Pictograph A graph that uses pictures to show information.

Place value The value of a digit depends on its position within a number. In 9,267, the 9 stands for 9,000, while in 91, the 9 stands for 90.

PRINT A command that tells a computer to print what is on a screen.

Probability The chance of something happening.

Product The answer in multiplication.
Example: $7 \times 6 = 42$
42 is the product.

Quotient The answer in division.
Example: $36 \div 9 = 4$
4 is the quotient.

Radius A line segment from the center of a circle to a point on the circle.

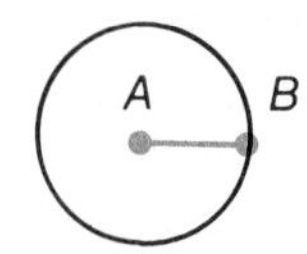

Rectangle A figure formed by four line segments. It has four right angles.

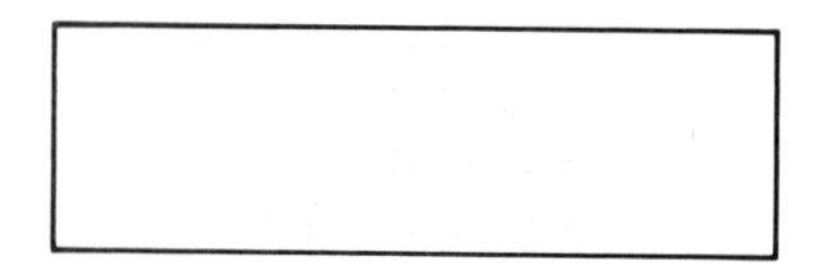

Rectangular prism A figure that has six faces, each of which is a rectangle.

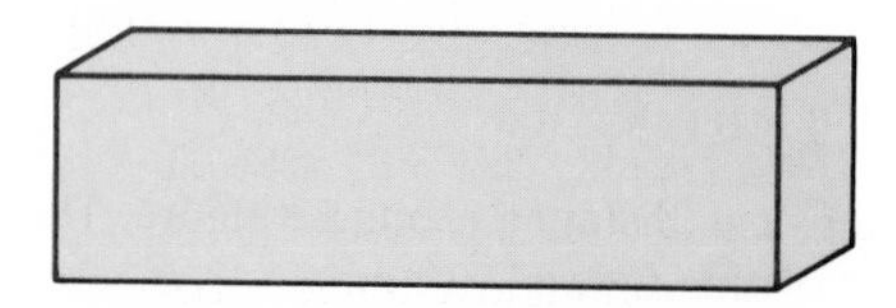

Related sentences Related sentences are the same numbers and the same or opposite operations.

$3 \times 8 = 24$
$8 \times 3 = 24$
$24 \div 8 = 3$
$24 \div 3 = 8$

Remainder In the division $19 \div 5$, the quotient is 3 and the remainder is 4.

```
   3 R4
 5)19
   15
    4
```

Right angle An angle that forms a square corner.

Sphere A space figure that has the shape of a ball.

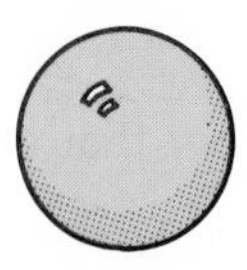

Square A figure that has all four sides the same length and four right angles.

Standard form The usual way of writing a number. In standard form, *twelve* is written as 12.

Sum The answer in addition.
Example: $4 + 5 = 9$. 9 is the sum.

Symmetry If the parts match when a figure is folded on a line, the figure is symmetrical.

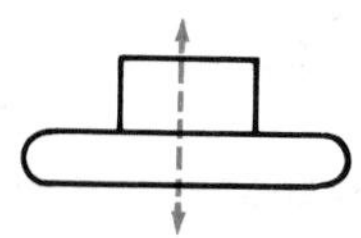

Triangle A figure formed by three line segments.

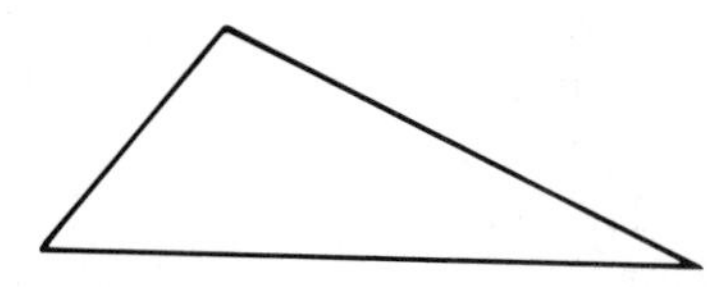

Vertex Two sides of a figure meet at a vertex.

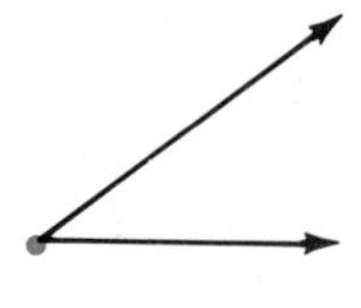

Volume The number of cubic units that it takes to fill a space. The volume of this figure is 24 cubic units.

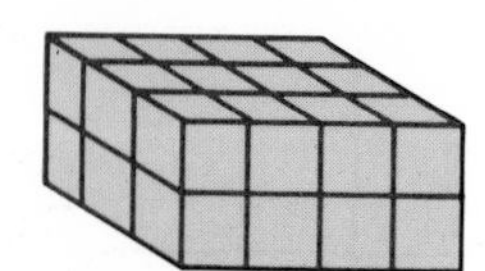

Whole number Any of the numbers such as 0, 1, 2, 3, ...

INDEX

E

F

G

N

O

P

Q

R

S

T

V

W

Y

Z